OCCUPATIONAL AND ENVIRONMENTAL SAFETY ENGINEERING AND MANAGEMENT

OCCUPATIONAL AND ENVIRONMENTAL SAFETY ENGINEERING AND MANAGEMENT

By

H. R. Kavianian
California State University, Long Beach

C. A. Wentz, Jr.
Argonne National Laboratory

VNR VAN NOSTRAND REINHOLD
New York

Library of Congress Catalog Card Number 89-22514

ISBN 0-442-23822-3

Printed in the United States of America

Van Nostrand Reinhold
115 Fifth Avenue
New York, New York 10003

Chapman & Hall
2-6 Boundary Row
London SE1 8HN, England

Thomas Nelson Australia
102 Dodds Street
South Melbourne, Victoria 3205, Australia

Nelson Canada
1120 Birchmount Road
Scarborough, Ontario M1K 5G4, Canada

16 15 14 13 12 11 10 9 8 7 6 5 4 3 2

Library of Congress Cataloging-in-Publication Data

Kavianian, Hamid R., 1949-
Occupational and environmental safety engineering and management / Hamid R. Kavianian, Charles A. Wentz.
p. cm.
Includes bibliographical references.
ISBN 0-442-23822-3
1. Industrial safety. I. Wentz, Charles A. II. Title.
T55.K38 1990
363.11—dc20

89-22514
CIP

Contents

Preface

Our primary purpose in writing this book was to bring together all legal, engineering, and scientific aspects of safety in a single, integrated treatment.

As the field of occupational and environmental safety has grown quite rapidly, changes in engineering and science, laws and regulatory demands, and the attitude of workers and the public have nearly outpaced technology's ability to respond to change.

This textbook has been written for students at the college level in engineering and science. In addition, managers and technical personnel should find it a useful reference in dealing with complex safety matters. Government officials and regulatory personnel as well as members of the legal profession and the general public also may find this book helpful in giving them a general overview of the subject.

Employees in the workplace have a growing need to be more knowledgeable about the hazards of their work environment. It is imperative that individuals and governments work together to better inform and protect these workers, as they are exposed to a variety of complex and potentially hazardous situations. Additionally, public relations aspects of safety are becoming increasingly important in the management and operation of industrial and manufacturing facilities, as the public has a right to know and understand the potential hazards of such operating facilities within or near their communities in order to anticipate problems that may arise.

The better management understands potential hazards and the implementation of measures to either eliminate or reduce the risk connected with these hazards, the better the relationship will be between the operating facility, the workers, and the community.

Historically, many aspects of safety have been relegated to personnel lacking the engineering and scientific knowledge required for a thorough understanding of the hazards of the work environment. As a result, engineers often have not participated in decisions related to occupational and environmental safety. It is imperative that engineers become more involved in both engineering and management decisions regarding safety. We hope that this book will provide insight and direction to facilitate this change. Safety management and engineering must be approached as a science with well-defined goals and objectives, not as an exercise in lip service and sloganeering.

H. R. Kavianian
C. A. Wentz

Acknowledgment

We wish to express our sincere gratitude to Mr. Jack H. Dobson, Jr., technical consultant to the American Society of Safety Engineers. We also would like to thank the Defense Reutilization and Marketing Service for the opportunity to develop the format for this textbook. We also wish to express our appreciation to Dr. J. K. Rao, Chairman and Professor of Civil Engineering at California State University, Long Beach. The assistance of the word processing and graphics personnel at Argonne National Laboratory is greatly appreciated. We also wish to express our thanks to Gabriel Brown, our chemical engineering student assistant.

The support provided by National Institute for Occupational Safety and Health, and in particular Mr. John T. Talty, for preparation of material on "System Safety Analysis in Process Design," is greatly appreciated.

OCCUPATIONAL AND ENVIRONMENTAL SAFETY ENGINEERING AND MANAGEMENT

Chapter

1

Introduction to Safety Engineering and Management

Members of the engineering profession are becoming increasingly aware of their legal liability and their responsibility for the public welfare, especially with respect to the public's tolerance for engineering and management failures and the need for safe working conditions. Putting a man on the moon was considered a scientific achievement, but the Challenger disaster was regarded as an engineering and management failure. Although engineers and managers can minimize the risk of being sued,[1] they cannot avoid their professional responsibility to protect human health and the environment with their expertise and knowledge. The law may keep us from doing wrong, but ethical constraints influence us to do whatever is right.

Accidents are the leading cause of death for younger workers (see Figure 1.1). Occupational accidents alone cost society an estimated 30 billion dollars a year, and workplace hazards pose a serious risk to approximately 100 million workers in the United States. According to the Bureau of Labor Statistics, 4,000 work-related fatalities and 4.9 million occupationally related injuries took place in the United States in 1982.[5]

The risk assessment, engineering, and management needed to guard against potential hazards and disasters have become critically important because of the enormity of their impact. Bhopal, Chernobyl, Challenger, Three Mile Island, and Seveso are among the disasters that might have been avoided through better engineering and operational safety. Engineering risk analysis and management must be a fundamental element of the engineering, scientific, and managerial components in chemical process, defense, space, nuclear power, and other industries.

Sometimes only a disaster will cause engineers and managers to acquire vital information about hazardous technology and its effect upon the public and industry.[2] Technological improvements, refinements in production methods, and market demand can outpace requisite advances in process safety equipment and procedures. A complacent attitude may develop as a result of the successful operation of a manufacturing facility involved with hazardous chemicals and processes, allowing certain coincidental events to occur that can lead to a disaster of catastrophic proportions. This was certainly the case at both Bhopal and Chernobyl, which have provided dramatic examples of the need for increased safety in engineering design and implementation.

Society is also showing an increasing interest in occupational safety and environmental health-related injuries caused by hazardous technologies. This concern was evidenced by a court ruling on June 13, 1985, when three company executives were sentenced to 25 years in prison and a $10,000 fine over the cyanide poisoning death of an employee.[3]

Occupational safety and environmental health issues can easily be accorded mere lip service unless engineers and managers have taken them seriously and have developed a scientific approach to solving safety and environmental problems. This scientific approach must be incorporated into all aspects of the engineering and management of hazardous technology, from conceptual design, feasibility studies, pilot plant operation, semiworks, and commercial operation to

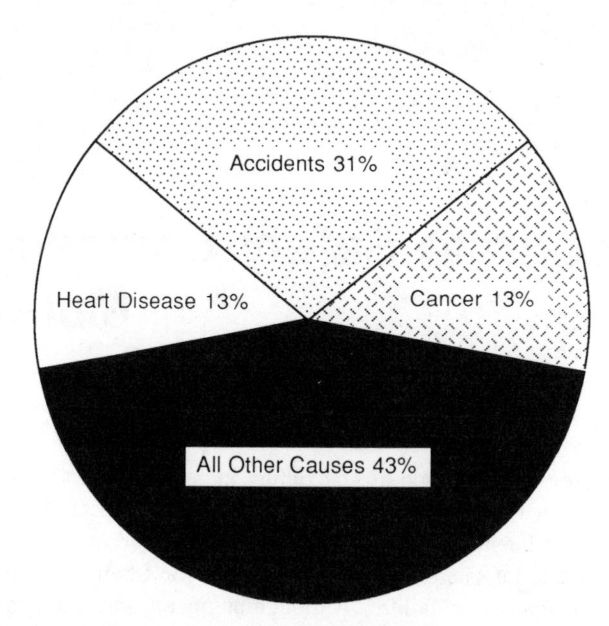

Figure 1.1. Accidents are the leading cause of death in the United States among younger workers. (Source: Reference 5)

shipment of products and disposal of wastes. This enormous task can be accomplished only if engineers and managers have the proper training and knowledge on occupational safety and environmental health issues.

In recent years the public has put a tremendous amount of pressure on industry to improve the quality of life by advancing technology safely, with minimum interference with public safety and the environment. This pressure has manifested itself in the form of strict occupational and environmental laws, regulations, and standards. In spite of its effects on industry, this social pressure has not been transmitted proportionately to our nation's engineering and business schools, which are responsible for the education and training of engineers and managers who ultimately will make crucial decisions regarding occupational safety and environmental health issues. As a result, there is a great need to incorporate the science of safety into the curricula of these schools.

Steps to integrate this concept into engineering education are exemplified by NIOSH (National Institute for Occupational Safety and Health) programs such as Project SHAPE (Safety/Health Awareness for Preventive Engineering). Project SHAPE focuses on integrating safety and health into the curricula of engineering schools. The management of occupational safety and health must bring together schools of business, industrial participants, and government agencies. The goals of all of these programs should be to produce managers and engineers who are attuned to occupational safety and health hazards.[3] Students of business schools need to understand the importance of *managing* safety and health, just like any other business function, and engineers need to stress the importance of sound engineering design to occupational safety and health.[3]

Management Styles and the Effect on Occupational and Environmental Safety

In his book *Managing*, Harold Geneen, former president and board chairman of ITT, stresses that the primary responsibility of management is to manage.[4] Programs now are emphasizing that occupational safety and health can be managed like any other aspect of industry. Indeed, since the passage of the Occupational Safety and Health Act in 1970, management has been held responsible for safety and health by government regulators as well as by others in the community.[3] Occupational safety and health engineering and management encompass such areas of concern as:

- Occupational injuries
- Occupational diseases
- Engineering control systems
- Loss control
- Risk assessment
- System safety
- Air containments
- Emission control
- Heat stress control
- Noise control
- Vibration control
- Electrical safety
- Mechanical guarding
- Work practices
- Materials handling
- Monitoring
- Industrial toxicology
- Product liability
- Waste disposal
- Radiation control

The safety and welfare of employees in the workplace is a matter of increasing concern to workers and often is a deciding factor in both job selection and worker satisfaction. Closely related to this is the issue of worker compensation for work-related injuries and illnesses. The increasing cost of workmen's compensation is of great concern to industry, as it is perceived to deplete funds that otherwise could be used to further profitability. Another significant area is that of product liability. According to *Market Facts*, about 500,000 product liability cases are filed annually, resulting in millions of dollars in court awards and out-of-court settlements.[5] Clearly, these financial concerns, as well as the safety and health issues themselves, provide strong incentives for management to seriously regard occupational safety and environmental health issues.

Hazard Evaluation

It is necessary to do more than just emphasize the importance of these issues to managers and engineers; practical measures must be designed for the management of safety and health. Hazard evaluations of existing or proposed facilities should be an integral part of industrial management. The American Institute of Chemical Engineers (AIChE) has published *Guidelines for Hazard Evaluation Procedures*, outlining a number of procedures intended primarily for use in the evaluation of hazards in chemical and petrochemical process plants.[6]

Most of the AIChE procedures can be applied to many industries. Primarily, these procedures were developed to identify hazards, their possible consequences, and the likelihood of occurrence of events that could cause an accident, as well as what, if any, safety and mitigating systems, emergency alarms, and evaluation plans could be used to properly eliminate or reduce the consequences. The *Guidelines* point out that proper corporate policy and management procedures are an essential component of hazard evaluation and control, which can be accomplished through adherence to good practice and predictive hazard evaluation. In the implementation of a predictive evaluation, a systematic approach will result in the most effective and efficient use of the procedure. Usually, a coarse screening evaluation precedes more detailed and costlier procedures, which are used for more complete evaluations once major problems are identified. This approach provides management with important information it then can use to further formulate or refine policy and implement procedures to assure occupational safety and environmental health.

Information

Of primary importance in any organization's work is the matter of how the information it deals with is handled. In the chapter on "Two Organizational Structures" in *Managing*, Geneen states that:

> The fundamental and basic job of management in any company is to manage. . . . Management manages by making decisions and by seeing that those decisions are implemented. *And the only way* management can do that *successfully* is to have full access to the facts of any situation affecting the welfare of the company.[4]

Writing in the October 1987 issue of the *Journal of Professional Issues in Engineering*, David Bella pointed out that what is at the bottom of many problems in today's technological organizations is the systematic distortion of information as it passes through the organization.[7] His thesis is that this systematic distortion is not necessarily due to unethical behavior, but rather is a function of the mishandling of information. Whether one fully agrees with Bella on this question, it is still fairly easy to see that there must be something fundamentally wrong for an organization to carry on systematic self-deception, and that steps need to be taken to correct the situation. If management wants only to hear and see favorable information, then that is the type of information management will receive. In the field of occupational safety and environmental health, distorted information can be substantially more than bad business—it can be disastrous, as the space shuttle Challenger's explosion and the accident in the nuclear reactor at Chernobyl tragically illustrate.[7]

In the implementation of safety and health policies and procedures, as in other areas of management, managers will be influenced by their modes of thinking and styles of decision-making. Management thinking and decision-making styles have been scrutinized at least as much as and perhaps more than safety. One of the lessons learned has been that effective management techniques can greatly help an organization move forward in all aspects of its endeavor, whereas ineffective management can harm the organization. The crucial question then becomes one of distinguishing good management from bad, or effective techniques from those that are counterproductive.

There is no simple answer. To begin with, there is no set formula for managing. Just as each person is an individual, with his or her own personal traits, each organization usually takes on characteristics that are unique to it, though it normally also will reflect the personality of its leader. Carrying the analogy even further, most people—however different they may be from each other—can benefit from the observations of management, and organizations also can benefit from analyzing and improving themselves.

Management Styles

Management must make sure that its style of thinking promotes full communication within the organization, and this means more than lip service. Studies have identified a variety of management styles, each with its strengths and weaknesses. Whatever the style, management activities include planning, organizing, motivating, and controlling.

For the effective communication of occupational safety and environmental health concerns, the best style is participative management, as shown in Table 1.1. Participative management provides a free flow of information, both up and down the organization; and this information—especially upward-flowing information—tends to be more accurate with the participative style. Also, in terms of decisions, goals, and leadership, the participative mode encourages full involvement by subordinates in decision-making as well as goal-setting, and employees generally feel free to discuss matters with their superiors, who in turn show considerable confidence in them.

Authoritative management, on the other hand, tends to stifle the flow of information; managers hear only what they want to hear and not what they need to hear. Clearly, managers employing an authoritative approach must seriously take stock of their situation if they are to avoid dire consequences in the field of safety. Of course, an organization may not necessarily fit into one of these two styles of management, but may be somewhere in between.

An organization needs to deal with complete facts in the management of occupational safety

Table 1.1. Management styles and their relationship to information access and communication.

Characteristic	Management Style: Participative	Management Style: Authoritative
Is there confidence in subordinates?	Yes	None exists
Are subordinates free to communicate upward?	Yes	No
Are different ideas sought?	Yes, always	Very seldom
Nature of communication	All directions	Only downward
Accuracy of upward communication	Very accurate	Wrong information
Decision-making process	Everyone within organization	Only top management
Nature of goal-setting	Performed by group action	Orders issued
Nature of control	Widely shared	Only at top
Informal organized resistance?	Does not exist	Always exists

and environmental health systems, and it behooves every organization to adopt a style of management and decision-making that permits and encourages free upward flow of vital information.[7,9]

Incorporation of Occupational and Environmental Safety and Health Issues Into the Curriculum of Engineering Schools

There is a great need for integration of safety within the standard engineering curriculum, especially with the laboratory and design orientation of engineering science and analysis. Engineering analysis, design, and management should be considered as a tetrahedron with: (1) life-cycle costs and performance, (2) functional and technical efficiency, (3) quality, and (4) safety as the four points, as shown in Figure 1.2, so that it will be adaptive to the public's increasing expectations of the engineering profession.[8]

Engineers should focus on the unique relationships among public expectations, industry, engineering, and the universities in the crucial area of environmental safety and occupational health. The complexity of our technology is directly proportional to our expectations regarding quality-of-life factors. It is inevitable that the public will put more and more pressure on engineers to improve the quality of life while minimizing interference with valuable natural resources. The engineers are responsible for the design, construction, and safe operation of the resulting complex technology.

The public has put a great deal of pressure on industry to ensure that the safety and health of the community will be considered in the design of community preparedness plans for disaster mitigation. However, this pressure has not been proportionately transmitted to the engineering schools responsible for the education and training of future engineers and for the continuing education of engineers already engaged in professional practice.

The lack of safety awareness in our engineering schools is really a two-fold problem. The engineering students' exposure to hazard engineering and safety systems is minimal, and there may be an administrative problem. The latter problem is exemplified by the fact that most schools of engineering have not incorporated any general safety courses as part of the *requirements* for an undergraduate degree in any engineering branch, although some schools do require it in their industrial engineering program or offer it as an elective. The graduating engineer, therefore, has either very little or no knowledge of safety-related topics.

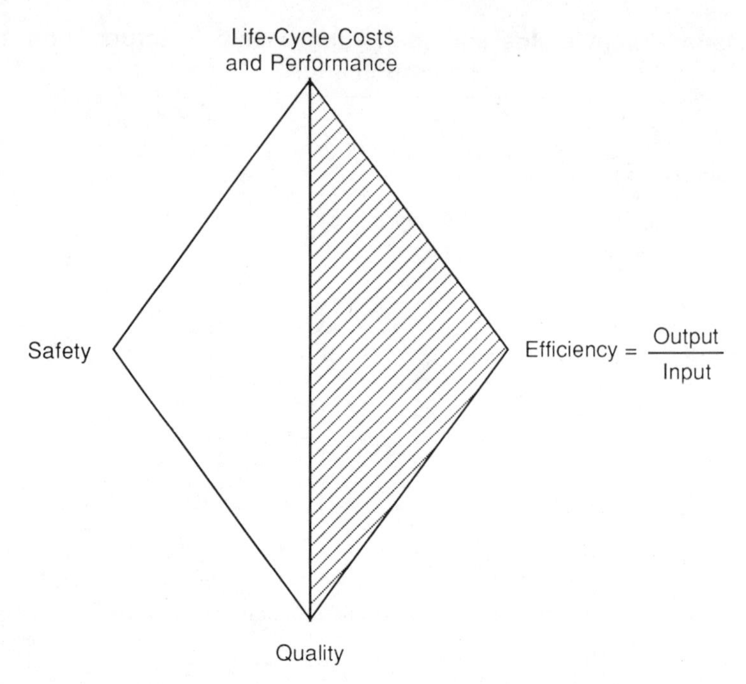

Figure 1.2. Tetrahedron of engineering for functional effectiveness.

The lack of safety awareness in most engineering schools also is exemplified in the operation of most undergraduate and graduate research laboratories. Although many schools require certain minimum safety standards in the laboratories, some of them lack effective enforcement of these regulations, protective equipment, proper chemical storage, and general safety rules. The importance of safety must be inculcated in engineering students, just like quality and efficiency, to complete the engineering tetrahedron of costs/performance–quality–efficiency–safety and to increase its effectiveness.

Graduating engineers who succumb to thinking of success only in terms of traditional technical and managerial project teamwork may experience a rude awakening when they are told upon employment that safety is as important as production, quality, and employee morale—and that if a job cannot be done safely, it will not be done.

Although the incorporation of safety and health issues in the crowded engineering curriculum is not an easy task, the rewards to the public and to industry are so great that every effort should be made to successfully train engineers who are capable of applying the laws of nature to the advancement of technology safely, with minimum interference to the environment.

References

1. Heuer, Charles R., No liability insurance? Limit your exposure to legal claims, *J. of Management in Engineering*, ASCE, Vol. 3, No. 1 pp. 3–7 (Jan. 1987).
2. Cowan, P. A., and J. K. Rao., Disaster abatement and control in the chemical process industry through comprehensive safety systems engineering and emergency management: Lessons from Flixborough and Bhopali, *Proc.*, World Conference on Chemical Accidents, Institute Supervisor di Sainta, Rome, pp. 7–10 (July 1987).
3. Talty, J. T., and J. B. Walters, Integration of safety and health into business and engineering school curriculum, Professional Safety, *Proc.* ASSE, pp. 26–32 (Sept. 1987).
4. Geneen, H., *Managing*, Doubleday, New York (1984).

5. U.S. Chamber of Commerce, Annual report (1982).
6. AIChE, *Guidelines for Hazard Evaluation Procedures* (1985).
7. Bella, D. A., Organizations and systematic distortion of information, *J. of Professional Issues in Engineering*, ASCE, 360, Vol. 113, No. 4 (Oct. 1987).
8. Hershey, P., and K. H. Blanchard, Life cycle theory of leadership, *Training and Development J.* (May 1969).
9. Hill, R., Classifying executives' management style, *IEEE Engineering Management Review* (June 1987).

Chapter

2

Legislation, Regulations, and Standards

The responsibility for occupational and environmental safety, for both workers and the public, has been assigned to employers by laws and regulations. The first three sections of this chapter discuss the provisions of the Occupational Safety and Health Act of 1970, OSHA's Hazard Communication Standard, and OHSA's Hazardous Waste Operations and Emergency Response Standard. Attention then focuses on provisions of Title III of the Superfund Amendments and Reauthorization Act (SARA), commonly known as the Community Right-to-Know Act, followed by an introduction to the Resource Conservation and Recovery Act (RCRA) of 1976.

Occupational Safety and Health Act of 1970

Background and History

The federal government entered the safety and health area in 1890 by passing legislation for safety standards in coal mines. However, no comprehensive effort was made to protect the safety and health of workers in their workplace until the late 1960s, when an estimated 14,000 workers were dying annually and 2.2 million were suffering disabling injuries as a result of work-related accidents.[1] At the same time, the annual cost of lost time injuries was estimated at $1.5 billion to industry and $8 billion in the gross national product. After over 20 years of congressional debate and controversy, on the afternoon of December 17, 1970 the Occupational Safety and Health Act was passed by the Congress and was made ready for President Nixon's signature. At the signing ceremony, the president said, "This bill represents in its culmination the American system at its best." The law, which went into effect on April 28, 1971, directs the secretary of labor to set forth and enforce safety and health standards for any employer who is involved in a business affecting commerce and has one or more employees. The term "employer" does not include the United States or any state or political subdivision of a state. The term "employee" means anyone who is employed by an employer in a business that affects commerce. The Department of Labor estimated that some 57 million workers were affected by the act.[1,2]

The act created the position of assistant secretary of labor for occupational safety and health to carry out the responsibilities of the secretary of labor under the act and to head the Occupational Safety and Health Administration (OSHA). OSHA's activities were organized and coordinated in a national office in Washington, D.C., and regional offices and area offices throughout the nation. OSHA's major responsibilities include enforcement of all provisions of the act, setting forth standards and regulations, overseeing state plans, and employer–employee training.

As a result of the creation of OSHA within the Department of Labor, two other agencies came into existence: the Occupational Safety and Health Review Commission (OSHRC) and the National Institute for Occupational Safety and Health (NIOSH).[2,3]

OSHRC is independent of the Department of Labor, and its major responsibility is to review and decide on disputes between OSHA and employers. The commission is comprised of three members, who are appointed by the president for six-year terms. The chairman of the commission is given the power to appoint judges who will hear and adjudicate any disputes.

NIOSH is an agency within the Department of Health and Human Services (HHS). The major responsibility of NIOSH is to develop research and experimental programs to promote the understanding and prevention of occupationally related diseases.

Major Objectives of the Act

A principal objective of the act was to ensure a safe and healthy working environment for all workers. This goal is accomplished by promulgation and enforcement of safety and health standards and regulations, by research conducted into occupationally related diseases, and by the training of employers and employees in creating a safe and healthy work environment. The major objectives of the act can be summarized as follows:[1–4]

- To encourage identification and elimination of safety and health problems in the workplace.
- To define employer and employee responsibilities and rights in creating a safe and healthy work environment.
- To keep track of work-related accidents, fatalities, and injuries by enforcing the recordkeeping and reporting provisions of the act.
- To set forth and enforce safety and health standards and regulations.
- To encourage states to create their own occupational safety and health programs, which must be at least as effective as the federal program.
- To make operations more cost-effective by eliminating industrial accidents and lost time injuries.

OSHA Standards

OSHA's standards fall predominantly into four major categories, as described below.

Interim or National Consensus Standards

Prior to OSHA's existence, several safety and health guidelines were developed by organizations such as the American National Standard Institute (ANSI) and the National Fire Protection Association (NFPA) and by such acts as the Fair Labor Standard Act. When OSHA came into existence, these national consensus standards were adopted and promulgated immediately.[1,4]

Permanent Standards

The U.S. Congress had realized that occupational safety and health information and knowledge are part of a fast-changing field; so it made provisions in the act by which permanent standards could replace the interim standards. In creating new permanent standards, Congress developed procedures to give affected employers and employees a voice in the standard-setting process.[1]

General Standards

These standards cover hazards that are common to many industries. For example, standards covering worker safety in fire prevention fall under this category.[3]

Emergency Temporary Standards

In emergency situations or when the health or safety of employees is in imminent danger, OSHA has the authority to create temporary emergency standards. These standards were not meant to take the place of permanent standards, and their use is intended for limited situations where the health or safety of employees is in grave danger, such as the possibility of exposure to a highly toxic chemical.[1,4]

Employer Responsibilities

Under OSHA regulations, an employer is responsible for:[1,3,5]

1. Creating and enforcing a safety and health program.
2. Complying with OSHA's regulations on recordkeeping, posting, and reporting.
3. Providing and maintaining any personal protective equipment (PPE) needed for the safe conduct of any operations in the workplace.
4. Ensuring that the concentration of any hazardous and toxic material in the work environment is monitored by appropriate monitoring devices.
5. Informing employees about any safety or health hazard in the workplace. The Hazard Communication Standard (discussed in the next section) puts stringent requirements on employers in terms of informing employees on chemical hazards.

Examples of OSHA's General Industry Standards

The Code of Federal Regulations Title 29, Part 1910 (29 CFR Part 1910) includes OSHA general industry standards.[3–5] These standards cover areas that are common to many industrial operations. The following are examples of the areas of coverage of these standards.[3]

Electrical Standards

The regulations that apply to the installation and use of electrical equipment are found in 29 CFR 1910.308 and 1910.309. Although these standards had been adopted by industry years before the creation of OSHA, noncompliance with them constitutes the bulk of citations issued by OSHA. The following is a summary of OSHA's electrical standard requirements:

- All installations must conform with the National Electric Code.
- All electrical equipment must be properly grounded.
- Access space should be provided around electrical equipment.
- All electrical equipment should be tested and approved by at least one nationally recognized laboratory.
- Outlets should be properly wired to provide adequate grounding.
- The switch box doors and control panels must be kept closed.
- No employee should be allowed to perform any electrical maintenance work that should be performed by a professional person familiar with electrical hazards.
- No electrical wiring should be exposed that could induce shock.
- Electrical wirings should not pose a tripping hazard.

Machinery and Machine Guarding Standards

The employer is responsible for ensuring that workers are protected from hazards posed by any machine in the work area. The standard requires that the worker be protected from injury by a machine regardless of his or her own behavior [29 CFR 1910.212(a)(1)]. Most machines can be made safe through engineering and administrative controls to protect employees against hazards.

The possibility of any of the following situations in a work area is a violation of OSHA standards for machine guarding:

- Contact with a moving part during normal operation of the machine.
- Clothing getting caught in any moving part of the machine.

- Random ejection of material during normal operation.
- The machine's being turned on accidentally.
- Controls not easily reached for emergency shutoff.
- Operator able to bypass the guards.
- Not enough space provided for operation and maintenance.
- Insufficient illumination in the area.
- Inadequate ventilation for machines that generate dusts, fumes, vapors, mists, or gases.
- Poor housekeeping in machine shop area.

Standards for Exits

OSHA's standard for exits is contained in 29 CFR 1910.35. This standard requires that all workplaces be constructed and arranged in such a way that in case of an emergency such as fire or explosion, its occupants are protected. Any one or a combination of the following conditions would constitute a violation of this standard:

- Exits that are not clearly marked.
- Doorways that do not lead to safety and are not marked "NOT AN EXIT."
- Stairways and elevator shafts that are not built according to fire resistance rating specifications.
- Emergency exits that are not marked and/or blocked.
- Exits that are covered by draperies, mirrors, and so on.
- Exits that are arranged in such a way that a single fire can block all exits.
- Exits that are kept locked during working hours.

Material Handling and Storage Standards

The standards in 29 CFR 1910.176 deal with the potential dangers involved in the handling of materials and the storage of all items. This standard requires that:

- Sufficient safe clearances be provided for aisles, at loading docks, and in turnaround areas.
- Aisles and passways be kept clear and in good condition.
- Storage of any material be done in such a way that it will not create a hazard.
- Good housekeeping practices be used in all storage areas.
- Clearance signs to warn of clearance limits be installed.
- Covers and/or guards be provided to protect personnel from hazards of open pits, tanks, ditches, and so on.

Walking and Working Surfaces

The standards in 29 CFR 1910.22 describe rules and regulations dealing with requirements for the purchase, construction, operation, inspection, and maintenance of both permanent and temporary facilities that are used as working or walking surfaces or provide access to such surfaces. OSHA states that:

- All places of employment should be kept clean, orderly, and in sanitary condition.
- Floors should be clean, dry, and free of nails, holes, boards, and so forth.
- When mechanical equipment is used, safe clearances must be provided.
- Permanent aisles should be appropriately marked.
- Covers and guardrails should be provided when necessary to protect personnel.
- Placement on any floor or roof of a building of a load greater than that approved by the building official is unlawful.

Personal Protective Equipment

The standard [29 CFR 1910.132(a)] deals with equipment needed to protect employees from potential hazards that cannot be eliminated by engineering or administrative control. Personal protective equipment—such as eyewear, ear muffs and plugs, protective footwear, face shields, headgear, clothing, and respiratory devices—must be provided in a sanitary manner and in good working condition. The employer is responsible for examining all PPE used on the job to ensure that it is of a safe design and in proper condition. The employer must make sure that:

- Employees use the protective equipment.
- Provisions exist for obtaining additional and replacement equipment.
- Protective equipment is inspected on a regular basis and is well maintained.
- Employee-owned equipment is of an approved type and in the proper condition.

Hazardous Materials

OSHA has adopted standards for materials that are toxic, flammable, or reactive. These standards set forth certain rules and regulations for the storage and handling of hazardous materials.

For compressed gas (29 CFR 1910.101), the employer is responsible for routine inspection of all gas cylinders to ensure their safe condition. Because oxygen cylinders can start or greatly contribute to the spread of fires, precautions must be taken in conjunction with the use and storage of compressed oxygen. Flammables should not be stored near oxygen tanks (a 20-foot minimum distance is required), and tank cylinder threads or pressure gage and pipe fittings should not be lubricated because most lubricating materials are petroleum-based and can start a fire.

Acids, alkalis, and solvents can cause severe burns; so the employer is responsible for training employees in their safe use and for providing eyewashes, safety showers, and any PPE necessary to protect the employees against these potential hazards. The employer also must comply with provisions of the Hazard Communication Standard (HCS) (29 CFR 1910.1200).

Fire Protection

The standards in 29 CFR 1910.156 through 165 describe the employer's responsibilities in fire protection. The employer is required to provide and appropriately install firefighting equipment suitable for fighting the class of fires that might be anticipated in a given area.

- *Class A* fires involve materials such as wood, paper, and cloth.
- *Class B* fires involve flammable gases, liquids, and greases, including gasoline and most hydrocarbon liquids.
- *Class C* fires involve fires in live electrical equipment.
- *Class D* fires involve combustible metals such as magnesium, zirconium, potassium, and sodium.

The employer is required to:

- Prepare a written statement summarizing the organization's policy for fire protection.
- Provide training and education for those personnel who might be expected to fight fires.
- Provide and maintain adequate firefighting equipment.
- Provide proper firefighting personal protective equipment.

Fire extinguishers must be properly labeled regarding the class(es) of fire they are capable of fighting. The employer must classify areas as light, ordinary, or extra hazardous.

The employer is also responsible for the selection and provision of fire extinguishers, as well as their installation, inspection, maintenance, and testing. The standard in 29 CFR 1910.157(d)(2) states that, in accordance with these steps, the employer must ensure that the extinguishers are:

- Not blocked.
- Not tampered with.
- Not physically damaged.
- Inspected and recharged at regular intervals, with the dates of inspections shown on a tag.
- Properly maintained.

OSHA also outlines certain general requirements in conjunction with the installation and use of standpipe systems, automatic sprinkler systems, and local fire-alarm signaling systems.[3,5]

Medical and First Aid

OSHA requires that, if the worksite is not within a reasonable distance of a medical facility, one or more employees be adequately trained in first aid procedures. The standard in 29 CFR 1910.151 also requires that medical personnel be available for advice and consultation on matters of plant health. The first aid supplies approved by the consulting physician should be readily available. Where there is a potential for injury to the eyes or body of any person by corrosive or toxic chemicals, suitable facilities such as eyewash stations and safety showers must be provided.

Environmental Controls for Occupational Health

OSHA places a strong emphasis on limiting the exposure of employees to air contaminants and noise through engineering and administrative controls. Engineering control reduces the concentration of contaminants and the level of noise in work areas, whereas administrative control make changes in personnel exposure such as reducing the amount of time that a worker is exposed to relatively high levels of contaminants or noise. If engineering and administrative controls cannot resolve the problem, the employer must provide personal protective equipment to facilitate full compliance.

OSHA has specific requirements regarding the control of such environmental hazards as noise, air contaminants, and radiation. It should be noted that OSHA's Hazard Communication Standard, discussed below, sets stringent requirements for worker exposure to chemical hazards.

Ventilation

OSHA requirements for ventilation for abrasive blasting, grinding, polishing, and buffing operations are contained in 29 CFR 1910.94. OSHA requires that if the air contaminant hazards cannot be eliminated by engineering control procedures, a ventilation system must be installed. The requirements for abrasive blastings are summarized below:

- Keep the concentration of respirable dust or fume in the breathing zone of the abrasive blasting operator below the specified levels (1910.1000).
- Blast-cleaning enclosures should be ventilated in such a way as to maintain a continuous inward flow of air at all openings in the enclosure.
- All exhaust and ventilation systems must conform to the principles and requirements set forth in American National Standard Institute fundamentals governing the design and operation of local exhaust systems.
- Adequate respiratory protection must be provided by the employer.
- Dusts are not permitted to accumulate outside the enclosure.

Table 2.1. Permissible noise exposure limits.

Duration of Exposure, Hours/Day	Sound Level, Decibels
8	90
6	92
4	95
3	97
2	100
1.5	102
1	105
0.5	110
0.5 or less	115

Source: Reference 5.

Noise

OSHA's requirements for occupational noise control are contained in 29 CFR 1910.95.[5] OSHA requires that hearing protection be provided when sound levels exceed those shown in Table 2.1. OSHA also requires that when employees are exposed to sound levels exceeding those in Table 2.1, practical engineering and administrative controls be exercised. If such controls fail, the employer is responsible for providing adequate hearing protection devices. The employer also must maintain a "hearing conservation program" whenever employee noise exposures equal or exceed an 8-hour time weighted average (TWA) of 85 decibels.

Radiation

OSHA has standards for both nonionizing radiation (29 CFR 1910.97(a)2) and ionizing radiation (29 CFR 1910.96b1). The term "nonionizing radiation" refers to the portion of the spectrum commonly known as the radio frequency range, which also includes the microwave frequency region. For normal environmental conditions, OSHA prescribes the following guidelines for electromagnetic energy of frequencies between 10 MHz and 100 GHz: power density—10 mW/cm^2 for periods of 0.1 hour or more; energy density—1 $mW\text{-}hr/cm^2$ (milliwatt hour per square centimeter) during any 0.1-hour period. This guide applies whether the radiation is continuous or intermittent. OSHA's guidelines also include installation of appropriate warning signs.

Ionizing radiation, covered in 29 CFR 1910.96(b)(1), includes alpha rays, beta rays, gamma rays, X rays, neutrons, high speed electrons, high speed protons, and other atomic particles. OSHA's requirements for ionizing radiation include the following:

- The employer must ensure that no individual in a restricted area receives higher levels of radiation than those summarized in Table 2.2.
- The employer is responsible for ensuring that no employee under 18 years of age receives, in one calendar year, a dose of ionizing radiation in excess of 10% of the values shown in Table 2.2.
- The employer is responsible for the provision and use of radiation monitoring devices such as film badges.
- Where there is a potential for exposure to radioactive materials, appropriate warning signs must be installed.

Variances

In certain situations employers may not be in compliance with specific OSHA standards either because they lack the means to do so, they have more stringent criteria than those specified by

Table 2.2. Radiation exposure limits.

Part of Body	Dose, Rems/Quarter
Whole body: head and trunk, active blood-forming organs, lens of eyes, or gonads	1.25
Hands and forearms; feet and ankles	8.75
Skin of whole body	0.5

Source: Reference 5.

the particular standards, or compliance would cause irreparable harm to their business or workplace. Under such circumstances an employer can request that the assistant secretary of labor issue a variance. There is no special format for the variance application, which must be typewritten. There are basically two types of variances for which an employer can apply, temporary and permanent.[1-4]

Temporary Variances

The intent of temporary variances is to give an employer additional time to comply with a standard. The employer must prove that the business is unable to comply with the standard for lack of technical people, structures, or equipment. The employer also must prove that there is an ongoing program that would bring the facility into compliance with the standard. Financial inability to comply with a standard does not constitute grounds for issuance of a temporary variance. This type of variance can be issued for a period of one year, and it can be extended twice, each time for a six-month period.[4]

Permanent Variance

An employer can make a request for a permanent variance from a standard when the business can prove, by a preponderance of the evidence, that the existing methods and procedures in the business render a working place as safe and healthful as the procedures mandated by the standard. A permanent variance can be issued only after a hearing when the affected employees are notified about the proposed variance. A permanent variance can be modified or revoked after six months from the date of its issuance upon an application from either the employees, the employer, or OSHA.

Enforcement of the Standards

The enforcement of the act, which was a point of great controversy during congressional debates, is performed by OSHA compliance officers, who are authorized to conduct inspection of any place of employment.[1,4]

OSHA Inspections

An OSHA compliance officer, upon showing proper credentials, can enter any place of employment covered by the act at any time during working hours for the purpose of conducting an inspection of the premises. The employer, his or her representative, or an authorized representative of the employees can accompany the officer to witness the inspection process. Management is not given an advance notice of the inspection; indeed, the act contains a specific prohibition against advance notification:[4]

> Any person who gives advance notice of any inspection to be conducted under this Act, without authority from the secretary or his designees, shall, upon conviction, be punished by a fine of not more than $1,000 or by imprisonment for not more than six months, or by both.

Under certain circumstances, however, the OSHA area director may determine that an advance notice of the inspection must be given to the management. These situations may arise where there is immediate danger to the safety and health of employees or when special arrangements for the inspection are required.

OSHA's inspections fall into two general categories: regular inspections and special inspections. A regular inspection takes place as a result of OSHA's normal inspection schedule for workplaces. With millions of work establishments in the United States, OSHA usually takes a "worst first" approach with this type of inspection; that is, establishments with a poor safety and health record usually receive top priority on OSHA's regular inspection list.

A special inspection, on the other hand, is an inspection triggered by the request of an employee or employee representative. Although such a request eventually must be put in writing, initially it may start with a phone call. The secretary or his or her designee will decide if there are sufficient grounds for inspection of the premises. If the employee's request regarding an inspection is turned down, the secretary must report this action in writing to the employee who made the request. The employee has a right to appeal the secretary's decision.

To ensure that an inspection has led to the rectification of hazards, the area director may schedule follow-up inspections of the premises. The follow-up inspections are normally scheduled when an employer has been cited for a serious, willful, or repeated violation. Upon a follow-up inspection, if the hazard has been eliminated, the inspector will mark the file accordingly; if the hazard still exists, a notice of failure is given, and additional penalties are imposed.

If an accident results in the death of one employee or the hospitalization of five or more employees, the Department of Labor must be so notified within 48 hours. Although inspections are not automatically ordered as a result of such accidents, usually the entire plant may receive an OSHA inspection.

Citations

At the completion of the inspection of a facility, the compliance officer holds a conference with the employer and states what standards, rules, or regulations are being violated. Based on these findings, the inspector prepares a report for the OSHA area director with recommendations for or against issuance of a citation.[1,4] After reviewing the inspector's report, the area director may issue a citation. The inspector may issue a citation on the spot only if the safety and health of employees is in immediate danger, and must obtain the verbal approval of the area director to issue an expedited citation.

After the issuance of a citation, the employer has a right to appeal it. However, if the employer fails to notify the secretary of an appeal within 15 days of the issuance, the citation is considered to be final. As part of the citation, the secretary specifies a reasonable time by which the violation must be corrected.

If, in the inspector's opinion, a violation is minor and does not have any direct effect on the safety or health of employees, a notice might be issued instead of a citation. The means of rectification of the problem must be identified, and there is no penalty associated with a notice.

The types of violations that OSHA might cite are classified according to the degree of severity:

- Imminent danger violations are those safety and/or health violations that can cause death or serious physical or health harm. For example, the failure to provide proper respiratory protection for personnel who work in a toxic atmosphere could constitute an imminent danger violation. On discovering an imminent danger violation during an inspection, the OSHA inspector asks the employer to eliminate the hazard voluntarily. Although inspectors do not have the authority to close down an operation, they can go to court and obtain a court order to stop operation in a hazardous area.
- Willful violations are classified by OSHA as those in which an employer willfully and knowingly violates a standard, or makes no effort to correct a hazardous situation that has been

found to be a violation of a standard. The penalties for willful violations are up to $10,000 for each violation.

- Repeated violations are regarded as less severe than willful violations. An employer who violates the same standard twice can be cited for a repeat violation. If the employer has been cited once for the violation of a standard, and during another inspection the same standard is violated in another area of the plant, a repeat violation may be issued. The penalties for repeat violations also can run up to $10,000 for each violation.
- Another category for which OSHA can issue a citation is that of serious violations. If, in the opinion of the inspecting officer, there is a substantial probability that death or serious physical or health impairment could result from a given condition, a citation for a serious violation may be issued. For example, the incompatible storage of chemicals, such as storing flammables with oxidizers, amounts to an invitation to a devastating fire that could cause a large number of fatalities. Such conditions may be cited by the inspecting officer as serious violations. The penalty for this type of violation can be up to $1,000 for each violation.
- Nonserious violations are defined as those in which a standard is violated, but the possible consequences of the violation are not death or serious injury. For example, poor housekeeping that creates tripping hazards in a work area is unlikely to cause deaths, and thus is a nonserious violation. The penalty for a nonserious violation can be up to $1,000 for each violation.
- Continuing violations are classified as those that were not corrected within the time frame set in the Labor Department citations. Continuing violations can carry a penalty of up to $1,000 for each day the violation continues.

It should be noted that, besides the civil penalties mentioned above, criminal charges can be brought against employers in certain situations:

- An employer who willfully violates a standard so that the death of an employee results can be fined $10,000 or get six months in prison for his first conviction, with a fine of $20,000 and one year in prison for the second conviction.
- An employee who files a ''false document'' can be fined up to $10,000 or receive six months in prison or both.
- Assaulting, or hampering the work of, an inspector carries a fine of up to $5,000 or a sentence to three years in prison or both.

If the penalties associated with the violations are not paid, the government can file a lawsuit in the federal district court to force payment.

Reports, Recordkeeping, and Posting

Prior to the Act of 1970, no comprehensive effort had been made to keep records and statistics of occupational fatalities and injuries. Such records, even when they were kept, were sketchy at best and were limited to a worker's compensation context. The act mandates that any employer with 11 or more employees must maintain and, upon request, make available to an inspection officer the following forms:[1,3,4]

- *The log and summary of occupational injuries and illnesses (OSHA Form 200)*: Each employer is required to record and maintain a log of recordable injuries and illness on OSHA Form 200, as shown in Figure 2.1. A recordable injury or illness is defined as one that is directly or indirectly related to the worker's occupational environment. If there are no recordable injuries, the employer does not need to maintain the log. All recordable injuries must be recorded within six days of the employer's learning about them.
- *The supplementary record (OSHA Form 101)*: In addition to OSHA Form 200, an employer with 11 or more employees must maintain a supplementary record for each recordable illness

Bureau of Labor Statistics
Log and Summary of Occupational
Injuries and Illnesses

NOTE: **This form is required by Public Law 91-596 and must be kept in the establishment for *5 years*. Failure to maintain and post can result in the issuance of citations and assessment of penalties. *(See posting requirements on the other side of form.)***

RECORDABLE CASES: You are required to record information about every occupational **death**; every nonfatal occupational **illness**; and those nonfatal occupational **injuries** which involve one or more of the following: loss of consciousness, restriction of work or motion, transfer to another job, or medical treatment (other than first aid). ***(See definitions on the other side of form.)***

Case or File Number	**Date of Injury or Onset of Illness**	**Employee's Name**	**Occupation**	**Department**	**Description of Injury or Illness**
Enter a nonduplicating number which will facilitate comparisons with supplementary records.	Enter Mo./day.	Enter first name or initial, middle initial, last name.	Enter regular job title, not activity employee was performing when injured or at onset of illness. In the absence of a formal title, enter a brief description of the employee's duties.	Enter department in which the employee is regularly employed or a description of normal workplace to which employee is assigned, even though temporarily working in another department at the time of injury or illness.	Enter a brief description of the injury or illness and indicate the part or parts of body affected. Typical entries for this column might be: Amputation of 1st joint right forefinger; Strain of lower back; Contact dermatitis on both hands; Electrocution--body.
(A)	(B)	(C)	(D)	(E)	(F)
					PREVIOUS PAGE TOTALS →
					TOTALS (Instructions on other side of form.) →

OSHA No. 200

FOLD

Figure 2.1. Log and summary of occupational injuries and illnesses.

U.S. Department of Labor

For Calendar Year 19 ______ Page ___ of ___

ompany Name	Form Approved O.M.B. No. 1220-0029
stablishment Name	
stablishment Address	

xtent of and Outcome of INJURY						Type, Extent of, and Outcome of ILLNESS												
atalities	Nonfatal Injuries					Type of Illness							Fatalities	Nonfatal Illnesses				
njury elated	Injuries With Lost Workdays				Injuries Without Lost Workdays	**CHECK** Only One Column for Each Illness *(See other side of form for terminations or permanent transfers.)*							Illness Related	Illnesses With Lost Workdays				Illnesses Without Lost Workdays
nter **DATE** of death. o./day/yr.	Enter a **CHECK** if injury involves days away from work, or days of restricted work activity, or both.	Enter a **CHECK** if injury involves days away from work.	Enter number of **DAYS** *away from work.*	Enter number of **DAYS** of *restricted work activity.*	Enter a **CHECK** if no entry was made in columns 1 or 2 but the injury is recordable as defined above.	Occupational skin diseases or disorders	Dust diseases of the lungs	Respiratory conditions due to toxic agents	Poisoning (systemic effects of toxic materials)	Disorders due to physical agents	Disorders associated with repeated trauma	All other occupational illnesses	Enter **DATE** of death. Mo./day/yr.	Enter a **CHECK** if illness involves days away from work, or days of restricted work activity, or both.	Enter a **CHECK** if illness involves days away from work.	Enter number of **DAYS** *away from work.*	Enter number of **DAYS** of *restricted work activity.*	Enter a **CHECK** if no entry was made in columns 8 or 9.
)	(2)	(3)	(4)	(5)	(6)	(7)							(8)	(9)	(10)	(11)	(12)	(13)
						(a)	(b)	(c)	(d)	(e)	(f)	(g)						

INJURIES

ILLNESSES

Certification of Annual Summary Totals By ______________ Title ______________ Date ______________

OSHA No. 200

POST ONLY THIS PORTION OF THE LAST PAGE NO LATER THAN FEBRUARY 1.

Figure 2.1 (Continued)

Bureau of Labor Statistics
Supplementary Record of
Occupational Injuries and Illnesses

U.S. Department of Labor

This form is required by Public Law 91-596 and must be kept in the establishment for *5 years.* Failure to maintain can result in the issuance of citations and assessment of penalties.	Case or File No.	Form Approved O.M.B. No. 1220-0029

Employer

1. Name

2. Mail address *(No. and street, city or town, State, and zip code)*

3. Location, if different from mail address

Injured or Ill Employee

4. Name *(First, middle, and last)* — Social Security No.

5. Home address *(No. and street, city or town, State, and zip code)*

6. Age — 7. Sex: *(Check one)* Male ☐ Female ☐

8. Occupation *(Enter regular job title, not the specific activity he was performing at time of injury.)*

9. Department *(Enter name of department or division in which the injured person is regularly employed, even though he may have been temporarily working in another department at the time of injury.)*

The Accident or Exposure to Occupational Illness

If accident or exposure occurred on employer's premises, give address of plant or establishment in which it occurred. Do not indicate department or division within the plant or establishment. If accident occurred outside employer's premises at an identifiable address, give that address. If it occurred on a public highway or at any other place which cannot be identified by number and street, please provide place references locating the place of injury as accurately as possible.

10. Place of accident or exposure *(No. and street, city or town, State, and zip code)*

11. Was place of accident or exposure on employer's premises? Yes ☐ No ☐

12. What was the employee doing when injured? *(Be specific. If he was using tools or equipment or handling material, name them and tell what he was doing with them.)*

13. How did the accident occur? *(Describe fully the events which resulted in the injury or occupational illness. Tell what happened and how it happened. Name any objects or substances involved and tell how they were involved. Give full details on all factors which led or contributed to the accident. Use separate sheet for additional space.)*

Occupational Injury or Occupational Illness

14. Describe the injury or illness in detail and indicate the part of body affected. *(E.g., amputation of right index finger at second joint; fracture of ribs; lead poisoning; dermatitis of left hand, etc.)*

15. Name the object or substance which directly injured the employee. *(For example, the machine or thing he struck against or which struck him; the vapor or poison he inhaled or swallowed; the chemical or radiation which irriatated his skin; or in cases of strains, hernias, etc., the thing he was lifting, pulling, etc.)*

16. Date of injury or initial diagnosis of occupational illness — 17. Did employee die? *(Check one)* Yes ☐ No ☐

Other

18. Name and address of physician

19. If hospitalized, name and address of hospital

Date of report	Prepared by	Official position

OSHA No. 101 (Feb. 1981)

Figure 2.2. Supplementary record for occupational injuries and illnesses.

or injury. A sample OSHA form 101 is shown in Figure 2.2. The supplementary record also must be filed within six days after the employer learns about a recordable illness or injury. This record provides details on each injury, such as how the accident occurred, and so on. Many companies have been using their own forms to record job-related illnesses and injuries for years. It should be noted that deviations from the Form 101 format are much less serious than deviations from the log format.

- *The annual summary*: An annual summary of occupational injuries and illnesses must be prepared from the last page of OSHA Form 200 by February 1 of each calendar year. The law mandates that the annual summary be posted from February 1 until March 1 in a conspicuous place or where employee notices usually are posted.
- *Records of any employee exposure to toxic or physical agents.*

State OSHA Plans

After passage of the Occupational Safety and Health Act, many states started to develop their own occupational safety and health plans. Some states enforce primarily federal OSHA standards, others primarily state OSHA standards, and some a combination of the two. A complete copy of a state plan must be kept at the office of the state agency administering the plan (29 CFR 1952.5).

The checklist in Table 2.3 may be used as guidance for compliance with OSHA standards.

Table 2.3. Checklist for compliance with OSHA standards.

Description	29 CFR Part
All electrical equipment and wiring conform to National Electric Code (NEC)	1910.309
Abrasive wheel machinery meet the general requirements	1910.215
Portable fire extinguishers are properly maintained, installed, and distributed	1910.157
Exits are clearly marked and meet the general requirements	1910.037
Machines are properly guarded	1910.212
Floors and wall opening holes are properly guarded	1910.023
General requirements for housekeeping are met	1910.022
Recordkeeping and posting requirements are met	1903.002
Hand and portable power tools are properly grounded or have double insulation	1910.242
Fire extinguishers are properly mounted	1910.157
General requirements for floor loading are met	1910.022
Flammable and combustible liquids are properly handled and stored	1910.106
Stairways have proper railing	1910.023
Log of occupational injuries and illnesses is properly prepared and maintained	1904.002
Annual summary report is properly prepared and posted	1904.005
General requirements for personal protective equipment are met	1910.132
Woodworking machinery requirements are met	1910.213
General requirements for first aid procedures are met	1910.151
General requirements for dimensions of portable wood ladders are met	1910.025
General requirements for material handling and storage are met	1910.176
Exits are not obstructed	1910.036
Toilet facilities meet general requirements for sanitation	1910.141
Aisles and passageways are properly marked	1910.022
Welding, cutting, and brazing operation and maintenance of ARC welding and cutting equipment are proper	1910.252
Approved containers are used for handling of flammable and combustible liquids	1926.152
Ladders meet the general requirements	1926.450
Overhead and gantry cranes meet inspection requirements	1910.179
General ventilation requirements are met	1910.094
Compressed gases have proper safety valve are stored properly	1910.101

Exercises

1. What are the major responsibilities of OSHRC and NIOSH?
2. Describe four types of OSHA standards? Briefly describe the major objectives of each.
3. What are the requirements for the issuance of temporary and permanent variances?
4. Name and briefly discuss the general categories of OSHA inspections.
5. What is the major difference between an OSHA citation and an OSHA notice?
6. List and briefly discuss different types of OSHA standard violations.
7. What specific forms must the employer maintain to comply with OSHA requirements for recordkeeping and posting?

OSHA's Hazard Communication Standard (29 CFR 1910.1200)

On November 25, 1983 the Federal Occupational Safety and Health Administration issued a new rule called Hazard Communication Standard (HCS), also known as the ''Workers' Right to Know.''[5,7–12] The major objective of the new standard was to ensure that workers would be protected from physical and health hazards of chemicals in their workplace. The standard requires that the manufacturers, importers, and distributors of the hazardous chemicals in Standard Industrial Classification (SIC) codes 20 through 39 (Table 2.4), must identify the physical and health hazards of their chemicals. This information must then be transmitted to employers in the form of container labels and material safety data sheets (MSDSs). Under provisions of the standard, employers must establish a program to properly communicate the information on hazards of chemicals to their employees.

Life without chemicals would be impossible; all the world is made of chemicals. Although properly handled chemicals can serve a useful, even vital, role in society, the mishandling of hazardous chemicals can cause fatalities, serious health problems, fires, and property damage. For example, the hazardous chemical vinyl chloride, when handled safely, can produce polyvinylchloride (PVC), which is used in a variety of applications. However, when vinyl chloride is mishandled, it can create serious health hazards, having been implicated in both cancer and birth defects.

Table 2.4. Standard industrial classification.

Code	Group
20	Food and kindred products
21	Manufacturers of tobacco
22	Textile mill products
23	Apparel and other textile products
24	Wood and lumber products
25	Fixtures and furniture
26	Paper and allied products
27	Publishing and printing
28	Chemicals and allied products
29	Petroleum and coal products
30	Rubber and plastic products
31	Leather and products of leather
32	Glass products, stone and clay
33	Primary metal industries
34	Fabricated metal products
35	Machinery, except electrical
36	Electrical equipment and supplies
37	Transportation equipment
38	Instruments and related products
39	Miscellaneous manufacturing products

As industry has moved to enhance the quality of life through advancing technology and the introduction of new products, questions have been raised about legal liability, as well as the professional responsibility of members of industry for the public welfare. The public and employee groups have pressured industry to enhance their quality of life by advancing technology safely; and in recent years, this public pressure has resulted in strict occupational and environmental laws and regulations for industry.

Years ago, no laws existed to protect workers from the dangers of chemicals they faced at their workplace. Container labels and information about chemicals, even when they were provided, were sketchy at best, lacking such critical safety information as what to do in an emergency or what type of personal protective equipment was needed.

OSHA's Hazard Communication Standard (HCS) clarified this confusion by defining what information people needed in order to work safely with hazardous chemicals, and how this information should be provided and communicated to the workers. Under provisions of the HCS, chemical manufacturers, importers, and distributors must identify the physical and health hazards of chemicals they manufacture, and they must communicate this information to the users of those chemicals in the form of container labels and MSDSs.

Coverage and Responsibilities for Compliance

The main objective of the HCS was to ensure that the hazards of all chemicals produced or imported were evaluated, and that information about those hazards was transmitted to employers and employees in the manufacturing sector. This goal must be accomplished by means of a comprehensive hazard communication program, which includes container labeling and other forms of warning, MSDSs, and employee training. The affected employers and employees are those in the Standard Classification Codes (SIC) 20 through 39 (Table 2.4), and the provisions of the standard apply to any hazardous chemical that is present in the workplace in such a manner that employees may be exposed either under conditions of normal use or in an emergency.

Supplier Responsibilities

The HCS mandated that manufacturers and importers of chemicals identify the hazards of the chemicals they produce. The employers are not required to evaluate hazards unless they choose not to rely on manufacturers' information. The standard requires that each container of a hazardous chemical be labeled by the manufacturer.

The employer must ensure that all containers of hazardous chemicals in the workplace are properly labeled. Although the label must be written in English, a second label in a different language may be added to the container, if appropriate.

The standard also mandated that manufacturers or importers of hazardous chemicals provide an MSDS for every hazardous chemical they manufacture or import. Although OSHA does not set a specific format for the MSDS, the following information must be included:

- Hazardous ingredients.
- Physical and chemical characteristics, such as boiling point and vapor pressure.
- Potential for hazards such as fire, explosion, or reaction.
- Possible health hazards and symptoms.
- Pathways into the body such as inhalation or skin exposure.
- Safe exposure limits and carcinogenic potential.

The employer is responsible for ensuring that copies of MSDSs are available to employees at all times for every hazardous chemical in the workplace. It is imperative that the chemical identity on the MSDS and on the hazardous materials list be the same to facilitate locating a hazardous chemical. The MSDS describes how to use the chemical safely, in terms of work procedures, hygiene, and protective equipment. Also, the MSDS explains what to do about spills, leaks, or emergencies, and it describes first aid measures for exposure or accidents.

The MSDSs must be updated by manufacturers within three months after they obtain new hazard information about chemicals. The supplier must send a revised MSDS to each customer with the next hazardous chemical shipment. Importers and distributors are permitted to rely on the original manufacturer for the initial hazard determination and preparation of labels and MSDSs, as well as the information from the updates.

The manufacturer also must determine the specific hazards of ingredients of a chemical mixture and make this information available to its users through the MSDS. In the evaluation of chemical hazards, manufacturers must establish, in writing, the procedures and/or test methods used in hazard identification. This written document may be included in a "Written Hazard Communication Program," which is required under the law. The written document must be made available, upon request, to employees or their representatives.

Employer and Employee Responsibilities

Under provisions of the HCS, employers are responsible for developing a "Written Hazard Communication Program," which must describe the following:

- How employees are informed about the provisions of the HCS.
- How the HCS is being put into effect in their workplace.
- Ways to train employees to recognize and understand labels and MSDSs and to work safely with chemicals.

The written program also must include:

- A list of hazardous chemicals in the workplace. This list must use the same identity for a chemical as appears on the MSDS.
- The methods and procedures that the employer uses to inform employees of the hazards involved in nonroutine tasks, such as the hazards present for chemicals flowing through pipes.
- Methods used to inform contractors of chemical hazards and suggestions for protective measures.

Every organization must establish a training program to ensure that its employees are provided information about the provisions of the HCS, and that they are properly trained to use and understand labels and the MSDS. The employees must be informed about hazards of chemicals at the time of their initial assignments, and whenever a new hazard is introduced in the work area. The employees also must be informed of:

- Any operation in their work area where hazardous chemicals are present.
- The location and availability of the written hazard communication program, including the required list of hazardous chemicals and MSDSs.

Employees must be trained on how to recognize hazardous chemicals and how to protect themselves against them. The training program must explain:

- How to detect the presence or release of a hazardous chemical in a workplace, such as the use of chemical monitoring devices.
- The physical and health hazards of chemicals in the workplace.

- Methods by which employees can protect themselves against the hazards of chemicals, such as the use of proper protective equipment and emergency procedures.
- The use of labels and MSDSs, and how valuable information about chemical hazards can be extracted from these sources.

Employees are responsible for reading the labels and the MSDSs, heeding any warnings, and following any handling instructions.

Labeling

The HCS requires that all containers of hazardous chemicals be labeled by the manufacturer, importer, or distributor. Although OSHA does not require any special format for labels, the following information must be provided:

- Identity of the hazardous chemicals.
- Appropriate hazard warnings such as CAUTION or DANGER.
- Name and address of the chemical manufacturer, importer, or other responsible party.

The chemical manufacturer must ensure that any container of hazardous chemical has an OSHA specification label that does not conflict with the requirements of the Hazardous Materials Transportation Act. The employer must not remove the labels on the incoming or existing containers of hazardous chemicals unless another label that satisfies OSHA requirements is attached to the same container. The label must be written in English and be legible. However, in areas where the employer has a large number of non–English-speaking workers, a second label in their language may be provided.

The American National Standards Institute (ANSI) published a voluntary labeling standard (ANSI Z129.1-1982) that suggests including the following nine items on the label:

1. Chemical identity.
2. A signal word such as CAUTION, WARNING, or DANGER.
3. A statement of the physical and health hazards.
4. Precautionary measures for using the product, such as wearing rubber gloves, goggles, or respirators.
5. Instructions in case of contact or overexposure, such as whether to induce vomiting in case of swallowing.
6. Antidotes for poisoning.
7. Notes to physicians as to emergency treatment recommended.
8. Instuctions in case of fire and spill or leak.
9. Instructions for container handling and storage, such as keeping it in a cool place, away from fires, and away from strong acids.

Another form of warning label in widespread use is the NFPA 704M system. A sample of the 704M label is illustrated in Figure 2.3. This system can give an indication of the severity of fire hazards in four different categories: flammability (red background), dangerous reactivity or explosivity (yellow background), general health hazard (blue background), and any special information about the chemical (white background).

The NFPA system does not classify the hazards to the extent required by the HCS, but it does provide at a glance a good visual warning of how hazardous a material is. The numerals used in each of the three hazard rating areas are 0, 1, 2, 3, and 4, with 0 representing no hazard, 4 a very serious hazard, and the other numbers intermediate degrees of hazard severity. In addition, a symbol representing special hazard problems, such as water reactivity, dangerous polymeriz-

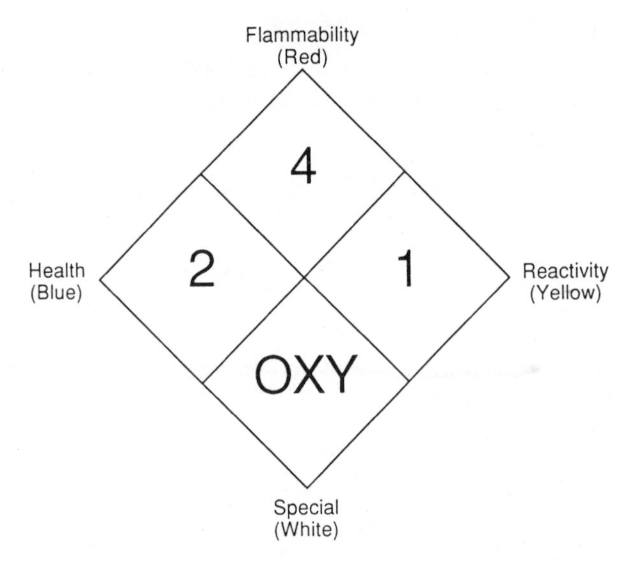

Figure 2.3. National Fire Protection Association (NFPA) 704 labeling system.

ability, oxidizer, or radioactivity, can be placed in the bottom section of the symbol. Department of Transportation (DOT) symbols such as those shown in Figure 2.4 are finding widespread use on containers of hazardous chemicals.

Labeling Exemptions

Items regulated by governmental agencies other than OSHA are exempt from OSHA's labeling requirements, including the following:[10]

- Foods, food additives, drugs, or cosmetics, including flavors and fragrances, if they are subject to the labeling requirements and regulations of the Food and Drug Administration under the Federal Food, Drug, and Cosmetic Act.
- Alcoholic beverages that are not intended for industrial use, if they are subject to the labeling requirements and regulations of the Treasury Department's Bureau of Alcohol, Tobacco, and Firearms.
- Consumer products or hazardous substances defined in the Consumer Product Safety Act and the Federal Hazardous Substances Act, if they are subject to a consumer product safety standard or the labeling requirements or regulations of the Consumer Product Safety Commission.
- Pesticides as defined in the Federal Insecticide, Fungicide, and Rodenticide Act (FIFRA), if they are subject to labeling requirements or regulations of the U.S. Environmental Protection Agency (EPA).
- Tobacco and tobacco products.
- Portable containers of hazardous chemicals that are transferred from a labeled container and are intended for the immediate use (i.e., same shift) of the employee who transfers the chemical. That is, a supervisor who transfers a hazardous chemical from a labeled container into a pail can use the pail without label as long as it is under his or her control at all times, and he or she empties the pail by the end of the workshift. However, if the same supervisor hands the pail to a worker, then it must be properly labeled.
- Hazardous wastes defined by RCRA and controlled under EPA regulations. The EPA reg-

Figure 2.4. Some Department of Transportation labels.

ulations for such wastes cover labeling, manifesting, emergency provisions, contingency plans, and employee training.

- Wood and wood products, even when the wood is treated with hazardous preservatives and other chemicals.

Protection of Trade Secrets

In order to protect the business interests of manufacturers, importers, and employers, the act provides for shielding a chemical's identity when it can support the manufacturer's trade secret claim. Under such circumstances the identity of the chemical need not be revealed on the MSDS. However, any physical or health hazards posed by the chemical must be mentioned along with means of protecting workers from the hazards of the chemical.

In medical emergencies when a health professional needs a chemical's identity to treat a patient, the manufacturer must immediately disclose the identity to the health professional without requesting a written confidentiality agreement. However, a confidentiality agreement may be executed later, when circumstances permit it. A health professional, as defined by OSHA, is a physician, an industrial hygienist, a toxicologist, or an epidemiologist. In nonemergency situations, a chemical manufacturer, importer, or employer must disclose the identity of a chemical that is protected by the trade secret provisions of the act if a written request is made, and the basis for the request is one of the following occupational health needs:[5]

- To determine the hazards of a chemical to which employees will be exposed.

- To conduct sampling of the workplace atmosphere to determine employee exposure levels.
- To conduct periodic surveillance of exposed employees or to provide treatment to exposed employees.
- To determine the proper personal protective equipment for exposed employees.
- To design or assess engineering controls for exposed employees.
- To conduct studies to determine the health effects of exposure.

Physical Hazard Identification

Chemical hazards fall into two broad categories: physical hazards and health hazards. Materials that present physical hazards and are covered by the HCS include combustible liquids, flammable materials, all compressed gases, explosives, organic peroxides, oxidizers, pyrophoric materials, unstable (dangerously reactive) materials, and water-reactive materials. Physical hazards are discussed in the following paragraphs.

Fire Hazards

Fire hazards are those chemicals that have the potential for creating a fire or aiding an ongoing fire. These materials are flammables, combustibles, oxidizers, pyrophoric materials, and organic peroxides.

Flammable Liquids

A flammable liquid, by definition, is a liquid whose flash point is below 100°F (38°C). The flash point is defined as the minimum temperature at which enough vapor is generated in an open container to support combustion in the presence of a source of ignition. Examples of flammable liquids are gasoline, acetone, turpentine, and ethyl alcohol. Flammable liquids with the lowest flash points pose the greatest fire threat. For example, a liquid with a flash point of −40°F is more dangerous than a flammable liquid with a flash point of +80°F. In general, special care must be exercised in handling flammable liquids, as open containers of flammable liquids can generate a fire at or near room temperature. Special precautions also must be taken for the storage of flammable liquids. Because these liquids can generate a fire at relatively low temperatures, all sources of ignition must be eliminated in a flammable-liquid storage area.[6] Also, because flammable liquids are capable of generating static electricity due to friction between different liquid layers, and the spark generated by the static electricity can act as a source of ignition, proper bonding and grounding techniques must be used where flammable liquids are transferred, stored, or handled.

Flammable Gases

In order to understand what a flammable gas is, we first must define the lower flammability limit (LFL) and the upper flammability limit (UFL). The LFL is defined as the minimum concentration, by volume, of a gas in air that will burn in the presence of a source of ignition, whereas the UFL is the maximum concentration, by volume, of a gas in air that will burn in the presence of a source of ignition. Any mixture of a gas and air below the LFL is said to be too lean to burn, and any mixture of a gas and air above the UFL is said to be too rich to burn. All concentrations of a gas and air that lie between the LFL and the UFL will support combustion when a source of ignition is present. Figure 2.5 shows a typical explosive range for an air–gas mixture.

A flammable gas can be defined in terms of the LFL and the UFL. A flammable gas would satisfy at least one of the following criteria: (a) a gas that at ambient temperature and pressure has an LFL of less than 13% by volume in air; (b) a gas that at ambient temperature and pressure has a UFL that exceeds its LFL by more than 12% by volume. For example, a gas with an LFL

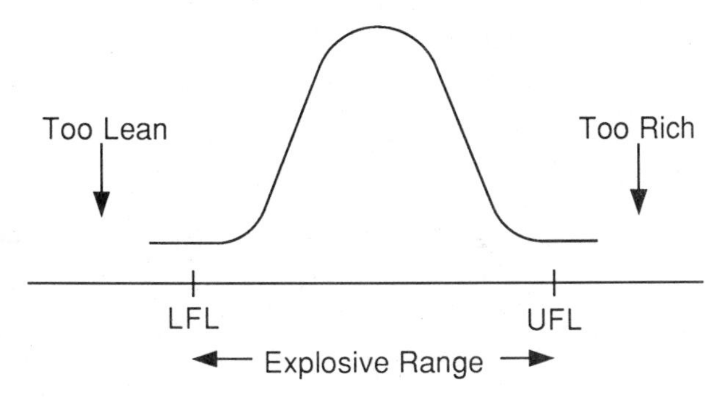

Figure 2.5. Explosive range for a gas.

of 9% is considered flammable because its LFL is less than 13%; and a gas with an LFL of 15% and a UFL of 30% also is considered flammable because the difference between its UFL and its LFL exceeds 12% although its LFL is above 13%.

It is extremely important to report both temperature and pressure when reporting gas flammability data, as the volume of a gas is a strong function of these two variables. Because flammability data are reported in terms of volume percent, the UFL and the LFL for a given gas can be quite different at different temperatures and pressures.

Table 2.5 shows flash points and limits of flammability in air for some common chemicals.[11]

Flammable Solids

A flammable solid ignites and burns with a self-sustained flame at a rate of at least 0.1 inch per second along its major axis. Examples of flammable solids include magnesium and nitrocellulose film. This category of hazard does not include blasting agents or explosives.

Combustible Liquids

A combustible liquid, by definition, is any liquid with a flash point between 100°F (38°C) and 200°F (93°C). Combustible liquids present a smaller fire hazard than flammable liquids simply because they ignite at higher temperatures. Combustible liquids must be handled and stored with

Table 2.5. Flash points and flammability data for some common hydrocarbons.

Substance	Flash Point, °F	LFL, Vol %	UFL, Vol %
Acetone	0	2.6	12.8
Acetylene	gas	2.5	81
Ammonia	gas	16	25
Benzene	12	1.4	7.1
Ethyl alcohol	55	4.3	19
Methane	gas	5.3	14
Methyl alcohol	52	7.3	36
Phenol	175	—	—
Toluene	40	1.4	6.7
Normal heptane	25	1.2	6.7
Ethylene oxide	0	3	100
Propane	gas	2.2	9.5

Source: National Fire Codes, vol. I, ''Flammable liquids and gases, NFPA.

caution. For example, an open container of a combustible liquid with a flash point of 105°F poses a severe fire hazard when the ambient temperature is above 105°F. Examples of combustible liquids are fuel oil, diesel fuel, and phenol.

Oxidizers

An oxidizer is a chemical compound that is capable of initiating or promoting a fire in other compounds through the release of oxygen or other gases. The most common oxidizer is oxygen, which is a necessary component of any fire. Other oxidizers such as hydrogen peroxide, potassium permanganate, nitric acid, and chlorine also can promote combustion reactions.

Pyrophoric Materials

A pyrophoric material is a substance that can be ignited as a result of contact with oxygen in the absence of an ignition source at temperatures below 130°F. Examples of pyrophoric materials are alkyls of group III metals such as trimethylaluminum, trimethylgallium, and trimethylindium.

Organic Peroxides

Organic peroxides contain both fuel, in the form of carbon, and excess oxygen and thus can pose a severe fire hazard. In a peroxide, two oxygen atoms are directly attached to each other, creating a potentially unstable compound.

Compressed Gases

All compressed gases pose an additional physical hazard. If the valve on a compressed gas cylinder ruptures or breaks, the cylinder easily can turn into a powerful rocket, endangering everything in its path. Therefore, all compressed gas cylinders must be properly chained and stored.

A compressed gas may be defined by any of the following criteria: (a) a gas or gas mixture in a container having an absolute pressure exceeding 40 psia at 70°F (21.1°C); (b) a gas or gas mixture in a container having an absolute pressure exceeding 104 psia at 130°F (54.4°C), regardless of the pressure at 70°F (21.1°C); (c) a liquid having a vapor pressure exceeding 40 psi at 100°F (37.8°C). Vapor pressure, which in a strong function of temperature, is the pressure exerted by the vapor on the surface of a liquid and the walls of a closed container when the liquid and vapor phases have reached a state of equilibrium.

Explosive Materials

Explosives are compounds that can be decomposed in a violent chemical reaction with the production of heat, pressure, and large quantities of gas. Pressure, ignition, or a high temperature can initiate such a chemical reaction. Examples of explosives are trinitrotoluene (TNT) and nitroglycerin.

Unstable Materials

Certain compounds in their pure form can undergo vigorous decomposition or polymerization under moderate conditions of pressure, shock, or temperature. These materials, which are called unstable compounds, pose a hazard by virtue of their instability. Some unstable compounds also pose a fire hazard. For example, polymerization is an exothermic reaction, generating large quantities of heat. The heat produced by such an exothermic reaction, if not removed, will raise the temperature and increase the rate of the polymerization reaction, causing even more heat to be generated. This condition, known as a runaway reaction, can raise the temperature above the flash point of the reactants and result in a severe fire or explosion.

Water-Reactives Compounds

These materials can react vigorously with water to produce a toxic or flammable gas. Water-reactives include such compounds, as sodium, potassium and acetic anhydride.

Health Hazard Characterization

Under the provisions of the HCS, chemical manufacturers have to determine the health hazards of any chemical they manufacture and communicate this information to the users of the chemical. The health hazard determination is complicated by a number of factors. A particular chemical may produce inconsistent health effects for members of any given population because of such factors as age, sex, and condition of health. Also, many occupationally related diseases may be due to nonoccupational exposures in the population, making it difficult to separate the effects of exposure. This matter may be further complicated by a lack of scientific data on the effects of human exposure compared to laboratory animal exposure for a given chemical.

Chemical health hazards can be divided into the two broad categories of acute and chronic hazards. Acute health hazards usually cause a measurable effect soon after a single exposure. Chronic health hazards, on the other hand, produce permanent health effects as a result of repeated exposures over a long period of time. Although it is not possible precisely to define and quantify occupational health effects, workers should be informed about known or potential chemical health hazards in the workplace.

Several excellent references are available for determining the occupational health hazards of chemicals, including the following:

- *The Condensed Chemical Dictionary* published by the Van Nostrand Reinhold Company.
- *The Merck Index, An Encyclopedia of Chemicals and Drugs*, published by Merck Co. Inc.
- *The Registry of Toxic Effects of Chemical Substances*, published by the U.S. Department of Health and Human Services.
- Appendix C of the 29 CFR 1910.1200, which lists additional references for health hazard determination.

Health hazards can be classified into the groups described in the following paragraphs.[10]

Corrosives

A corrosive chemical is capable of causing visible destruction or irreversible alteration in a living tissue by chemical action at the site of contact. The corrosivity of a chemical usually is determined by conducting skin tests on laboratory animals. Examples are sulfuric acid and sodium hydroxide.

Irritants

An irritant chemical can cause reversible inflammation upon contact with living tissue. An example is nitric oxide.

Toxic Chemicals

Toxicity normally is determined by conducting tests on laboratory animals. In these tests, the animals are exposed to a given chemical through either inhalation, ingestion, or skin absorption. The amount of chemical administered per unit weight of the organism is known as the dose. The dose of a chemical that can kill half of the test animals is called the lethal dose fifty, or LD_{50}. When the route of exposure is by inhalation, the concentration of chemical in air that kills half of the test animals is called the lethal concentration fifty, or LC_{50}.

A chemical is defined as a highly toxic agent if it satisfies any of the following criteria:

- An oral LD_{50} of less than 50 milligrams per kilogram (mg/kg) of body weight in rats.
- A skin LD_{50} of 200 mg/kg of body weight, or less, in rabbits.
- An LC_{50} of 200 parts per million (ppm), or less, in rats.

A chemical is classified as a toxic agent if it satisfies any of the following criteria:

- An oral LD_{50} in rats of more than 50 mg/kg but less than 500 mg/kg of body weight: 50 mg/kg < oral rats LD_{50} < 500 mg/kg.
- A skin LD_{50} of more than 200 mg/kg but less than 1,000 mg/kg of body weight in rabbits: 200 mg/kg < skin rabbit LD_{50} < 1,000 mg/kg.
- A lethal concentration in air of more than 200 ppm but less than 2,000 ppm administered to rats for one hour: 200 ppm < rats LC_{50} < 2,000 ppm.

Skin Hazards

Skin hazards are chemicals that can affect the dermal layer of the body by defatting the skin, resulting in rashes and irritation. Examples of this class of chemicals are the ketone family and chlorinated hydrocarbon compounds.

Eye Hazards

An eye hazard is a chemical that can adversely affect the eye or diminish the visual capacity of a human. Examples of eye hazards are organic solvents, acids, and bases.

Blood or Hematopoietic System Hazard

Hazards of the blood system are caused by chemicals that decrease the hemoglobin function; as a result, body tissues are deprived of oxygen. Symptoms are usually cyanosis and loss of consciousness. Examples are carbon monoxide and the cyanides.

Carcinogens

Carcinogens are chemicals that have been known to cause cancer or have the potential of causing cancer in humans, as recognized by any of the following:

- The International Agency for Research on Cancer (IARC).
- The annual report published by the National Toxicology Program (NTP) of the U.S. Public Health Service.
- Regulation by OSHA as a carcinogen (29 CFR 1910 Subpart Z).

Examples of carcinogens are benzene and vinyl chloride.

Sensitizers

Sensitizers do not create an adverse effect on first exposure, but on subsequent repeated exposures they can cause an allergic reaction to them to develop in living tissue.

Target Organ Chemicals

Target organ chemicals can enter the bloodstream and adversely affect vital organs. For example, halogenated hydrocarbons can enter the bloodstream through skin absorption or through broken skin, and having thus entered the body can adversely affect the function of the kidney.

Neurotoxins

Neurotoxins are chemicals that produce their toxic effects primarily on the central nervous system. Their symptoms include narcosis, behavioral changes, and a decrease in normal functions. Mercury and carbon disulfide are examples of these chemicals.

Nephrotoxins

Nephrotoxins are chemicals that can produce toxic effects on kidneys, with symptoms that include edema and proteinuria. Uranium and halogenated hydrocarbons are examples of these agents.

Reproductive Toxins

Reproductive toxins are chemicals, such as lead and vinyl chloride, that have the potential of adversely affecting the reproductive system by the creation of mutations or teratogenesis, which can produce adverse effects on fetuses, sperm, or egg cells.

Hepatotoxins

Hepatotoxins are chemicals that can adversely affect the liver. Examples include carbon tetrachloride and ethyl alcohol.

Lung Hazards

Among agents that damage the lung are such chemicals as asbestos and silica, which can irritate or damage the pulmonary tissue resulting in coughing, tightness in the chest, and shortness of breath.

Mandatory Requirements for Hazard Evaluation

The provisions of the HCS do not require chemical manufacturers and importers to follow any specific methods or procedures in determining the health hazards of chemicals they manufacture or import. The standard does, however, mandate that such determination be thorough and scientifically defensible. OSHA states that any hazard determination conducted by manufacturers or importers of chemicals must at least satisfy the following requirements:[5]

- A chemical should be classified as a carcinogen if a determination by the National Toxicology Program, the International Agency for Research on Cancer, or OSHA has revealed that the chemical is a known or potential carcinogen.
- Human data such as epidemiological studies and case reports of adverse health effects, where available, must be considered in hazard determination.
- Because comprehensive toxicological data on humans are not generally available, the results of tests on laboratory animals must be considered in the hazard determination of chemicals.

Material Safety Data Sheet (MSDS)

The MSDS, which is the worker's guide to the safe use of a chemical, contains valuable information on the physical and health hazards of the chemical and means of protecting against them. It contains safe procedures to be followed in case of a spill or an emergency. The MSDS should be consulted before one works with any chemical, not just during an emergency.

The HCS requires that a copy of the MSDS must be available for every hazardous chemical in the workplace. The law also requires that copies of the MSDS be available to employees at all times. The employees must be trained in reading and understanding labels and the MSDS.

The Hazard Communication Standard has put a considerable burden of responsibility on employers to ensure that workers are protected from the hazards of chemicals in the workplace. It is up to the employees, however, to read the label and the MSDS before working with any chemical, and to follow safe work practices. The best way for a worker to find out whether a chemical is hazardous is to check the hazardous chemical list. If the name of the chemical appears on that list, it is hazardous, and an MSDS must be available for that chemical. (The standard requires that the chemical's name be the same on the hazardous chemical list and on the MSDS to facilitate the worker's obtaining information about the hazardous chemical.) Next the worker obtains a copy of the MSDS and studies it prior to working with the hazardous chemical.

Figure 2.6 shows a typical MSDS for acetone, which will now be reviewed.

Section I of the MSDS in Figure 2.6 contains the following information about acetone:

- Name, address, and telephone number of the manufacturer, which can be used to obtain additional information about acetone.
- Chemical name, formula, and formula weight: Although this information may not be directly related to safety, it provides a general understanding of how the chemical would react with other chemicals and what kind of reaction products can be produced. Also, members of a given chemical family often pose similar hazards.
- Chemical abstract number (CAS No.) and common synonyms: The CAS No. can be used to locate and extract any additional technical information that might be needed. The synonyms can help those workers who are familiar with the chemical under a different name. For example, sodium hydroxide is known by many workers as caustic soda.
- Chemical hazard summary: The NFPA 704 labeling system divides the hazards of chemicals into four categories: health, flammability, reactivity, and special (reserved for any special information about the chemical). In each category a rating system from 0 to 4 is used to indicate the severity of the hazard, with 0 indicating no hazard and 4 extreme hazard. Studying this part of the MSDS for acetone reveals that the health hazard associated with acetone is slight, the flammability hazard is severe, the reactivity hazard is moderate, and there is a slight hazard when acetone contacts the skin.
- Laboratory protective equipment: Study of this part of the MSDS indicates that safety glasses, a laboratory coat, adequate ventilation, proper gloves, and a Class B fire extinguisher are required for working with acetone. It should be noted that in working with any chemical, it is imperative that the personal protective equipment (PPE) be matched to the specific hazard. Use of the wrong kind of PPE can create a false sense of security and is an invitation to an accident.
- Precautionary label statement: The HCS requires that labels have adequate hazard warnings such as caution, warning, and danger, in order of increasing severity. This part of the MSDS suggests that containers of acetone should carry a label that emphasizes the flammability dangers of this chemical.

Section II of the MSDS identifies the hazardous components of the chemical, in this case 90 to 100% acetone.

Section III of the MSDS summarizes the important physical data. Study of this section for acetone provides the following information:

- Boiling point: The boiling point of a chemical, at a given pressure, is the temperature at which the chemical changes phase from liquid to vapor. This information is important in the safe storage of chemicals. If a liquid were stored at a temperature at or near its boiling point, it could generate enough vapor to burst a closed container or fill the space in a storage area with flammable or toxic vapor. It should be noted that the boiling point is a function of

J. T. Baker Chemical Co.

222 Red School Lane Phillipsburg, N.J. 08865
24-Hour Emergency Telephone -- (201) 859-2151

Chemtrec # (800) 424-9300
National Response Center # (800) 424-8802

A0446 -01 Acetone Page: 1
Effective: 10/11/85 Issued: 10/14/85

SECTION I - PRODUCT IDENTIFICATION

Product Name:	Acetone
Formula:	$(CH_3)_2CO$
Formula Wt:	58.08
CAS No.:	00067-64-1
NIOSH/RTECS No.:	AL3150000
Common Synonyms:	Dimethyl Ketone; Methyl Ketone; 2-Propanone
Product Codes:	9010,9006,9254,9009,9001,9004,5356,A134,9007,9005,9008,9002

PRECAUTIONARY LABELLING

BAKER SAF-T-DATA™ System

HEALTH	FLAMMABILITY	REACTIVITY	CONTACT
1	3	2	1
SLIGHT	SEVERE	MODERATE	SLIGHT

Laboratory Protective Equipment

SAFETY GLASSES LAB COAT VENT HOOD PROPER GLOVES EXTIN-GUISHER

Precautionary Label Statements

DANGER!
EXTREMELY FLAMMABLE
HARMFUL IF SWALLOWED OR INHALED
CAUSES IRRITATION

Keep away from heat, sparks, flame. Avoid contact with eyes, skin, clothing. Avoid breathing vapor. Keep in tightly closed container. Use with adequate ventilation. Wash thoroughly after handling. In case of fire, use water spray, alcohol foam, dry chemical, or carbon dioxide. Flush spill area with water spray.

SECTION II - HAZARDOUS COMPONENTS

Component	%	CAS No.
Acetone	90-100	67-64-1

SECTION III - PHYSICAL DATA

Boiling Point: 56°C (133°F) Vapor Pressure(mmHg): 181

Continued on Page: 2

Figure 2.6. Material safety data sheet. (Source: J. T. Baker Chemical Company, 1985).

J. T. Baker Chemical Co.

222 Red School Lane Phillipsburg, N.J. 08865
24-Hour Emergency Telephone -- (201) 859-2151

Chemtrec # (800) 424-9300
National Response Center # (800) 424-8802

A0446 -01 Acetone Page: 2
Effective: 10/11/85 Issued: 10/14/85

SECTION III - PHYSICAL DATA (Continued)

Melting Point: -95°C (-139°F) Vapor Density(air=1): 2

Specific Gravity: (H_2O=1) 0.79 Evaporation Rate: (Butyl Acetate=1) 5.6

Solubility(H_2O): Complete (in all proportions) % Volatiles by Volume: 100

Appearance & Odor: Clear, colorless liquid with fragrant sweet odor.

SECTION IV - FIRE AND EXPLOSION HAZARD DATA

Flash Point: -18°C (0°F) NFPA 704M Rating: 1-3-0

Flammable Limits: Upper - 13 % Lower - 2 %

Fire Extinguishing Media
Use alcohol foam, dry chemical or carbon dioxide.
(Water may be ineffective.)

Special Fire-Fighting Procedures
Firefighters should wear proper protective equipment and self-contained (positive pressure if available) breathing apparatus with full facepiece. Move exposed containers from fire area if it can be done without risk. Use water to keep fire-exposed containers cool.

Unusual Fire & Explosion Hazards
Vapors may flow along surfaces to distant ignition sources and flash back. Closed containers exposed to heat may explode. Contact with strong oxidizers may cause fire.

SECTION V - HEALTH HAZARD DATA

Threshold Limit Value (TLV/TWA): 1780 mg/m^3 (750 ppm)

Short-Term Exposure Limit (STEL): 2375 mg/m^3 (1000 ppm)

Toxicity: LD_{50} (oral-rat)(mg/kg) - 9750
LD_{50} (ipr-mouse)(g/kg) - 1297

Effects of Overexposure
Contact with skin has a defatting effect, causing drying and irritation. Overexposure to vapors may cause irritation of mucous membranes, dryness of mouth and throat, headache, nausea and dizziness.

Continued on Page: 3

Figure 2.6. (*Continued*)

J. T. Baker Chemical Co.

222 Red School Lane Phillipsburg, N.J. 08865
24-Hour Emergency Telephone -- (201) 859-2151

Chemtrec # (800) 424-9300
National Response Center # (800) 424-8802

A0446 -01 Acetone Page: 3
Effective: 10/11/85 Issued: 10/14/85

SECTION V - HEALTH HAZARD DATA (Continued)

Emergency and First Aid Procedures

If swallowed, if conscious, immediately induce vomiting.
If inhaled, remove to fresh air. If not breathing, give artificial respiration. If breathing is difficult, give oxygen.
In case of contact, immediately flush eyes with plenty of water for at least 15 minutes. Flush skin with water.

SECTION VI - REACTIVITY DATA

Stability: Stable Hazardous Polymerization: Will not occur

Conditions to Avoid: heat, flame, sources of ignition

Incompatibles: sulfuric acid, nitric acid, strong oxidizing agents

SECTION VII - SPILL AND DISPOSAL PROCEDURES

Steps to be taken in the event of a spill or discharge

Wear suitable protective clothing. Shut off ignition sources; no flares, smoking, or flames in area. Stop leak if you can do so without risk. Use water spray to reduce vapors. Take up with sand or other non-combustible absorbent material and place into container for later disposal. Flush area with water.

J. T. Baker Solusorb[R] solvent adsorbent is recommended for spills of this product.

Disposal Procedure

Dispose in accordance with all applicable federal, state, and local environmental regulations.

EPA Hazardous Waste Number: U002 (Toxic Waste)

SECTION VIII - INDUSTRIAL PROTECTIVE EQUIPMENT

Ventilation: Use general or local exhaust ventilation to meet TLV requirements.

Respiratory Protection: Respiratory protection required if airborne concentration exceeds TLV. At concentrations up to 5000 ppm, a gas mask with organic vapor cannister is recommended. Above this level, a self-contained breathing apparatus with full face shield is advised.

Eye/Skin Protection: Safety glasses with sideshields, polyvinyl acetate gloves are recommended.

Continued on Page: 4

Figure 2.6. ***(Continued)***

J. T. Baker Chemical Co.
222 Red School Lane Phillipsburg, N.J. 08865
24-Hour Emergency Telephone -- (201) 859-2151
Chemtrec # (800) 424-9300
National Response Center # (800) 424-8802

```
A0446 -01                               Acetone                              Page: 4
Effective: 10/11/85                                                   Issued: 10/14/85
=====================================================================================
                 SECTION IX - STORAGE AND HANDLING PRECAUTIONS
=====================================================================================
SAF-T-DATA(TM) Storage Color Code:   Red

Special Precautions
     Bond and ground containers when transferring liquid.  Keep container
     tightly closed.  Store in a cool, dry, well-ventilated, flammable liquid
     storage area.
=====================================================================================
          SECTION X - TRANSPORTATION DATA AND ADDITIONAL INFORMATION
=====================================================================================
DOMESTIC (D.O.T.)

Proper Shipping Name        Acetone
Hazard Class                Flammable liquid
UN/NA                       UN1090
Labels                      FLAMMABLE LIQUID

INTERNATIONAL (I.M.O.)

Proper Shipping Name        Acetone
Hazard Class                3.1
UN/NA                       UN1090
Labels                      FLAMMABLE LIQUID
=====================================================================================
N/A = Not Applicable or Not Available
---
The information published in this Material Safety Data Sheet has been compiled
from our experience and data presented in various technical publications. It is
the user's responsibility to determine the suitability of this information for
the adoption of necessary safety precautions. We reserve the right to revise
Material Safety Data Sheets periodically as new information becomes available.
```

Figure 2.6. (*Continued*)

pressure. As the pressure goes up, so does the boiling point. The boiling point of a chemical at atmospheric pressure is called the normal boiling point. The boiling point reported on the MSDS is the normal boiling point unless specified otherwise.

- Vapor pressure: In order to understand vapor pressure, we have to define the concept of phase equilibria. When a liquid is placed in a closed container, some molecules from the liquid phase will vaporize into the vapor phase, and some vaporized molecules will condense back into the liquid phase. If the system is given enough time, a condition will be achieved where the rate of vaporization of the liquid is equal to the rate of condensation of the vapor—a condition called equilibrium, in which the liquid and the vapor phase are said to be at equilibrium with each other. Under conditions of equilibrium the pressure that the vapor phase exerts on the surface of the liquid and on the walls of container is called the liquid vapor pressure at the temperature of the system. It should be noted that vapor pressure is a function of temperature and increases with an increase in temperature.

The concept of vapor pressure is extremely important in the safe handling and storage of chemicals. Generally, the higher the vapor pressure is at a given temperature, more care one should exercise in container selection, ventilation equipment, and storage requirements. The higher vapor pressure means that at a given temperature more of the liquid can vaporize and exert a force on the walls of the container. Because vapor pressure always increases with an increase in temperature, it is important to understand that a given liquid on a cool morning can have a much lower vapor pressure than it has on a hot afternoon. One of the approximate ways to estimate vapor pressure at different temperatures is the Antoine equation:

$$\log P^* = A - \frac{B}{T + C}$$

where P^* is the vapor pressure in mm Hg, T is temperature in °C, and A, B, and C are constants for a given material. Table 2.6 summarizes Antoine constants for some common substances.[12]

- Melting point: The melting point of a substance is the temperature at which the solid phase changes into the liquid phase. Acetone has a melting point of −95°C (−139°F), which indicates that it is a liquid at room temperature.
- Specific gravity: The specific gravity of a liquid, by definition, is the ratio of the density of the liquid to the density of water. Specific gravity indicates whether a given liquid is heavier or lighter than water. A specific gravity greater than one means that the liquid is heavier than water, whereas a specific gravity less than one indicates that the liquid is lighter than water. Considering the effects of solubility, specific gravity can determine the proper firefighting procedures. For example, most hydrocarbons have a specific gravity less than one and are insoluble in water; so most hydrocarbons when mixed with water would float on the surface of the water phase. This is why putting water on a hydrocarbon fire may not be very effective.
- Vapor density: Vapor density, by definition, is the ratio of the density of a vapor or a gas to the density of air. A vapor density greater than one indicates that the vapor or gas is heavier than air and would collect in low areas such as near the floor surface or the surface of laboratory cabinets. A vapor density less than one indicates that the vapor or gas is lighter than air, and thus tends to rise in air and collect near the ceiling. The concept of vapor density is important in selecting the type and location of ventilation systems. For example, acetone has a vapor density of two and is heavier than air, which suggests that the local ventilation system must be placed in low areas.
- The physical data section of the MSDS for acetone also provides information on the solubility of the compound in water, percent volatiles, and appearance and odor.
- Evaporation rate is an indication of the volatility of a liquid. Evaporation rates are normally compared to that of butyl acetate, which is assumed to be one. A liquid that has an evapo-

Table 2.6. Antoine equation constants.

Substance	A	B	C
Acetaldehyde	6.81089	992.0	230
Acetone	7.02447	1161.0	224
Ethyl alcohol	8.04494	1554.3	222.65
Nomal heptane	6.90240	1268.115	216.9
Methyl ethyl ketone	6.97421	1209.6	216

Source: *Lange's Handbook of Chemistry*, 9th Ed., Handbook Publishers, Inc. Sandusky, Ohio, 1956.

ration rate of less than one vaporizes more slowly than butyl acetate. Acetone has an evaporation rate of 5.6, which means that it vaporizes 5.6 times faster than butyl acetate.

Section IV of the MSDS for acetone provides previously discussed fire and explosion hazard data such as flash point, lower and upper flammability limits, and the NFPA 704 hazard rating. This information is extremely valuable in the safe handling and storage of the material. For example, acetone has a flash point of −18°C (0°F), which suggests that it is a flammable liquid (flash point <100°F) and should be stored in a properly designed flammable storage area. This section also identifies the proper fire extinguishing media, special firefighting procedures, and any unusual fire and explosion hazards.

Section V of the MSDS contains important health hazard information. This section indicates that acetone has a threshold limit value (TLV), based on a time weighted average (TWA) of 8 hours, equal to 1780 mg/m^3 (750 ppm). The TLV is the concentration of a chemical that a worker may be exposed to 8 hours of a day for the rest of his or her working life without creating any adverse health effects. This information is vital because any exposure above the TLV value is considered to be a health hazard. In situations when workers are not sure whether they are exposed to concentrations above the TLV, monitoring devices should be installed to ensure that exposure concentrations are below the TLV values. This section also specifies the short term exposure limits (STEL) for acetone. As can be noted from Figure 2.6, the value of the STEL for acetone is 2375 mg/m^3, which is higher than the TLV value. The STEL sets the exposure limits for workers who are exposed to a chemical for a relatively short period of time, whereas the TLV sets the safe limit, assuming that a worker is exposed to a given chemical on every day of his or her working life.

The health hazard section also presents toxicity data. For acetone, the oral LD_{50} in rats is 9,750 mg/kg. This means that acetone is not a toxic agent. However, it should be noted that any chemical, including water, is toxic when the dose is large enough. This section also describes the effect of overexposure to the chemical, along with emergency and first aid procedures.

Section VI summarizes the reactivity data. As mentioned earlier, some chemicals can undergo hazardous polymerization or decomposition. The heat effects associated with these reactions or the products of reaction could create a physical or health hazard. For acetone, the MSDS specifies that hazardous polymerization or decomposition does not occur under normal conditions, and that the compound is stable. However, the MSDS indicates that because of its high flammability, acetone should not be stored with oxidizers, and it should be kept away from all sources of ignition.

Section VII covers spill response and proper disposal procedures. All hazardous wastes must be disposed of according to federal, state, and local regulations; and there are still civil and criminal penalties for the illegal disposal of hazardous wastes.

Section VIII covers the industrial protective equipment required when working with acetone. PPE comes in a variety of designs and materials, and one must consult the MSDS to ensure that PPE that matches the workplace hazards is selected.

Section IX discusses storage and handling precautions. Because of its high flammability, acetone must be stored in a well-ventilated, cool, flammable storage area away from all sources of ignition. The container must be properly bonded or grounded to prevent static electricity sparks during the transfer of the liquid.

The last section of the MSDS summarizes important transportation data and any additional information that can help in the safe handling of the chemical.

OSHA does not require a special format for the MSDS; so a variety of formats are used. However, they all provide comparable and valuable information that is vital for working safely with chemicals. Before using any chemical, one should study the MSDS to understand the associated hazards and proper means of protection against them.

Exercises

1. What are the major objectives of OSHA's Hazard Communication Standard?
2. Under HCS, what information must be provided on container labels of hazardous chemicals?
3. Under HCS, what information must be provided on the MSDS?
4. Under HCS, what are the chemical manufacturer's responsibilities when new hazard information becomes available about a hazardous chemical?
5. What are the issues that a "Written Hazard Communication Program" must address?
6. What are the employee training requirements?
7. List five chemicals that are exempt from OSHA labeling requirements.
8. What are OSHA's labeling requirements for transfer containers?
9. What is flash point? How is flash point used to classify liquids into flammables and combustibles?
10. Define a flammable gas.
11. What is the major difference between an oxidizer and an organic peroxide?
12. What is a runaway reaction, and how can it create a physical hazard?
13. What are LD_{50} and LC_{50}, and how do they relate to chemical toxicity?
14. Describe how a hematopoietic system hazard can affect body function.
15. A 55-gallon drum is half-filled with acetone. The drum is sealed and placed in a flammable liquid storage. Calculate the amount of force exerted on the drum by the acetone vapor on a hot afternoon when the temperature is 133°F. Perform your calculation using the vapor pressure data presented on the MSDS. The drum diameter is 2.5 ft.
16. Can water be used to extinguish a fire in a flammable liquid with specific gravity of 3 and no solubility in water?

OSHA's Hazardous Waste Operations and Emergency Response Standard (29 CFR 1910.120)

OSHA's hazardous waste operations and emergency response standard[13] covers employers and employees who are engaged in hazardous substance response and cleanup operations; operations involving hazardous waste storage, disposal, and treatment; operations at sites designated for cleanup; and emergency operations for release of hazardous substances. The standard defines and requires compliance in specific areas such as:

- Site characterizationand analysis
- Site control
- Training
- Medical surveillance
- Engineering controls, work practices, and PPE
- Monitoring
- Informational programs
- Material handling
- Decontamination
- Emergency response
- Illumination
- Sanitation

Site Characterization

The standard requires an evaluation of hazardous waste sites to determine appropriate safety and health control measures that might be needed to protect employees from hazards. The standard mandates that a preliminary site evaluation be conducted prior to site entry to determine the appropriate protective measures to be utilized. An additional evaluation of site hazards must be performed, by a trained person, during site entry to identify the appropriate engineering control measures and any personal protective equipment needed for the tasks to be performed.

The major objective of the evaluation is to identify all conditions or suspected conditions that might be immediately dangerous to life or health. The secondary evaluation must identify these hazardous conditions and recommend appropriate protective measures. Examples of such hazards may include, but are not limited to, entry into confined spaces, potentially explosive or flammable situations, visible vapor clouds, or areas where a biological indicator, such as dead animals or vegetation, is located.

The employer is responsible for obtaining the following information, to the extent available, prior to asking employees to enter the hazardous waste site:

- Location and approximate size of the site.
- Description of the nature of tasks to be performed on-site.
- Expected period of time that employees will spend on-site.
- Description of the site topography.
- Information regarding the accessibility of the site by air or roads.
- Possible pathways for hazardous substance dispersion.
- Emergency personnel to provide assistance on-site in case of an emergency.
- Health or physical hazards posed by hazardous substances on-site.

The employer is also responsible for identifying and providing personal protective equipment (PPE) to protect employees from site-specific hazards. The following requirements apply to the selection and use of PPE at hazardous waste sites:

- PPE must be capable of providing protection within the permissible exposure limit of the hazardous substance.
- An appropriate self-contained breathing apparatus must be provided.
- If complete information on the hazards is not available, level B PPE (see Chapter 4) along with monitoring devices must be provided.

The standard requires that, where ionizing radiation information is not available, the employer must provide monitoring devices for the following conditions:

- Hazardous levels of ionizing radiation.
- Dangerous levels of air contaminants.

Once the concentrations of hazardous substances have been identified, the employer must assess the risks associated with the hazards, and must inform employees of those risks and any information on physical, chemical, and toxicological properties that is relevant to the tasks to be performed.

Site Control

The standard requires that, prior to any cleanup operation at a hazardous waste site, a site control program be developed by the employer. The major objective of this program is to prevent contamination of employees by the hazardous substances. This program must, at a minimum, include the following:

- Site map.
- Site work zones.
- Use of a buddy system to provide emergency assistance to affected employees.
- Site communication.
- Standard operating procedures for safe work practices.
- Identification of nearest medical assistance.

Training

All employees who have a potential for being exposed to hazardous substances must be trained by the employer in the following areas:

- Names of personnel responsible for site safety and health.
- Safety, health, and other hazards present on the site.
- Proper use of PPE.
- Safe work practices.
- Safe use of engineering controls and equipment.
- Signs and symptoms of overexposure to hazardous substances that might be present on the site.

The employer is required to provide at least 40 hours of training for employees off the site and three days of field experience on the site under the supervision of a trained person. The standard mandates that employees can refuse to participate in field activities requested by the employer until such time when they have received the appropriate level of training. The instructor must issue a certificate to those employees who have successfully completed the training program. Any person who has not been certified is prohibited from taking part in any hazardous waste operation. Employees who are responsible for responding to emergency situations must be trained in how to respond to expected emergencies, and supervisors and managers who direct hazardous waste operations also must be properly trained.

Medical Surveillance

The employer is responsible for organizing a medical surveillance program for employees potentially exposed to hazardous substances or those who wear a respirator for 30 days or more a year. The medical program must provide a medical examination and consultation to covered employees at least once a year and at the termination of employment. The medical examination also should be provided as soon as possible when the employer is informed that an employee has developed signs and symptoms of overexposure to a hazardous substance.

The employer must disclose the following information to the licensed physician conducting the examination:

- A copy of OSHA's hazardous waste and emergency response standard (29 CFR 1910.120).
- A description of the employee's duties as they relate to overexposure.
- The employee's exposure levels or anticipated exposure levels.
- A description of any personal protective equipment used by the employee.
- Information on previous medical examinations that is not available to the examining physician.

At the completion of the medical examination, the employer must obtain the written opinion of the examining physician and furnish this document to the employee. This document must contain the results of the medical examination and tests, the physician's opinion on the medical condition of the employee as it relates to working at the hazardous site, the physician's recommended limitations upon the employee's assigned work, and a statement that the employee has been informed of the results of the medical examination.

The employer is responsible for keeping a file of employees' medical records. An individual's record must contain the following information:

- Name and social security number of the employee.

- Physician's written opinion.
- Any employee medical complaints related to exposure to hazardous substances.
- A copy of the information provided to the examining physician by the employer.

Engineering Control, Work Practices, and PPE

Every attempt should be made to reduce employee exposure to hazardous substances through work practices and engineering controls. However, if changes in work practices or engineering controls are not successful in reducing employee exposure to such substances to values below permissible exposure limits, the employer is responsible for providing adequate PPE. The employer must identify the site hazards and select the proper type of PPE, which must match the site hazards. Specifically, when there is danger of skin contact by hazardous substances, the employer must provide fully encapsulating chemical protective suits (level A protection). As more information become available on the site condition, the employer must upgrade or downgrade the level of PPE. When site conditions are such that employees must use PPE to reduce their exposure to below values specified by permissible exposure limits, the employer must develop a PPE program that addresses the following:

- Site hazards.
- PPE selection.
- PPE use.
- Work mission duration.
- PPE maintenance and storage.
- PPE decontamination.
- PPE training and proper fitting.
- PPE donning and doffing procedures.
- PPE inspection.
- PPE in-use monitoring.
- Evaluation of the effectiveness of the PPE program.
- Limitations on equipment during temperature extremes.

Monitoring

The standard requires that the concentration of hazardous substances in air be measured to help workers in selecting the proper level of protection needed. As a first step, an air analysis must be conducted to identify any situations that might be immediately dangerous to life or health, such as the presence of toxic or oxygen-deficient atmospheres. Also, periodic monitoring should be conducted when:

- Work begins on a new portion of the site.
- Contaminants other than those previously identified are being handled.
- A new operation has been initiated (such as drum opening or sampling).
- Employees are handling leaking drums or spills.

Informational Programs

As part of the overall site safety and health program, the employer is responsible for the development and implementation of a safety and health plan. This plan must be available at the site for employee and OSHA representatives' inspection. The plan must describe the safety and health hazards of each phase of the site operation and must include the requirements and procedures for employee protection. The site safety and health plan, at a minimum, must address the following:

- The names of key personnel and alternates responsible for site safety and health and the appointment of a site safety officer.
- A safety and risk analysis for each site task and operation.
- Employee training assignments.
- PPE to be used by employees for each site task and operation.
- Medical surveillance requirements.
- Frequency and types of air monitoring, personnel monitoring, and environmental sampling techniques and instrumentation.
- Site control measures.
- Decontamination procedures.
- The site's standard operating procedures.
- The site's contingency plan.
- Confined space entry procedures.

Materials Handling

The standard also sets forth requirements for handling drums and other containers of hazardous wastes, which are covered in Chapter 10.

Decontamination

The employer is responsible for developing a decontamination procedure and communicating it to employees before any attempt is made to enter a hazardous area. Standard operating procedures must be developed to minimize employee contact with contaminated tools or equipment. The decontamination must include procedures for properly decontaminating employees, clothing, and equipment. The site safety and health officer must assess the effectiveness of the decontamination procedures and improve them as necessary.

Emergency Response

The standard requires that the employer develop and implement an emergency response plan to handle anticipated emergencies prior to the commencement of hazardous waste operations. The emergency response plan for on-site and off-site emergencies must, at a minimum, address the following:

- Pre-emergency planning.
- Personnel roles, lines of authority, training, and communication.
- Emergency recognition and prevention.
- Safe distances and places of refuge.
- Site security and control.
- Evacuation routes and procedures.
- Decontamination.
- Emergency medical treatment and first aid.
- Critique of response and follow-up.
- PPE and emergency equipment.

Besides the requirements outlined above, there are additional requirements for handling on-site and off-site emergencies. These requirements, which are discussed in detail in 29 CFR 1910.120, Paragraph L, have been summarized in Chapter 10 of this book.

Table 2.7. Minimum illumination requirements in foot-candles.

Foot-Candles	Area of Operations
5	General site areas.
3	Excavation and waste areas, accessways, active storage areas, loading platforms, refueling, and field maintenance areas.
5	Indoors; warehouses, corridors, hallways, and exitways.
5	Tunnels, shafts, and general underground work areas; exception: minimum of 10 foot-candles is required at tunnel and shaft heading during drilling, mucking, and scaling. Bureau of Mines approved cap lights shall be acceptable for use in the tunnel heading.
10	General shops (e.g., mechanical and electrical equipment rooms, active storerooms, barracks or living quarters, locker or dressing rooms, dining areas, and indoor toilets and workrooms.
30	First aid stations, infirmaries, and offices.

Source: Reference 13.

Table 2.8. Toilet facilities required for hazardous waste sites.

Number of Employees	Minimum Number of Facilities
20 or fewer	One.
More than 20, fewer than 200	One toilet seat and one urinal per 40 employees.
More than 200	One toilet seat and one urinal per 50 employees.

Source: Reference 13.

Illumination

The standard requires that work areas be lighted to not less than the minimum illumination intensities listed in Table 2.7 while work is in progress.

Sanitation

The employer must provide an adequate supply of drinking water on-site, with the following in mind:

- Portable containers of drinking water must have the capability of being tightly closed. These containers must have a tap. Water should not be dipped from containers.
- Containers of drinking water must be marked as to their contents, and they should not be used for any other purpose.
- When disposable cups are used, a sanitary container for unused cups and a receptacle for used cups must be provided.

The outlets for nonpotable water must be clearly marked with indications that the water is unsafe for drinking, washing, or cooking.

The employer is responsible for providing toilet facilities according to Table 2.8.

The employer must provide food service facilities, temporary sleeping quarters, and washing facilities that comply with the law.

Exercises

1. What are OSHA requirements for the selection and use of PPE at hazardous waste sites?
2. What is the major objective of a site control program at a hazardous waste site?

3. What are the major areas of training required for workers at hazardous waste sites?
4. What are OSHA requirements for the monitoring of hazardous substances at hazardous waste sites?
5. List at least five areas that must be covered in a hazardous waste site contingency plan.

Title III of Superfund Amendment and Reauthorization Act (SARA); Emergency Planning and Community Right-to-Know

Overview

As a result of the Bhopal tragedy, in which more than 2,000 people lost their lives and more than 200,000 were injured, Congress examined the possibility of similar accidents occurring in the United States. The legislative response to the Bhopal tragedy was SARA, a comprehensive system of information gathering, reporting, and sharing on the storage, release, and handling of hazardous and toxic materials. SARA was enacted into law on October 17, 1986.[14,15]

With thousands of new chemicals being developed each year, both citizens and public officials are concerned about accidents (e.g., train derailments, industrial incidents) happening in their communities. Recent evidence shows that many people consider hazardous materials incidents to be the most significant threat facing local jurisdictions. Of the more than 3,100 localities completing the Federal Emergency Management Agency's (FEMA's) Hazard Identification Capability Assessment and its Multi-Year Development Plan during fiscal year 1985, some 93% identified one or more hazardous materials risks (e.g., on highways and railroads, at fixed facilities) as a significant threat to their community. Communities need to prepare themselves to prevent such incidents and to respond to any accidents that do occur.[15]

Because of the risk of hazardous materials incidents and because local governments are completely on their own in the first stages of almost any such incident, communities must maintain a continuing preparedness capacity. A specific, tangible indication of such readiness is an emergency plan. However, even the most sophisticated and detailed written plans must be regularly tested and revised, or the communities they cover might be less well prepared than they should be for possible incidents.[15]

Title III of SARA, known as the Emergency Planning and Community Right-to-Know Act of 1986, establishes requirements for federal, state, and local governments and industry regarding emergency planning and reporting on hazardous and toxic chemicals. This legislation builds upon EPA's Chemical Emergency Preparedness Program (CEPP) and numerous state and local programs aimed at helping communities to better meet their responsibilities in regard to potential chemical emergencies. The community right-to-know provisions of Title III will help increase the public's knowledge and access to information on the presence of hazardous chemicals in their communities and the release of these chemicals into the environment. Title III is divided into three subtitles: (1) Subtitle A, Emergency Planning and Notification; (2) Subtitle B, Reporting Requirements; and (3) Subtitle C, General Provisions.[14]

Subtitle A, which includes Sections 301 through 305, deals with procedures for emergency planning and notification requirements. Sections 301 through 304 describe the emergency planning and notification requirements and procedures, and Section 305 deals with emergency training and systems.

Subtitle B, which includes Sections 311 through 313, is designed to provide information to the public on the nature, amount, location, disposal, and release of any hazardous and toxic materials. In many respects, Subtitle B can be considered to be the core of the community right-to-know provisions, as it requires companies to provide information to state and local emergency personnel, as well as to the public, on any chemicals used in the workplace. Section 311 deals with the availability and submission of MSDSs, Section 312 is concerned with emergency and hazardous chemical inventory requirements, and Section 313 describes the toxic chemical release reporting requirements.[14,18,19]

Subtitle C, which includes Sections 321 through 327, sets forth a system for the handling and processing of confidential or trade secrets. Section 321 describes the relationship between provisions of Subtitle C and other laws; Section 322 sets forth and explains procedures for the protection of trade secrets; Section 323 discusses the requirement for disclosure of the chemical identity of a trade secret chemical to a health professional; whereas Section 324 describes the need for public availability of plans, data sheets, and emergency notices. Sections 325, 326, and 327 describe enforcement of the law, citizens' suits, and transportation exemptions, respectively.[14,16]

Subtitle A: Emergency Planning and Notification

Subtitle A is devoted to the two-fold task of establishing a system to (a) ensure that members of the public have information about chemicals present in their communities and (b) ensure compliance with emergency planning and notification requirements.

Section 301: State and Local Government Responsibilities

State Emergency Response Commission. The law requires that each state governor designate a state emergency response commission (SERC). The commission must have broad-based representation. The important roles in state commission activities are played by public agencies and public and private sector groups concerned with issues relating to the environment, natural resources, emergency services, public health, occupational safety, and transportation.[14]

The state commission designates local emergency planning districts and appoints local emergency planning committees (LEPCs). The state commission is also responsible for:

- Supervising and coordinating the activities of the local emergency planning committees.
- Establishing procedures for receiving and processing public requests for information collected under other sections of Title III.
- Reviewing local emergency plans.

Local Emergency Planning Committee. The local committee must include:

- Elected state and local officials.
- Police, fire, civil defense, and public health professionals.
- Environmental, hospital, and transportation officials.
- Representatives of community groups and facilities subject to the emergency planning requirements.

Facilities subject to the emergency planning requirements must designate a representative to participate in the planning process. The local committee must establish rules, give public notice of its activities, and establish procedures for handling public requests for information.

The local committee's primary responsibility is to develop an emergency response plan. This plan must include:

- Identification of facilities and transportation routes for extremely hazardous substances.
- Emergency response procedures, on-site and off-site.
- Designation of a community coordinator and facility coordinator(s) to implement the plan.
- Emergency notification procedures.
- Methods for determining the occurrence of a release and the probable affected area and population.

- Description of community and industry emergency equipment and facilities and the identity of persons responsible for them.
- Evacuation plans.
- Description and schedules of a training program for emergency response personnel.
- Methods and schedules for exercising emergency response plans.

Section 302: Emergency Planning Notification

A key objective of Title III is to make adequate information available for the local committees so they can determine which facilities to include in emergency planning.

Any facility that produces, uses, or stores any EPA-listed extremely hazardous substance in an amount greater than its threshold planning quantity is subject to the emergency planning requirements. Appendix A lists the chemicals that are subject to reporting under Title III of SARA. The reportable quantity, the Chemical Abstract (CAS) No., and the threshold planning quantity are also included. A state commission or governor can designate additional facilities, after public comment, to be subject to these requirements. If a facility begins to produce, use, or store any of the extremely hazardous substances in threshold quantity amounts, it must so notify the state commission within 60 days.[14,16]

Section 303: Comprehensive Emergency Response Plan

Under Section 303, each local committee has the responsibility for developing a comprehensive emergency response plan for the district under its jurisdiction.[17,20,21] This plan must provide for:

- Identification of facilities subject to the requirements of this subtitle that are within the emergency planning district.
- Identification of routes that are likely to be used for the transportation of extremely hazardous materials.
- Identification of strategic facilities susceptible to high risk due to their proximity to facilities subject to the requirements of this subtitle. Examples are hospitals, natural gas facilities, water reservoirs, and so on.
- Methods and procedures to be followed by facility owners and local emergency and medical personnel in the event of an emergency due to the release of a toxic substance.
- Designation of a community emergency coordinator and a facility emergency coordinator responsible for executing the plan in case of an emergency.
- Effective and reliable procedures for timely notification of the public and persons designated in the emergency plan that a release of toxic substances has occurred.
- Methods for the determination of the occurrence of a release as well as the estimated area and population affected by such a release.
- Description of emergency equipment available and persons who are responsible for such equipment.
- Evacuation plans.
- Training programs, including schedules for the training of emergency and medical personnel.
- Methods and schedules for exercising the emergency plan.

Section 304: Emergency Notification

Facilities must immediately notify the local emergency planning committee and the state emergency response commission if there is a release of a listed hazardous substance that exceeds the reportable quantity for that substance. Substances subject to this requirement are substances on the list of 402 extremely hazardous substances as published in the *Federal Register* on 11/17/86 and substances subject to the emergency notification requirements under CERCLA Section 103(a).[14]

The initial notification can be done by telephone, by radio, or in person. Emergency notification requirements involving transportation incidents can be satisfied by dialing 911, or in the absence of a 911 emergency number, calling the telephone operator.

This emergency notification needs to include:

- The substance's chemical name.
- An indication of whether the substance is extremely hazardous.
- An estimate of the quantity released into the environment.
- The time and duration of the release.
- The medium into which the release occurred.
- Any known or anticipated acute or chronic health risks associated with the emergency and, where appropriate, advice regarding medical attention necessary for exposed individuals.
- Proper precautions, such as evacuation.
- Name and telephone number of a contact person.

Section 304 also requires a follow-up written emergency notice after the release. The follow-up notice or notices shall:

- Update information included in the initial notice.
- Provide information on:
  - Actual response actions taken.
  - Any known or anticipated data or chronic health risks associated with the release.
  - Advice regarding medical attention necessary for exposed individuals.

Until state commissions and local committees are formed, releases should be reported to appropriate state and local officials.

Section 305: Emergency Training

U.S. government officials who are responsible for emergency training in existing federal programs are authorized to provide training and education programs in emergency preparedness and other hazard mitigation procedures.[17]

Subtitle B: Reporting Requirements

Subtitle B of SARA III includes Sections 311, 312, and 313, dealing with the requirements for Material Safety Data Sheet (MSDS) submission, reporting requirements and the hazardous chemical inventory, and toxic chemical release reporting requirements.

Section 311: Material Safety Data Sheet Submissions

Under Section 311, a facility that must maintain MSDSs under OSHA regulations must either submit copies of the MSDSs to, or file a list of on-site hazardous substances with the local committee, state commission, and local fire department. In addition, the facility is required to submit the MSDS for any chemical on the list upon the request of the local committee or state commission.[14,16,17]

If the facility owner or operator chooses to submit a list of MSDS chemicals, the list must include the chemical name or common name of each substance and any hazardous component as provided on the MSDS. This list must be organized in categories of health and physical hazards as set forth in OSHA regulations.

If a list is submitted, the facility must submit the MSDS for any chemical on the list upon the request of the local planning committee. Under Section 311, EPA may establish threshold quantities for hazardous chemicals below which no facility must report.

The local emergency planning committee must submit a copy of an MSDS to any person who requests it. If the committee does not have a copy of the MSDS readily available, it must obtain one from the owner or operator of the facility and submit it to the person who requested it.

Section 312: Emergency and Hazardous Chemical Inventory Reporting Requirements

Section 312 requires the reporting of hazardous chemical inventories by a two-tiered approach: Tier I, the reporting of aggregate information by hazard type, and Tier II, the reporting of specific information by chemical.[14,17,22,23] Facilities must submit an emergency hazardous chemical inventory form on March 1 of each year to the local committee, state commission, and local fire department. The facility may use either the Tier I or the Tier II form. Hazardous chemicals to be included on the form are those reported under Section 311. Under Section 312, EPA has established threshold quantities for hazardous chemicals, below which no facility must be subject to the requirement.

Under Tier I, facilities must submit the following aggregate information for each applicable OSHA category of health and physical hazard:

- An estimate (giving ranges) of the maximum amount of chemicals for each category present at the facility at any time during the preceding calendar year.
- An estimate (giving ranges) of the average daily amount of chemicals in each category.
- The general location of hazardous chemicals in each category.

Upon the request of a local committee, state commission, or local fire department, the facility must provide the following Tier II information for each substance subject to the request:

- The chemical name or the common name as indicated on the MSDS.
- An estimate (giving ranges) of the maximum amount of the chemical present at any time during the preceding calendar year.
- A brief description of the manner of storage of the chemical.
- The location of the chemical at the facility.
- An indication of whether the owner elects to withhold location information from disclosure to the public.

The public may also request Tier II information from the state commission and the local committee. The information submitted by facilities under Sections 311 and 312 generally must be made available to the public by local and state governments during normal working hours.

Figures 2.7 and 2.8 show Tier I and Tier II forms developed by EPA.

Section 313: Toxic Chemical Release Reporting

Section 313 of Title III requires EPA to establish an inventory of toxic chemical emissions from certain facilities. The facilities subject to this reporting requirement are required to complete a toxic chemical release form for specified chemicals. The form must be submitted to EPA and those state officials designated by the governor on July 1 of each year, reflecting releases during the preceding calendar year.

The purpose of this reporting requirement is to inform government officials and the public about releases of toxic chemicals in the environment. It will also assist in research and the development of regulations, guidelines, and standards.

Page ____ of ____ pages
Form Approved OMB No. 2050-0072

Tier One

EMERGENCY AND HAZARDOUS CHEMICAL INVENTORY
Aggregate Information by Hazard Type

FOR OFFICIAL USE ONLY
ID #
Date Received

Important: Read instructions before completing form

Reporting Period From January 1 to December 31, 19____

Facility Identification

Name ____
Street Address ____
City ____ State ____ Zip ____
SIC Code ☐☐☐☐ Dun & Brad Number ☐☐-☐☐☐-☐☐☐☐

Owner/Operator

Name ____
Mail Address ____
Phone (   ) ____

Emergency Contacts

Name ____
Title ____
Phone (   ) ____
24 Hour Phone (   ) ____

Name ____
Title ____
Phone (   ) ____
24 Hour Phone (   ) ____

☐ Check if site plan is attached

	Hazard Type	*Max Amount**	*Average Daily Amount**	*Number of Days On-Site*	*General Location*
Physical Hazards	Fire	☐☐	☐☐	☐☐☐	____
	Sudden Release of Pressure	☐☐	☐☐	☐☐☐	____
	Reactivity	☐☐	☐☐	☐☐☐	____
Health Hazards	Immediate (acute)	☐☐	☐☐	☐☐☐	____
	Delayed (Chronic)	☐☐	☐☐	☐☐☐	____

Certification *(Read and sign after completing all sections)*

I certify under penalty of law that I have personally examined and am familiar with the information submitted in this and all attached documents, and that based on my inquiry of those individuals responsible for obtaining the information, I believe that the submitted information is true, accurate and complete.

Name and official title of owner/operator OR owner/operator's authorized representative

Signature ____ Date signed ____

* Reporting Ranges

Range Value	Weight Range in Pounds From...	To...
00	0	99
01	100	999
02	1000	9,999
03	10,000	99,999
04	100,000	999,999
05	1,000,000	9,999,999
06	10,000,000	49,999,999
07	50,000,000	99,999,999
08	100,000,000	499,999,999
09	500,000,000	999,999,999
10	1 billion	higher than 1 billion

Figure 2.7. Tier I information reporting form.

Form Approved OMB No. 2050-0072

Tier Two
EMERGENCY AND HAZARDOUS CHEMICAL INVENTORY
Specific Information by Chemical

Facility Identification

Name ______
Street Address ______
City ______ State ______ Zip ______
SIC Code ______ Dun & Brad Number ______

FOR OFFICIAL USE ONLY ID # ______ Date Received ______

Owner/Operator Name

Name ______ Phone ( )
Mail Address ______

Emergency Contact

Name ______ Title ______
Phone ( ) 24 Hr. Phone ( )

Name ______ Title ______
Phone ( ) 24 Hr. Phone ( )

Important: Read all instructions before completing form

Reporting Period From January 1 to December 31, 19____

Chemical Description	Physical and Health Hazards (check all that apply)	Inventory: Max. Daily Amount (code)	Inventory: Avg. Daily Amount (code)	Inventory: No. of Days On-site (days)	Storage Codes and Locations (Non-Confidential): Storage Code	Storage Locations
CAS ______ Trade Secret ☐ Chem. Name ______ *Check all that apply:* ☐ Pure ☐ Mix ☐ Solid ☐ Liquid ☐ Gas	☐ Fire ☐ Sudden Release of Pressure ☐ Reactivity ☐ Immediate (acute) ☐ Delayed (chronic)					
CAS ______ Trade Secret ☐ Chem. Name ______ *Check all that apply:* ☐ Pure ☐ Mix ☐ Solid ☐ Liquid ☐ Gas	☐ Fire ☐ Sudden Release of Pressure ☐ Reactivity ☐ Immediate (acute) ☐ Delayed (chronic)					
CAS ______ Trade Secret ☐ Chem. Name ______ *Check all that apply:* ☐ Pure ☐ Mix ☐ Solid ☐ Liquid ☐ Gas	☐ Fire ☐ Sudden Release of Pressure ☐ Reactivity ☐ Immediate (acute) ☐ Delayed (chronic)					

Certification *(Read and sign after completing all sections)*

I certify under penalty of law that I have personally examined and am familiar with the information submitted in this and all attached documents, and that based on my inquiry of those individuals responsible for obtaining the information, I believe that the submitted information is true, accurate, and complete.

Name and official title of owner/operator OR owner/operator's authorized representative ______ Signature ______ Date signed ______

Optional Attachments *(Check one)*

☐ I have attached a site plan
☐ I have attached a list of site coordinate abbreviations

Figure 2.8. Tier II information reporting form.

Page ______ of ______ pages
Form Approved OMB No. 2050-0072

Tier Two

EMERGENCY AND HAZARDOUS CHEMICAL INVENTORY

Specific Information by Chemical

Facility Identification

Name ______
Street Address ______
City ______ State ______ Zip ______
SIC Code [ ][ ][ ][ ] Dun & Brad Number [ ][ ]-[ ][ ][ ]-[ ][ ][ ][ ]

FOR OFFICIAL USE ONLY: ID # ______ Date Received ______

Owner/Operator Name

Name ______ Phone ( )
Mail Address ______

Emergency Contact

Name ______ Title ______
Phone ( ) ______ 24 Hr. Phone ( ) ______

Name ______ Title ______
Phone ( ) ______ 24 Hr. Phone ( ) ______

Important: Read all instructions before completing form

Reporting Period From January 1 to December 31, 19____

Confidential Location Information Sheet

	Storage Codes and Locations (Confidential)
	Storage Codes / *Storage Locations*
CAS # [ ][ ][ ][ ][ ][ ] [ ][ ] [ ] Chem. Name	
CAS # [ ][ ][ ][ ][ ][ ] [ ][ ] [ ] Chem. Name	
CAS # [ ][ ][ ][ ][ ][ ] [ ][ ] [ ] Chem. Name	

Certification *(Read and sign after completing all sections)*

I certify under penalty of law that I have personally examined and am familiar with the information submitted in this and all attached documents, and that based on my inquiry of those individuals responsible for obtaining the information, I believe that the submitted information is true, accurate, and complete.

Name and official title of owner/operator OR owner/operator's authorized representative ______ Signature ______ Date signed ______

Optional Attachments *(Check one)*

- [ ] I have attached a site plan
- [ ] I have attached a list of site coordinate abbreviations

The reporting requirement applies to owners and operators of facilities that have ten or more full-time employees, that are in Standard Industries Classification Codes 20 through 39 (i.e., manufacturing facilities), and that have manufactured, processed, or otherwise used a listed toxic chemical in excess of specified threshold quantities.

The list of toxic chemicals subject to reporting consists initially of chemicals listed for similar reporting purposes by the states of New Jersey and Maryland, there being over 300 chemicals and categories on these lists. EPA can modify this combined list. In adding a chemical to the combined Maryland and New Jersey lists, EPA must consider the following questions:

- Is the substance known to cause cancer or serious reproductive or neurological disorders, genetic mutations, or other chronic health effects?
- Can the substance cause significant adverse acute health effects outside the facility as a result of continuous or frequently recurring releases?
- Can the substance cause an adverse effect on the environment because of its toxicity, persistence, or tendency to bioaccumulate?

Chemicals can be deleted if there is insufficient evidence to satisfy any of these questions. State governors may petition the EPA administrator to add or delete a chemical from the list on the basis of any of the above factors.

In toxic chemical release reporting, the following information must be supplied:

- The name, location, and type of business.
- Whether the chemical is manufactured, processed, or otherwise used, and the general categories of use of the chemical.
- An estimate (giving ranges) of the maximum amounts of the toxic chemical present at the facility at any time during the preceding year.
- Waste treatment/disposal methods and efficiency of methods for each waste stream.
- Quantity of the chemical entering each environmental medium annually.
- A certification by a senior official that the report is complete and accurate.

Figure 2.9 shows the EPA form R for reporting a toxic chemical release. A summary of reporting requirements under SARA Title III is presented in Table 2.9.

Subtitle C: General Provisions

Subtitle C[16] of Title III of SARA contains seven sections (321–327), dealing with the relationship between Title III and other laws, the requirements and procedures for trade secret protection (an area that probably is of great concern to companies that must comply with provisions of Title III), requirements for the disclosure of information to health professionals, the public availability of information, enforcement provisions of the law, citizen suits, and transportation exemptions.

Section 321: General Provisions

This section discusses the relationship between Title III and other laws.[17] It should be noted that Title III does not preempt any state or local law and does not affect the authority of any state or local official to enforce any state or local law. It also does not affect the obligations or liabilities of any person under provisions other than federal laws.

Section 322: Trade Secrets

This section of Title III addresses trade secrets and applies to emergency planning, community right-to-know, and toxic chemical release reporting.[14] Any person may withhold the specific

Form Approved OMB No.: 2070-0093

Approval Expires: 01/91

(Important: Type or print; read instructions before completing form.)

Page 1 of 5

U.S. Environmental Protection Agency

EPA TOXIC CHEMICAL RELEASE INVENTORY REPORTING FORM

EPA FORM **R**

Section 313, Title III of The Superfund Amendments and Reauthorization Act of 1986

(This space for EPA use only.)

PART I. FACILITY IDENTIFICATION INFORMATION

1. 1.1 Does this report contain trade secret information? ☐ Yes (Answer 1.2) ☐ No (Do not answer 1.2)
   1.2 Is this a sanitized copy? ☐ Yes ☐ No
   1.3 Reporting Year

2. CERTIFICATION (Read and sign after completing all sections.)

I hereby certify that I have reviewed the attached documents and that, to the best of my knowledge and belief, the submitted information is true and complete and that the amounts and values in this report are accurate based on reasonable estimates using data available to the preparers of this report.

Name and official title of owner/operator or senior management official

Signature | Date signed

3. FACILITY IDENTIFICATION

3.1 Facility or Establishment Name

Street Address

City | County

State | Zip Code _ _ _ _ _ - _ _ _ _

3.2 This report contains information for: (check one)

a. ☐ An entire covered facility.

b. ☐ Part of a covered facility.

3.3 Technical Contact | Telephone Number (include area code) ( ) -

3.4 Public Contact | Telephone Number (include area code) ( ) -

3.5 a. SIC Code | b. | c.

3.6 Latitude: Deg. Min. Sec. | Longitude: Deg. Min. Sec.

3.7 Dun & Bradstreet Number(s) a. | b.

3.8 EPA Identification Number (RCRA I.D. No.) a. | b.

3.9 NPDES Permit Number(s) a. | b.

Where to send completed forms:

U.S. Environmental Protection Agency
P.O. Box 70266
Washington, DC 20024-0266
Attn: Toxic Chemical Release Inventory

3.10 Name of Receiving Stream(s) or Water Body(s)
a.
b.
c.

3.11 Underground Injection Well Code (UIC) Identification No.

4. PARENT COMPANY INFORMATION

4.1 Name of Parent Company

4.2 Parent Company's Dun & Bradstreet No.

EPA Form 9350-1 (1-88)

Figure 2.9. EPA form R for reporting toxic chemical release.

(Important: Type or print; read instructions before completing form.)

(This space for EPA use only.)

EPA FORM R
PART II. OFF-SITE LOCATIONS TO WHICH TOXIC CHEMICALS ARE TRANSFERRED IN WASTES

1. PUBLICLY OWNED TREATMENT WORKS (POTW)

Facility Name

Street Address

City | County

State | Zip ___ - ___

2. OTHER OFF-SITE LOCATIONS - Number these locations sequentially on this and any additional page of this form you use.

☐ **Other off-site location**

EPA Identification Number (RCRA ID. No.)

Facility Name

Street Address

City | County

State | Zip ___ - ___

Is location under control of reporting facility or parent company? ☐ Yes ☐ No

☐ **Other off-site location**

EPA Identification Number (RCRA ID. No.)

Facility Name

Street Address

City | County

State | Zip ___ - ___

Is location under control of reporting facility or parent company? ☐ Yes ☐ No

☐ **Other off-site location**

EPA Identification Number (RCRA ID. No.)

Facility Name

Street Address

City | County

State | Zip ___ - ___

Is location under control of reporting facility or parent company? ☐ Yes ☐ No

☐ Check if additional pages of Part II are attached.

EPA Form 9350-1(1-88)

Figure 2.9. (*Continued*)

(Important: Type or print; read instructions before completing form.)

Page 3 of 5

(This space for EPA use only.)

EPA FORM **R**

PART III. CHEMICAL SPECIFIC INFORMATION

1. CHEMICAL IDENTITY	
1.1	☐ Trade Secret (Provide a generic name in 1.4 below. Attach substantiation form to this submission.)
1.2	CAS # ☐☐☐☐☐☐ - ☐☐ - ☐ (Use leading zeros if CAS number does not fill space provided.)
1.3	Chemical or Chemical Category Name
1.4	Generic Chemical Name (Complete only if 1.1 is checked.)

2.	MIXTURE COMPONENT IDENTITY (Do not complete this section if you have completed Section 1.) Generic Chemical Name Provided by Supplier (Limit the name to a maximum of 70 characters (e.g., numbers, letters, spaces, punctuation)).

3. ACTIVITIES AND USES OF THE CHEMICAL AT THE FACILITY (Check all that apply.)				
3.1	Manufacture:	a. ☐ Produce	b. ☐ Import	c. ☐ For on-site use/processing
		d. ☐ For sale/ distribution	e. ☐ As a byproduct	f. ☐ As an impurity
3.2	Process:	a. ☐ As a reactant	b. ☐ As a formulation component	c. ☐ As an article component
		d. ☐ Repackaging only		
3.3	Otherwise Used:	a. ☐ As a chemical processing aid	b. ☐ As a manufacturing aid	c. ☐ Ancillary or other use

4. MAXIMUM AMOUNT OF THE CHEMICAL ON SITE AT ANY TIME DURING THE CALENDAR YEAR

☐☐ (enter code)

5. RELEASES OF THE CHEMICAL TO THE ENVIRONMENT

You may report releases of less than 1,000 lbs. by checking ranges under A.1.		A. Total Release (lbs/yr) A.1 Reporting Ranges 0	1–499	500–999	A.2 Enter Estimate	B. Basis of Estimate (enter code)	C. % From Stormwater
5.1 Fugitive or non-point air emissions	5.1a					5.1b ☐	
5.2 Stack or point air emissions	5.2a					5.2b ☐	
5.3 Discharges to water 5.3.1 ☐ (Enter letter code from Part I Section 3.10 for streams(s).)	5.3.1a					5.3.1b ☐	5.3.1c
5.3.2 ☐	5.3.2a					5.3.2b ☐	5.3.2c
5.3.3 ☐	5.3.3a					5.3.3b ☐	5.3.3c
5.4 Underground injection	5.4a					5.4b ☐	
5.5 Releases to land 5.5.1 ☐☐☐ (enter code)	5.5.1a					5.5.1b ☐	
5.5.2 ☐☐☐ (enter code)	5.5.2a					5.5.2b ☐	
5.5.3 ☐☐☐ (enter code)	5.5.3a					5.5.3b ☐	

☐ (Check if additional information is provided on Part IV–Supplemental Information.)

EPA Form 9350-1(1-88)

Figure 2.9. (*Continued*)

EPA FORM **R**, Part III (Continued)

6. TRANSFERS OF THE CHEMICAL IN WASTE TO OFF-SITE LOCATIONS

You may report transfers of less than 1,000 lbs. by checking ranges under A.1.	A. Total Transfers (lbs/yr)				B. Basis of Estimate (enter code)	C. Type of Treatment/ Disposal (enter code)
	A.1 Reporting Ranges			A.2 Enter Estimate		
	0	1–499	500–999			
6.1 Discharge to POTW					6.1b ☐	
6.2 Other off-site location (Enter block number from Part II, Section 2.) ☐					6.2b ☐	6.2c ☐☐☐
6.3 Other off-site location (Enter block number from Part II, Section 2.) ☐					6.3b ☐	6.3c ☐☐☐
6.4 Other off-site location (Enter block number from Part II, Section 2.) ☐					6.4b ☐	6.4c ☐☐☐

☐ (Check if additional information is provided on Part IV-Supplemental Information)

7. WASTE TREATMENT METHODS AND EFFICIENCY

A. General Wastestream (enter code)	B. Treatment Method (enter code)	C. Range of Influent Concentration (enter code)	D. Sequential Treatment? (check if applicable)	E. Treatment Efficiency Estimate	F. Based on Operating Data? Yes	No
7.1a ☐	7.1b ☐☐☐	7.1c ☐	7.1d ☐	7.1e %	7.1f ☐	☐
7.2a ☐	7.2b ☐☐☐	7.2c ☐	7.2d ☐	7.2e %	7.2f ☐	☐
7.3a ☐	7.3b ☐☐☐	7.3c ☐	7.3d ☐	7.3e %	7.3f ☐	☐
7.4a ☐	7.4b ☐☐☐	7.4c ☐	7.4d ☐	7.4e %	7.4f ☐	☐
7.5a ☐	7.5b ☐☐☐	7.5c ☐	7.5d ☐	7.5e %	7.5f ☐	☐
7.6a ☐	7.6b ☐☐☐	7.6c ☐	7.6d ☐	7.6e %	7.6f ☐	☐
7.7a ☐	7.7b ☐☐☐	7.7c ☐	7.7d ☐	7.7e %	7.7f ☐	☐
7.8a ☐	7.8b ☐☐☐	7.8c ☐	7.8d ☐	7.8e %	7.8f ☐	☐
7.9a ☐	7.9b ☐☐☐	7.9c ☐	7.9d ☐	7.9e %	7.9f ☐	☐
7.10a ☐	7.10b ☐☐☐	7.10c ☐	7.10d ☐	7.10e %	7.10f ☐	☐
7.11a ☐	7.11b ☐☐☐	7.11c ☐	7.11d ☐	7.11e %	7.11f ☐	☐
7.12a ☐	7.12b ☐☐☐	7.12c ☐	7.12d ☐	7.12e %	7.12f ☐	☐
7.13a ☐	7.13b ☐☐☐	7.13c ☐	7.13d ☐	7.13e %	7.13f ☐	☐
7.14a ☐	7.14b ☐☐☐	7.14c ☐	7.14d ☐	7.14e %	7.14f ☐	☐

☐ (Check if additional information is provided on Part IV-Supplemental Information.)

8. OPTIONAL INFORMATION ON WASTE MINIMIZATION

(Indicate actions taken to reduce the amount of the chemical being released from the facility. See the instructions for coded items and an explanation of what information to include.)

A. Type of modification (enter code)	B. Quantity of the chemical in the wastestream prior to treatment/disposal			C. Index	D. Reason for action (enter code)
	Current reporting year (lbs/yr)	Prior year (lbs/yr)	Or percent change		
☐☐	______	______	______ %	☐.☐	☐☐

EPA Form 9350-1(1-88)

Figure 2.9. (*Continued*)

(Important: Type or print; read instructions before completing form.)

EPA FORM **R**

PART IV. SUPPLEMENTAL INFORMATION

Use this section if you need additional space for answers to questions in Parts I and III. Number or letter this information sequentially from prior sections (e.g., D,E, F, or 5.54, 5.55).

(This space for EPA use only.)

ADDITIONAL INFORMATION ON FACILITY IDENTIFICATION (Part I - Section 3)

3.5	SIC Code		
3.7	Dun & Bradstreet Number(s)		
3.8	EPA Identification Number(s) RCRA I.D. No.)		
3.9	NPDES Permit Number(s)		
3.10	Name of Receiving Stream(s) or Water Body(s)		

ADDITIONAL INFORMATION ON RELEASES TO LAND (Part III - Section 5.5)

Releases to Land		A. Total Release (lbs/yr)				B. Basis of Estimate (enter code)
		A.1 Reporting Ranges			A.2 Enter Estimate	
		0	1-499	500-999		
5.5___ (enter code)	5.5___a					5.5___b
5.5___ (enter code)	5.5___a					5.5___b
5.5___ (enter code)	5.5___a					5.5___b

ADDITIONAL INFORMATION ON OFF-SITE TRANSFER (Part III - Section 6)

		A.Total Transfers (lbs/yr)				B. Basis of Estimate (enter code)	C. Type of Treatment/ Disposal (enter code)
		A.1 Reporting Ranges			A.2 Enter Estimate		
		0	1-499	500-999			
6.___ Discharge to POTW	6.___a					6.___b	
6.___ Other off-site location (Enter block number from Part II, Section 2.)	6.___a					6.___b	6.___c.
6.___ Other off-site location (Enter block number from Part II, Section 2.)	6.___a					6.___b	6.___c.

ADDITIONAL INFORMATION ON WASTE TREATMENT (Part III - Section 7)

A. General Wastestream (enter code)	B. Treatment Method (enter code)	C. Range of Influent Concentration (enter code)	D. Sequential Treatment? (check if applicable)	E. Treatment Efficiency Estimate	F. Based on Operating Data? Yes	No
7.___a	7.___b	7.___c	7.___d	7.___e %	7.___f	
7.___a	7.___b	7.___c	7.___d	7.___e %	7.___f	
7.___a	7.___b	7.___c	7.___d	7.___e %	7.___f	
7.___a	7.___b	7.___c	7.___d	7.___e %	7.___f	
7.___a	7.___b	7.___c	7.___d	7.___e %	7.___f	

EPA Form 9350-1(1-88)

Figure 2.10. (*Continued*)

Table 2.9. Summary of reporting requirements under SARA Title III.

Required Action	Title III Section	Regulation	Final Rule Published in *Federal Register*
Emergency planning	302	40 CFR 355	April 22, 1987
Emergency notification	304	40 CFR 355	April 22, 1987
Submit MSDS or list of chemicals	311	40 CFR 370	Oct. 15, 1987
Submit Tier I inventory	312	40 CFR 370	Oct. 15, 1987
Submit Tier II inventory	312	40 CFR 370	Oct. 15, 1987
Submit toxic chemical release form (EPA Form R)	313	40 CFR 372	Feb. 16, 1988

Source: Reference 14.

chemical identity of a hazardous chemical for specific reasons. Even if the chemical identity is withheld, the generic class or category of the chemical must be provided. The withholder must show each of the following:

- That the information has not been disclosed to any other person other than a member of the local planning committee, a government official, an employee of such person, or someone bound by a confidentiality agreement; that measures have been taken to protect confidentiality; and that the withholder intends to continue to take such measures.
- That the information is not required to be disclosed to the public under any other federal or state law.
- That the information is likely to cause substantial harm to the competitive position of the person.
- That the chemical identity is not readily discoverable through reverse engineering.

Section 323: Provision of Information to Health Professionals

Under circumstances in which chemical identity information can be legally withheld from the public, Section 323 provides for disclosure under certain circumstances to health professionals who need the information for diagnostic purposes or to local health officials who need the information for assessment activities.[14] In these cases, the person receiving the information must be willing to sign a confidentiality agreement with the facility.

Information claimed as a trade secret and substantiation for that claim must be submitted to EPA. This includes information that otherwise would be submitted only to state or local officials, such as emergency and hazardous material inventory information. People may challenge a trade secret claim by petitioning EPA, which then must review the claim and rule on its validity.

Section 324: Public Availability of Plans, Data Sheets, Forms, and Follow-up Notices

The law requires that each emergency response plan, MSDS, chemical inventory form, and toxic chemical release form and follow-up emergency notices be made available to the general public during normal working hours.[17] The EPA administrator, governor, or state emergency response commission must designate locations where the public can inspect these documents. The local emergency planning committee must publish each year, in a local newspaper, a notice that the above-mentioned documents are available for public examination. This notice also must mention that follow-up emergency notices may be issued.

Section 325: Enforcement

Lack of compliance with the provisions of Title III can be punished by assessment of criminal, civil, and administrative penalties as well as citizen suits.[16,17]

Penalties for Lack of Compliance with Emergency Notification (Section 304)

- Class I administrative penalty: A civil penalty of not more than $25,000 per violation may be assessed in the case of violation of the requirements of Section 304. This civil penalty may be assessed only after the accused person is given a notice and an opportunity for a hearing.
- Class II administrative penalty: A penalty of not more than $25,000 per day for each day that the violation of any requirements of Section 304 continues may be assessed. In the case of a second or subsequent violation, the amount of this penalty can be increased to $75,000 per day for each day that the violation continues.
- Criminal penalties: Any person who knowingly and willfully fails to provide notice in accordance with Section 304 can be punished by a fine not to exceed $25,000 or imprisonment for not more than two years, or by both. In the case of a second or subsequent conviction, the person can be fined up to $50,000 or imprisoned for not more than five years or both.

Penalties for Reporting Requirements

- Any person who violates any requirement of Section 312 or 313 shall be liable to the United States for a civil penalty in an amount not to exceed $25,000 for each violation.
- Any person who violates any requirement of Section 311 or 323, and any person who fails to furnish to the EPA administrator information required under Section 322 shall be fined in an amount not to exceed $10,000 for each violation.[17]

Penalties for Disclosure of Trade Secret Information

- Any person who knowingly and willfully discloses any information entitled to protection under Section 322 shall be subject to a fine of not more than $20,000 or to imprisonment not to exceed one year, or both.

Section 326: Civil Actions

Any person may file a civil action suit in U.S. District Court against:

- An owner or an operator of a facility for failure to do any of the following:
  - Submit a follow-up emergency notice under Section 304.
  - Submit an MSDS or a list under Section 311.
  - Complete and submit an inventory form under Section 312.
  - Complete and submit a chemical release form under Section 313.
- The EPA administrator for failure to do any of the following:
  - Publish inventory forms under Section 312.
  - Respond to a petition to add or delete a chemical under Section 313 within 180 days after receipt of the petition.
  - Publish a toxic chemical release form under Section 313.
  - Establish a computer data base under Section 313.
  - Promulgate trade secret regulations under Section 322.
  - Render a decision in response to a petition under Section 322 within nine months after receipt of the petition.

- The EPA administrator, a state governor, or a state emergency response commission, for failure to provide a mechanism for public availability of information in accordance with Section 324.
- A state governor or a state emergency response commission for failure to respond to a request for Tier II information under Section 312 within 120 days after the date of receipt of the request.

Section 327: Transportation Exemption

The transportation of regulated chemicals is exempt from the requirements of Title III. However, any releases of toxic chemicals must be reported in accordance with provisions of Section 304.

Exercises

1. Under Section 301 of Subtitle A, what are the major responsibilities of a state emergency response commission (SERC)?
2. Which industrial facilities are required to comply with emergency planning requirements of Subtitle A?
3. Which industrial facilities must submit copies of MSDSs or a list of hazardous substances to local committees, the state commission, and the local fire department?
4. How can a citizen obtain a copy of the MSDS for a hazardous substance that is used in an industrial facility within his or her community?
5. What aggregate information must facilities provide under Tier I reporting requirements?
6. Which facilities are subject to EPA toxic chemical release reporting?
7. What are the requirements for withholding information on the chemical identify of a substance under the provisions of Subtitle C?
8. What are the responsibilities of the EPA administrator, the governor, or the SERC in terms of making the information on hazardous chemicals available to the public?
9. What are the provisions of Section 304 of Subtitle A of SARA III?
10. What are the penalties for lack of compliance with the provisions of Section 311 or 312 of Subtitle C?

Introduction to the Resource Conservation and Recovery Act (RCRA)

The Resource Conservation and Recovery Act (RCRA),[24-30] an amendment to the Solid Waste Disposal Act, was passed in 1976 to address a problem of enormous magnitude—how to safely dispose of the huge volumes of municipal and industrial solid waste generated nationwide. The goals set by RCRA are:[24]

- To protect human health and the environment.
- To reduce waste and conserve energy and natural resources.
- To reduce or eliminate the generation of hazardous waste as expeditiously as possible.

To achieve these goals, three distinct yet interrelated programs were developed under RCRA. The first program, outlined under Subtitle D of RCRA, encourages states to develop comprehensive plans for the management of nonhazardous solid wastes, such as household waste. The second program, outlined under Subtitle C of the act, deals with hazardous waste management.

From the time that a hazardous waste is generated to the point when it is neutralized, destroyed, or disposed of, the government must know its location and what is being done with it. The last of three programs established under RCRA, outlined in Subtitle I of the act, sets standards and regulates the performance of certain underground storage tanks.

It should be noted that although RCRA sets standards for the management of hazardous and nonhazardous wastes, it does not deal with inactive or abandoned hazardous waste sites. These sites are covered by the Comprehensive Environmental Response, Compensation, and Liability Act (CERCLA), better known as Superfund.

Under provisions of the act, the Environmental Protection Agency (EPA) was required to develop a comprehensive set of regulations to implement the three RCRA programs discussed above. RCRA regulations are being developed continuously. When a regulation is developed, it first is published in the *Federal Register* to allow members of the public to express their opinions of it. RCRA regulations can be found in Volume 40 of the *Code of Federal Regulations* (CFR Parts 240–271). This section will present brief summaries of Subtitle D (solid wastes management) and Subtitle C (hazardous waste management) and discussions of regulations that apply to hazardous waste generators, to transporters of hazardous waste, to TSD facilities, and to underground storage tanks (Subtitle I).

Subtitle D: Solid Waste

Subtitle D of the act establishes a voluntary program through which participating states receive federal financial and technical support to develop and implement solid waste management plans. These plans are, among other things, intended to promote the recycling of solid wastes and to require the closing or upgrading of all environmentally unsound dumps. EPA's role in the Subtitle D program has been to establish regulations for states to follow in developing and implementing their plans, in approving those state plans that comply with such regulations, and in providing grant money for implementing the plans. EPA also has issued minimum technical standards that all solid waste disposal facilities must meet when disposing of solid wastes.

Subtitle D program goals can be summarized as follows:[24,25]

- To encourage environmentally sound solid waste management practices.
- To maximize the recycling of recoverable resources.
- To foster resource conservation.

Definition of Solid Waste

The term "solid waste" is very broad, including not only the traditional nonhazardous solid wastes, such as municipal garbage, but also hazardous solid wastes. The act defines solid waste as:[24,25,26]

- Garbage (e.g., milk cartons and coffee grounds).
- Refuse (e.g., metal scrap, wall board, and empty containers).
- Sludges from a waste treatment plant, water supply treatment plant, or pollution control facility (e.g., scrubber sludges).
- Other discarded material, including solid, semisolid, liquid, or contained gaseous material resulting from industrial, commercial, mining, agricultural, and community activities (e.g., boiler slag, fly ash).

In RCRA's definition of the term "solid waste," it is important to realize that a solid waste is not necessarily found in solid form. By the above definition, a solid waste can exist in a liquid or a gaseous form. Although the term includes both hazardous and nonhazardous waste, Subtitle D of RCRA deals only with nonhazardous wastes. The above definition of solid waste is broad enough to encompass any kind of waste produced by humans. However, the following materials are not considered solid waste under RCRA:[24]

- Domestic sewage (defined as untreated sanitary wastes that pass through a sewer system).
- Industrial wastewater discharges regulated under the Clean Water Act.
- Irrigation return flows.
- Nuclear materials, or by-products, as defined by the Atomic Energy Act of 1954.
- Mining materials that are removed from the ground during the extraction process.

Management of Solid Waste under Subtitle D

The Subtitle D program establishes a solid waste management framework that has two main components:[24]

- Regulations applicable to the development and implementation of state plans.
- Criteria used as minimum technical standards for solid waste disposal facilities and to identify open dumps.

State Plan Regulations

The regulations for state plan development are contained in 40 CFR Part 256. The objective of these regulations is to help states to develop an environmentally sound solid waste management program that can gain EPA approval. Although all states must comply with regulatory requirements, individual states can develop plans which tailored to their own needs. The minimum regulatory requirement for a state plan are summarized below:[24,29]

- Identifying the responsibilities of state, local, and regional authorities in implementing the plan.
- Describing a regulatory scheme that prohibits the establishment of new open dumps, provides for the closing or upgrading of all open dumps, and establishes any state regulatory powers required for the implementation of the plan.
- Ensuring that no state or local government within the state be prohibited from:
  - Establishing long-term contracts for the supply of solid waste to resource recovery facilities or for the operation of such facilities.
  - Securing long-term markets for material and energy recovered from resource recovery facilities.
  - Conserving materials or energy by reducing the waste volume.
- Detailing the combination of practices necessary to use or dispose of solid waste in an environmentally sound manner.

Minimum Technical Standards and Open Dump Criteria

The final component of the Subtitle D program is entitled Criteria for Classification of Solid Waste Disposal Facilities and Practices, commonly referred to as the Subtitle D Criteria (40 CFR Part 257). These criteria are used as:[24]

- A set of minimum technical standards with which all federal and nonfederal solid waste disposal facilities must comply.
- A means of determining if a solid waste disposal facility is an open dump.

The criteria cover eight areas: floodplains, endangered species, surface water, ground water, waste application limits for land used in the production of food chain crops, disease transmission, air, and safety. Under each of these areas, specific requirements are set. For example, 40 CFR Section 257.3-2 (a) and (b) states that:

> Facilities or practices shall not cause or contribute to the taking of any endangered or threatened species of plants, fish, or wildlife . . . [or] result in the destruction or adverse modification of the critical habitat of endangered or threatened species.

The details of minimum technical standards and open dump criteria can be found in 40 CFR Part 257.

Subtitle D has set minimum technical standards for all nonhazardous solid waste disposal facilities. Any facility that does not comply with the minimum standards is considered to be an open dump, and its operation must be upgraded or stopped. It should be noted that the provisions of Subtitle D apply to all nonhazardous solid waste disposal facilities whether or not the state has an approved plan with EPA.

Open Dump Criteria

An open dump is a disposal facility that does not comply with one or more of the criteria set by Subtitle D. A facility identified as an open dump either must be upgraded, or its operation must be stopped.

Certain solid waste disposal practices and facilities are exempt from having to comply with the Subtitle D criteria, including:[24]

- The use of agricultural wastes as fertilizers or soil conditioners.
- The land application of domestic sewage.
- Hazardous waste disposal facilities regulated under Subtitle C of RCRA.
- Industrial discharges that are point sources subject to permits under Section 402 of the Clean Water Act.

Subtitle C of RCRA: Management of Hazardous Waste

Overview

The improper management of hazardous waste is one of the most serious problems confronting the nation. This mismanagement has, in part, polluted the environment, and, if it continues, it has the potential of creating grave environmental consequences. It can result in the poisoning of our drinking water and the air we breathe, the pollution of our rivers and lakes, the destruction of wildlife, and the stripping of vegetation from land areas.

Subtitle C of RCRA was developed to ensure that the mismanagement of hazardous waste would not continue. Under the provisions of Subtitle C, the government must be informed when a hazardous waste is generated, and must be informed of its treatment from the point of generation to its ultimate disposal (''cradle to grave'' reporting). Subtitle C sets forth statutory and regulatory requirements for:[24–26]

- Identifying hazardous waste.
- Regulating generators of hazardous waste.
- Regulating transporters of hazardous waste.
- Regulating owners and operators of facilities that treat, store, or dispose of hazardous wastes.
- Issuing operating permits to owners or operators of treatment, storage, and disposal facilities.
- Enforcing the Subtitle C program.
- Transferring the responsibilities of the Subtitle C program under the federal government to the states.
- Requiring public participation in the Subtitle C program.

Subtitle C was created to assign responsibility for hazardous waste handling and to require that everyone who works with hazardous waste receive proper training in:

- General hazardous waste regulations.
- Safety practices.
- Chemical hazard recognition.
- Use of protective clothing and equipment.

- Use of respiratory protection.
- Emergency response procedures.
- Facility operation and maintenance procedures.

Any employer may have to report to EPA on how the company is meeting its RCRA responsibilities.

Subtitle C also requires that hazardous wastes be identified, handled, stored, treated, and disposed of correctly. The law requires companies that handle hazardous waste to:

- Obtain an identification number and permit from EPA.
- Identify and analyze new hazardous wastes.
- Provide a secure facility that keeps unauthorized people out.
- Inspect the facility regularly.
- Have a contingency plan for fire, explosion, and spills.
- Practice emergency response for fire, explosion, and spills.
- Provide proper protective clothing and equipment.
- Maintain EPA-required records.

The penalties for failure to comply with regulatory provisions range from administrative actions (e.g., warning letters) to criminal prosecution (including jail terms and fines).

Definition of Hazardous Waste

The RCRA definition of a hazardous waste is simple: "It is a solid waste that is either ignitable, corrosive, reactive, or toxic." Unfortunately, this definition of a hazardous waste does not reflect the complexity inherent in identifying a hazardous waste, or the complexity of the rules and the scope of their coverage. Some materials are mentioned by name, and some by the industrial processes in which they are used or produced; also, many substances are excluded. Note that RCRA defines hazardous wastes in terms of the properties of a solid waste; so if a waste is not a solid waste, it cannot be a hazardous waste. The amount of hazardous waste generated by U.S. industries is staggering. A 1985 EPA Survey identified 2959 facilities, regulated under RCRA, which managed a total of 247 million tons of wastes.[25] This translates into over 1 ton of hazardous waste generated per person, per year, in the United States. The vast majority of the waste comes from the chemical and petroleum industries, which alone generate 71% of all hazardous wastes produced. The remainder comes from a wide range of other industries, including metal finishing, general manufacturing, and transportation.

Hazardous waste, as defined in 40 CFR Part 261, is a solid waste that meets one or more of the following conditions:

- Exhibits, on analysis, any of the characteristics of a hazardous waste (ignitability, corrosivity, reactivity, toxicity).
- Has been named as a hazardous waste and listed.
- Is a mixture containing a listed hazardous waste and a nonhazardous solid waste (unless the mixture is specifically excluded or no longer exhibits any of the characteristics of a hazardous waste).
- Is not excluded from regulation as a hazardous waste.

Furthermore, the by-products of the treatment of any hazardous waste are also considered hazardous unless specifically excluded.

The primary responsibility in determining whether a solid waste exhibits any of the characteristics of a hazardous waste rests with the generators.[24–26]

Ignitability

A solid waste that falls in any of the following categories is considered a hazardous waste because of its ignitability:

- A liquid, except for aqueous solutions containing less than 24% alcohol, that has a flash point less than 60°C (140°F).
- A nonliquid that is capable, under normal conditions, of spontaneous and sustained combustion.
- An ignitable compressed gas per Department of Transportation (DOT) regulation.
- An oxidizer per DOT regulation.

Examples of ignitable wastes include waste oils and solvents. The main reason for defining ignitability is to identify those hazardous wastes that are capable of creating a fire during handling, transportation, or disposal.

Corrosivity

A solid waste that falls into one of the following categories is considered a hazardous waste because of its corrosivity:

- An aqueous material with pH less than or equal to 2 or greater than or equal to 12.5.
- A liquid that corrodes steel at a rate greater than ¼ inch per year at a temperature of 55°C (130°F).

The pH was selected as an indicator of corrosivity because most strong acids (pH < 2) and strong caustics (pH > 12.5) can react violently with other wastes, producing flammable or toxic products.

Steel corrosivity is an important consideration in dealing with a corrosive waste, as the waste might be capable of escaping from a corroded steel container and liberating other wastes. Examples of corrosive wastes include acidic wastes and used pickle liquor employed for cleaning steel.

Reactivity

A solid waste that behaves in any of the following ways is considered a hazardous waste because of its reactivity:

- Is normally unstable and reacts violently without detonating.
- Reacts violently with water.
- Forms an explosive mixture with water.
- Generates toxic gases, vapors, or fumes when mixed with water.
- Contains cyanide or sulfide and generates toxic gases, vapors, or fumes at a pH between 2 and 12.5.
- Is capable of detonation if heated under confinement or subjected to a strong initiating source.
- Is capable of detonation at standard temperature and pressure.
- Is listed by DOT as a Class A or B explosive.

Reactivity was chosen as a characteristic to identify unstable wastes that can pose a problem at any stage of the waste management cycle (e.g., an explosion). Examples of reactive wastes include water from TNT operations and used cyanide solvents.

EP Toxicity

Toxicity in a waste stream is determined by an extraction leaching procedure that simulates the natural conditions of landfills. The Subtitle C regulations contain a list of substances that, if their concentration exceeds a certain value, will characterize a stream as a hazardous waste. Substances that would render a waste hazardous by virtue of its toxicity, along with their threshold concentrations, are listed in Table 2.10.

Testing for Waste Characterization

Every generator of waste, of whatever description, has the burden of determining whether the waste is hazardous through sampling and analysis with an EPA-approved procedure. In addition, any other facility that handles waste (e.g., for treatment or disposal) may have to make this determination for an unidentified waste. The generator is also responsible for conducting tests on a representative sample of the waste to obtain results that properly describe its nature.

Listing of Hazardous Wastes

A solid waste is considered to be a hazardous waste if it is listed in any of EPA's lists. EPA has established these lists by considering substances' toxic effects. There are thousands of hazardous chemicals, with varying degrees of associated risk, and it is impossible to expose people to every substance to see how toxic it is. Scientists try to get an idea of poison risks by testing chemicals on laboratory animals. Some substances are toxic if inhaled, some if they get on the skin, and some if they are swallowed. Although not all the effects of toxic chemicals are known, such testing provides the basis for "listing" those chemicals that should be isolated from the environment.

The substances that are considered toxic are specified in the regulations by their generic chemical names. Any solid waste that contains one or more of these substances must be handled as a hazardous waste. EPA summarizes the list of toxic substances on two different chemical rosters. Its "P" list contains "acutely toxic" chemicals; special procedures must be followed in discarding this class of hazardous wastes. The other chemical list, known as the "U" list, contains other toxic chemicals (i.e., those that are not acutely toxic). Less stringent regulations govern this class of hazardous wastes.

Table 2.10. Constituents and concentrations for EP toxicity.

Constituent	Concentration, mg/1
Arsenic	5.0
Barium	100.0
Cadmium	1.0
Chromium	5.0
Lead	5.0
Mercury	0.2
Selenium	1.0
Silver	5.0
Endrin	0.02
Lindane	0.4
Methoxychlor	10.0
Toxaphene	0.5
2,4-D	10.0
2,4,5-TP	1.0

Source: Reference 24.

The Subtitle C regulations also list a number of wastes from specific industrial processes that are believed to produce hazardous waste, with a ''K'' list that describes wastes from about 75 industrial processes. A fourth list describes wastes generated in a variety of nonspecific industrial processes; these waste streams are characterized as ''F wastes.'' The ''F'' list includes spent halogenated and nonhalogenated solvents (F001–F005), which are common industrial hazardous wastes.[24–26]

Mixtures

The provisions of Subtitle C provide that any mixture of a solid waste that contains a hazardous ingredient should be treated and managed as a hazardous waste. The concentration of the hazardous ingredient in the mixture is not a factor. This rule was set forth by EPA to prevent generators from creating a major loophole in Subtitle C by commingling listed wastes with nonhazardous solid wastes. The exceptions to the mixture rule are:[24]

- If a wastewater discharge subject to regulation by the Clean Water Act is mixed with low concentrations of a listed waste, as specified in 40 CFR Section 261.3, the resultant mixture is not considered a listed hazardous waste. Of course, if such a mixture exhibited one of the hazardous waste characteristics, it would be deemed hazardous.
- Mixtures of nonhazardous wastes and listed wastes that were listed for exhibiting a characteristic are not considered hazardous if the mixture no longer exhibits any hazardous waste characteristics.

Solid Waste Exclusions

Congress has excluded from regulation a variety of waste streams that might otherwise be categorized as solid waste. The following categories of waste are excluded either because they already are regulated or because they are not considered hazardous enough to warrant regulation:[24]

- Domestic sewage.
- Industrial wastewater licensed under the National Pollutant Discharge Elimination System (NPDES).
- Irrigation waste or return water.
- Nuclear materials covered by the Atomic Energy Act.
- Mining wastes.
- Household wastes such as trash.
- Wastes that are returned to the earth as fertilizer.
- Products of combustion of a fossil fuel.
- Energy and mineral wastes such as drilling fluids and wastes from oil, gas, and geothermal energy exploration.

Small Quantity Generators (SQG)

Small quantity generators of hazardous waste are exempt from most of the hazardous waste regulatory requirements.[24,25,30] A small quantity generator was defined in Subtitle C as:

- A generator who produced less than 1,000 kg of hazardous waste at a site per month (or accumulated less than 1,000 kg at any one time).
- A generator who produced less than 1 kg of acutely hazardous waste per month (or accumulated less than 1 kg at any one time).

These small quantity generators were required to meet some minimum management requirements, including testing their waste, storing the waste properly, and disposing of the waste at approved facilities.

After 1980, concern developed that hazardous wastes exempted from regulation due to the SQG exclusion could be causing environmental harm. Therefore, Congress amended the definition of a SQG, reducing the cutoff point from 1,000 kg to 100 kg. Thus, the present definition of a SQG is:

- A generator who produces less than 100 kg of hazardous waste at a site per month (or accumulates less than 100 kg at any one time).
- A generator who produces less than 1 kg of acutely hazardous waste per month (or accumulates less than 1 kg at any one time).

Regulations That Apply to Hazardous Waste Generators

Any generator of hazardous waste not classified as a small quantity generator must comply with regulations under Subtitle C that are contained in 40 CFR Part 262. The generators of waste are responsible for conducting appropriate, EPA-approved tests to determine whether the solid waste that is generated is hazardous. If the waste is determined to be hazardous, it must be properly transported to an RCRA treatment, storage, or disposal facility (TSD).

Regulatory Requirements

The regulatory requirements for hazardous waste generators include:[24,25,28]

- Obtaining an EPA ID number.
- Handling of hazardous waste before transport.
- Manifesting of hazardous waste.
- Recordkeeping and reporting.

EPA ID Number

One way that EPA monitors and tracks generators is by assigning each generator a unique identification number. Without this number the generator is barred from treating, storing, disposing of, transporting, or offering for transportation any hazardous waste. Furthermore, the generator is forbidden from offering hazardous waste to any transporter or treatment, storage, and disposal facility that does not have an EPA ID number.

Pre-transport Regulations

Pre-transport regulations are designed to ensure the safe transportation of a hazardous waste from its origin to ultimate disposal. In developing these regulations, EPA adopted those used by the Department of Transportation (DOT) for transporting hazardous waste (49 CFR Parts 172, 173, 178, and 179). The DOT regulations require:

- Proper packaging to prevent leakage of hazardous waste, both during normal transport conditions and in potentially dangerous situations.
- Identification of characteristics and dangers associated with the wastes being transported through labeling, marking, and placarding of the packaged waste.

The above regulations apply to generators who ship containers of hazardous waste. Any generator can store a container of hazardous waste on site for a period of 90 days as long as he or she meets the following requirements:

- Proper storage: The waste is properly stored in containers or tanks marked with the words "Hazardous Waste" and the date on which accumulation began.
- Emergency plan: A contingency plan and procedures to use in an emergency must be developed.
- Personnel training: Facility personnel must be trained in the proper handling of the hazardous waste.

A generator who decides to store hazardous waste for a period greater than 90 days will be considered the operator of a hazardous waste storage facility and must comply with provisions for such facilities under Subtitle C.

The Manifest

As was mentioned before, the government must be informed of the fate of any hazardous waste from cradle to grave. The uniform hazardous waste manifest, shown in Figure 2.10, enables the government and generators to keep track of the movement of hazardous wastes. The following information is included in a manifest:

- Name and EPA identification number of the generator, the transporter(s), and the facility where the waste is to be treated, stored, or disposed of.
- U.S. DOT description of the waste being transported.
- Quantities of the waste being transported.
- Address of the treatment, storage, or disposal facility to which the generator is sending the waste (the designated facility).

In addition each manifest must certify that:

- The generator has in place a program to reduce the volume and toxicity of the waste to the degree economically practicable, as determined by the generator.
- The treatment, storage, or disposal method chosen by the generator is that currently available practicable method that minimizes the risk to human health and the environment.

Manifest Routing

Each time the hazardous waste changes hands, the manifest must be signed by the party who takes control of it. Under normal circumstances, the generator prepares the manifest and signs it. The hazardous waste and the manifest then are handed to the transporter, who also signs the manifest and gives a copy to the generator. When the hazardous waste with the manifest is delivered to the treatment, storage, and disposal facility, the operator of the facility must return a copy of the manifest to the generator. With this system, the generator is assured that the waste has reached its site of ultimate disposal. Figure 2.11 shows the manifest routing and the cradle-to-grave tracking system for a hazardous waste.

Recordkeeping and Reporting

The recordkeeping and reporting requirements for generators provide EPA and the individual states with a method for tracking the quantities of waste generated and the movement of hazardous

Please print or type. *(Form designed for use on elite (12-pitch) typewriter.)* *Form Approved. OMB No. 2050-0039. Expires 9-30-88*

UNIFORM HAZARDOUS WASTE MANIFEST

1. Generator's US EPA ID No. Manifest Document No.

2. Page 1 of

Information in the shaded areas is not required by Federal law.

3. Generator's Name and Mailing Address

A. State Manifest Document Number

B. State Generator's ID

4. Generator's Phone ( )

5. Transporter 1 Company Name

6. US EPA ID Number

C. State Transporter's ID

D. Transporter's Phone

7. Transporter 2 Company Name

8. US EPA ID Number

E. State Transporter's ID

F. Transporter's Phone

9. Designated Facility Name and Site Address

10. US EPA ID Number

G. State Facility's ID

H. Facility's Phone

GENERATOR

11. US DOT Description *(Including Proper Shipping Name, Hazard Class, and ID Number)*	12. Containers No.	12. Containers Type	13. Total Quantity	14. Unit Wt/Vol	I. Waste No.
a.					
b.					
c.					
d.					

J. Additional Descriptions for Materials Listed Above

K. Handling Codes for Wastes Listed Above

15. Special Handling Instructions and Additional Information

16. **GENERATOR'S CERTIFICATION:** I hereby declare that the contents of this consignment are fully and accurately described above by proper shipping name and are classified, packed, marked, and labeled, and are in all respects in proper condition for transport by highway according to applicable international and national government regulations.

If I am a large quantity generator, I certify that I have a program in place to reduce the volume and toxicity of waste generated to the degree I have determined to be economically practicable and that I have selected the practicable method of treatment, storage, or disposal currently available to me which minimizes the present and future threat to human health and the environment; **OR**, if I am a small quantity generator, I have made a good faith effort to minimize my waste generation and select the best waste management method that is available to me and that I can afford.

Printed/Typed Name | Signature | *Month Day Year*

TRANSPORTER

17. Transporter 1 Acknowledgement of Receipt of Materials

Printed/Typed Name | Signature | *Month Day Year*

18. Transporter 2 Acknowledgement of Receipt of Materials

Printed/Typed Name | Signature | *Month Day Year*

FACILITY

19. Discrepancy Indication Space

20. Facility Owner or Operator: Certification of receipt of hazardous materials covered by this manifest except as noted in Item 19.

Printed/Typed Name | Signature | *Month Day Year*

EPA Form 8700-22 (Rev. 9-86) Previous editions are obsolete.

Figure 2.10. The uniform hazardous waste manifest.

wastes. Subtitle C contains three primary recordkeeping and reporting requirements:

- Biennial reporting.
- Exception reporting.
- Three-year retention of reports, manifests, and test records.

Please print or type. (Form designed for use on elite (12-pitch) typewriter.)

Form Approved. OMB No. 2050-0039. Expires 9-30-88

UNIFORM HAZARDOUS WASTE MANIFEST *(Continuation Sheet)*	21. Generator's US EPA ID No.	Manifest Document No.	22. Page	Information in the shaded areas is not required by Federal law.
23. Generator's Name			L. State Manifest Document Number	
			M. State Generator's ID	
24. Transporter ____ Company Name	25. US EPA ID Number		N. State Transporter's ID	
			O. Transporter's Phone	
26. Transporter ____ Company Name	27. US EPA ID Number		P. State Transporter's ID	
			Q. Transporter's Phone	

28. US DOT Description *(Including Proper Shipping Name, Hazard Class, and ID Number)*	29. Containers No.	29. Containers Type	30. Total Quantity	31. Unit Wt/Vol	R. Waste No.
a.					
b.					
c.					
d.					
e.					
f.					
g.					
h.					
i.					

GENERATOR

S. Additional Descriptions for Materials Listed Above	T. Handling Codes for Wastes Listed Above

32. Special Handling Instructions and Additional Information

TRANSPORTER

33. Transporter ____ Acknowledgement of Receipt of Materials		Date
Printed/Typed Name	Signature	Month Day Year
34. Transporter ____ Acknowledgement of Receipt of Materials		**Date**
Printed/Typed Name	Signature	Month Day Year

FACILITY

35. Discrepancy Indication Space

EPA Form 8700-22A (Rev. 9-86) Previous edition is obsolete.

Figure 2.10. (*Continued*)

Biennial Reporting

Generators who transport hazardous waste off-site must submit a biennial report to the regional administrator of the EPA by March 1 of each even-numbered year. The report details the gen-

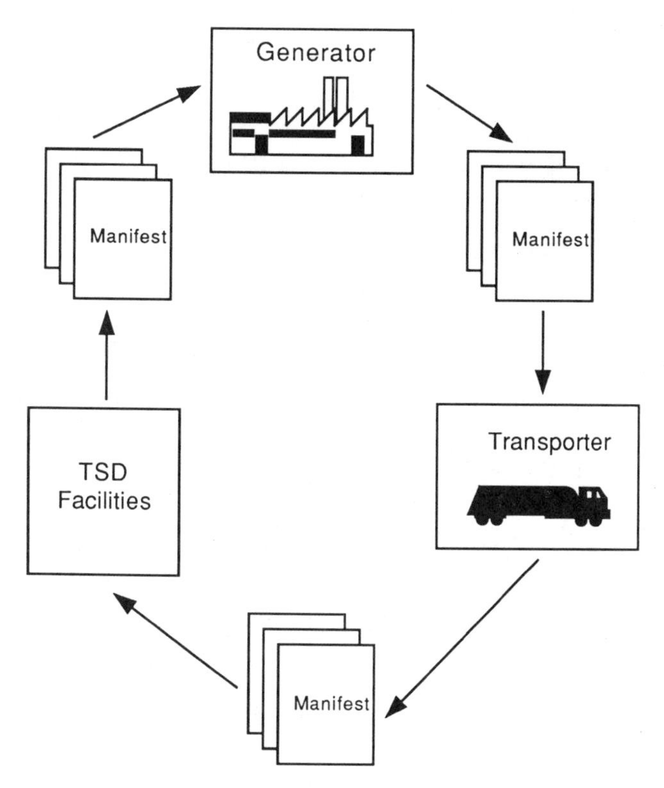

Figure 2.11. The cradle-to-grave tracking system for a hazardous waste. (Source: Reference 24)

erator's activities during the previous calendar year, including:

- The EPA identification number and the name of each transporter used during the year.
- The EPA identification number, name, and address of each off-site treatment, storage, or disposal facility to which waste was sent during the year.
- Quantities and nature of the hazardous waste generated.

The biennial report also should include a description of:

- Efforts taken to reduce the volume and toxicity of the wastes generated.
- Changes in volume or toxicity that were actually achieved, as compared with those achieved in previous years.

Generators who treat, store, or dispose of their hazardous waste on-site also must submit a biennial report that contains a description of the type and quantity of hazardous waste the facility handled during the year, and the method(s) of treatment, storage, or disposal used.

Exception Reports

In addition to the biennial report, generators who transport waste off-site must submit an exception report to the regional administrator of the EPA if they do not receive a copy of the manifest signed and dated by the owner or operator of the designated facility within 45 days of the date on which the initial transporter accepted the waste. The exception report must describe efforts taken to locate the waste and the results of those efforts.

Three-Year Retention of Reports, Manifests, and Test Records

The generator must keep a copy of each biennial report and any exception reports for a period of at least three years from the date the report was submitted. The generator also is required to keep a copy of all manifests for three years, or until he or she receives a copy of the manifest signed and dated from the owner or operator of the designated facility. The manifest from the facility then must be kept for at least three years from the date on which the hazardous waste was accepted by the initial transporter. Finally, the records of the waste analyses and determinations undertaken by the generator must be kept for at least three years from the date that the waste was last sent to an on-site or off-site TSD.

Regulations That Apply to Transporters of Hazardous Waste

EPA and DOT have jointly developed regulations that goven the transportation of hazardous waste.[24,25,28] These regulations, however, are not included in the same act, but are contained in DOT regulations under 49 CFR Parts 171 through 179 and in EPA regulations in 40 CFR Part 263 (Subtitle C of RCRA). A transporter of hazardous waste must be sure to comply with both EPA and DOT regulations. A transporter under Subtitle C is defined as ''any person engaged in off-site transportation of manifested hazardous waste, by air, rail, highway or water.''

Regulatory Requirements

A transporter is responsible for:

- Obtaining an EPA ID number: This would enable EPA to keep track of the transportation of hazardous waste by assigning a unique ID number to each transportation company. A transporter without an EPA ID number is forbidden from handling any hazardous waste.
- Complying with the manifest system: The major responsibility of the transporter is to deliver the waste to the TDS facility indicated on the manifest. Before the waste is handed over to the TDS facility, the manifest must be signed by the TSD facility, and the transporter must retain a copy of it for three years.
- Dealing with hazardous waste discharges: Transportation of hazardous waste is a dangerous activity, and it is always possible that an accident will occur. To deal with this possibility, the regulations require that transporters take immediate action to protect public health and the environment if a release occurs (e.g., notifying local authorities and/or diking off the discharge area).

Regulations That Apply to Treatment, Storage, and Disposal (TSD) Facilities

TSD facilities are the last link in the cradle-to-grave chain of the hazardous waste management system. Provisions of Subtitle C require that owners or operators of such facilities obtain a permit from EPA and abide by the regulations governing the operation of these facilities. The major objective of these regulations is to minimize the release of hazardous substances into the environment.[24–28]

Definition of a TSD Facility

A TSD, according to 40 CFR Part 260.10, encompasses three different functions:

- Treatment: Any method, technique, or process, including neutralization, designed to change the physical, chemical, or biological character or composition of any hazardous waste so as to neutralize it, or render it nonhazardous or less hazardous, or to recover it, make it safer to transport, store, or dispose of, or amenable for recovery, storage, or volume reduction.
- Storage: The holding of hazardous waste for a temporary period, at the end of which the hazardous waste is treated, disposed of, or stored elsewhere.
- Disposal: The discharge, deposit, injection, dumpling, spilling, leaking, or placing of any solid waste or hazardous waste into or on any land or water so that any constituent thereof may enter the environment or be emitted into the air or discharged into any waters, including ground waters.

Regulations Governing TSD Facilities

The act establishes two categories of regulations based on a facility's status regarding a permit: interim standards and permit standards. In developing the regulations, EPA had realized that it would take many years to issue or deny permits for all TSD facilities. Therefore, all TSDs in existence on or before November 19, 1980 were given ''interim'' status and were allowed to operate without a permit until the approval or denial of their permit application by EPA.

Generally, the requirements for interim status are much less stringent than those for permit status. Both interim and permit standards consists of two parts:[24,28]

- Administrative and nontechnical requirements.
- Technical requirements.

The administrative and nontechnical requirements for interim and permit facilities are almost identical. However, there is a substantial difference between the technical requirements of the two types of TSD facilities. These requirements are briefly discussed below.

Administrative and Nontechnical Requirements for Interim and Permit TSD Facilities

The objective of these regulations is to ensure that the owners and operators of TSD facilities have set forth adequate methods and procedures for the operation of their facilities. The term ''operation'' also includes the capability of dealing with any emergency situations that might arise as a result of handling hazardous wastes. These regulations fall into the following categories:

- Waste analysis.
- Security.
- Inspections.
- Training.
- Ignitable, reactive, or incompatible wastes.
- Location standards (permitted facilities).
- Preparedness and prevention.
- Contingency plans and emergency procedures.
- Manifest system, recordkeeping, and reporting.

All owners or operators of TSD facilities must meet these regulations except for the following:

- A farmer disposing of pesticides obtained for his or her own use.
- The owner or operator of a total enclosed treatment facility.
- The owner or operator of a neutralization unit or a waste water treatment unit. (Publicly owned treatment works that mix hazardous waste with other wastes are regulated.)
- A person responding to a hazardous waste spill or discharge.
- Facilities that reuse, recycle, or reclaim hazardous waste. (Persons who produce, burn, and distribute hazardous waste–derived fuel and used oil recyclers are regulated as a result of HSWA.
- Generators (including small quantity generators) accumulating wastes within the time periods specified in Sections 3001 and 3002 of the act.
- A transporter storing manifested shipments less than 10 days.
- A facility regulated by an authorized state program. (Such facilities are regulated by the state program instead of the federal program.)

General Facility Standards

Before handling any hazardous waste, owners or operators must obtain an ID number from EPA. They also must ensure that their wastes are properly identified and handled, and that their personnel have adequate training in the management of hazardous waste. To meet these requirements, the owners or operators of TSD facilities must:

- Conduct a waste analysis to ensure that they understand the hazards posed by the particular waste.
- Install security measures to minimize the unauthorized entry of people or livestock onto the facility.
- Conduct an inspection to detect any potential problem areas within the facility. Inspection procedures must be in writing, and any problems found during inspections must be remedied and recorded. The record of any problems identified must be kept for three years.
- Conduct training of personnel to minimize environmental damage due to mistakes in handling the hazardous waste. The personnel must be trained no later than six months after they are assigned to a job, and the records of such training must be kept by the owner or operator of the facility.
- Properly manage ignitable, reactive, or incompatible wastes. Doing so would necessitate the elimination of sources of ignition where ignitibles or reactives are handled, and the proper storage of incompatible wastes.

Emergency Procedures

The regulations requires that owners or operators of TSD facilities prepare for emergency situations by taking proper measures such as installing firefighting equipment, alarms, and so on. Every TSD facility also must have a contingency plan with a designated emergency coordinator who is responsible for directing personnel and equipment in an emergency situation.

Manifest System, Recordkeeping, and Reporting

These regulations mandate that the manifest be returned to the generator to complete the manifest loop. In addition to the manifest requirements, this section emphasizes requirements for recordkeeping and reporting, including operating records, biennial reports, unmanifested waste reports, and reports on releases, groundwater contamination, and closure.

Technical Requirements for Interim Status TSD Facilities

The main objective of these technical requirements is to minimize the environmental impact of hazardous wastes from operating facilities waiting to obtain a permit from EPA. There are two groups of requirements: (1) general standards that apply to several types of facilities, and (2) specific standards that apply to a waste management method.

General Standards. The general standards cover three areas, as described in the following paragraphs.

Groundwater Monitoring Requirements. The owners or operators of a surface impoundment, landfill, and land treatment facilities are mandated to develop procedures to adequately determine the concentration of hazardous substances in groundwater systems. This monitoring must continue for the life of the facility except for landfills, where it must continue for 30 years after the facility has closed.

Closure, Post-closure Requirements. Closure is the period when wastes are no longer accepted, during which owners or operators of TSD facilities complete treatment, storage, and disposal operations, apply final covers to or cap landfills, and dispose of or decontaminate equipment, structures, and soil. Post-closure, which applies only to disposal facilities, is the 30-year period after closure during which owners or operators of disposal facilities conduct monitoring and maintenance activities to preserve and look after the integrity of the disposal system. The purpose of the closure and post-closure requirements is to ensure that all facilities are closed in a manner that (1) minimizes the need for care after closure, and (2) controls, minimizes, or eliminates the escape of waste, leachate, contaminated rainfall, or waste decompostion products to ground or surface waters and the atmosphere. An owner or operator must develop a plan for closing the facility and keep it on file at the facility until closure is completed and certified. This plan must include:

- A description of how the facility will be closed.
- An estimate of the maximum amount of waste the facility will handle.
- A description of the steps needed to decontaminate equipment during closure.
- An estimate of the year of closure.
- A schedule for closure.

Post-closure is required for disposal facilities. When a disposal facility is closed, it must be monitored for 30 years to ensure the integrity of any waste containment systems and to detect contamination. Post-closure care consists of at least the following:

- Groundwater monitoring and reporting.
- Maintenance and monitoring of waste containment systems.
- Security.

Financial Requirements. Financial requirements were established to assure that funds are available to pay for closing a facility, for rendering post-closure care at disposal facilities, and to compensate third parties for bodily injury and property damage due to sudden and nonsudden accidents related to the facility's operation. (States and the federal government are exempted from abiding by these requirements.) There are two kinds of financial requirements: (1) financial assurance for closure/post-closure, and (2) liability coverage for injury and property damage.

The first step owners and operators must take in meeting the financial assurance requirements is to prepare written cost estimates for closing their facilities. If post-closure care is required, a cost estimate for providing this care also must be prepared. These cost estimates must reflect the actual cost of conducting all the activities outlined in the closure and post-closure plans, and they are adjusted annually for inflation. The cost estimate for closure is based on the point in the facility's operating life when closure would be the most expensive. Cost estimates for post-closure monitoring and maintenance are based on projected costs for the entire post-closure period.

Following the preparation of the cost estimates, the owner or operator must demonstrate to EPA the ability to pay the estimated amounts. This is known as financial assurance.

An owner or operator is financially responsible or liable for bodily injury and property damage to third parties caused by a sudden or a nonsudden accidental occurrence due to operations at a facility. Sudden occurrences usually are due to an accident, such as an explosion or fire. Nonsudden occurrences take place over a long period of time and include groundwater and surface-water contamination. Separate liability coverage for each of these two types of occurrences must be obtained.

Specific Standards. These regulations cover and set minimum technical standards for ten specific waste management methods:

- Containers
- Tanks
- Surface impoundments
- Waste piles
- Land treatment
- Landfills
- Incinerators
- Thermal treatment
- Chemical, physical, biological treatment
- Underground injection

Although the requirements are facility-specific, there are common elements in each of them:

- Waste analysis
- Monitoring and inspection
- Closure/post-closure
- Recordkeeping
- Requirements for ignitable, reactive, and incompatible wastes
- General operating requirements

The description of standards for each of the ten waste management techniques is beyond the scope of this book, but a complete description can be found in Subparts I through R of Volume 40 of the *Code of Federal Regulations*, Part 265. The method-specific requirements for the five elements outlined above can found in 40 CFR Part 265, Subparts I through P.

Technical Requirements for Permit TSD Facilities

The standards for a permit facility differ from those of an interim facility in the following areas:

- Permitting standards are more extensive and stringent than interim standards.

- Permitting standards require an owner or operator to take corrective action if groundwater contamination is detected.
- These standards compel owners and operators using the various waste management methods to design their units to prevent the release of hazardous waste to the environment.
- The specific requirements that an owner or operator of a permit facility must comply with are developed for each facility by permit writers, based on their best engineering judgment on a case-to-case basis.

As with interim standards, the technical requirements of permit standards are divided into two groups: (1) general standards and (2) specific standards.

General Standards. The general standards cover three areas: groundwater monitoring requirements (Subpart F), closure/post-closure requirements (Subpart G), and financial requirements (Subpart H). A summary of the groundwater monitoring requirements is presented below. Closure/post-closure and financial requirements for permit facilities are similar to the corresponding requirements under interim status.

Groundwater Monitoring Requirements. The groundwater protection requirements are more specific than those found under interim status. They also differ by requiring the owner or operator to clean up any groundwater contamination. There are three parts to the groundwater requirements: a detection monitoring program, a compliance monitoring program, and a corrective action program.

Detection monitoring is conducted to determine if hazardous wastes are leaking from a TSD. Monitoring is conducted at a compliance point specified in the permit. This point is located at the edge of the waste management area, best envisioned as an imaginary line on the outer limit of one or a group of disposal units.

The objective of the compliance monitoring program is to evaluate the concentration of certain hazardous constituents in the groundwater in determine if groundwater contamination is occurring. Each permit specifies constituents, and their concentration limits, that owners or operators must monitor. If compliance monitoring indicates any statistically significant increase in the concentration limits for those hazardous constituents specified in the permit, then corrective action must be instituted.

The objective of the corrective action program is to bring facility-contamininating groundwater into compliance. The owner or operator achieves this by removing the hazardous waste constituents from the groundwater or treating the groundwater in place. The permit details specific actions required to remove the contamination.

Specific Standards. Facility-specific standards cover the following waste management methods:

- Containers
- Tanks
- Surface impoundments
- Waste piles
- Land treatment
- Landfills
- Incinerators

Notes that there are no facility-specific standards for underground injection, thermal treatment, or chemical, physical, and biological treatment. The details of regulations covering each of the areas outlined above can be found in 40 CFR Part 264.

Regulations That Apply to Underground Storage Tanks (Subtitle I)

As many as 1.5 million underground storage tanks are used in the United States to contain hazardous substances or petroleum products. An estimated 100,000 to 3000,000 of these tanks now are leaking and polluting underground water supplies, and more may begin to leak in the next few years. In addition to causing groundwater contamination, leaking tanks can damage sewer lines and buried cables, poison crops, and contribute to fires and explosions. To address this problem, Congress created a new program to control and prevent leakage from underground storage tanks (USTs). The UST program breaks new ground in that, for the first time, the RCRA program applies to products as well as wastes. Subtitle I regulates underground tanks storing petroleum products (including gasoline and crude oil) and any substance defined as hazardous under Superfund.[24] It is important to note that Subtitle I does not regulate tanks storing hazardous wastes as defined by RCRA. Such tanks are already regulated under Subtitle C, which states:

> Tanks, which are stationary devices designed to contain an accumulation of hazardous waste and constructed primarily of non-earthen materials, are regulated in much the same way as containers. Persons using tanks, either to store or treat wastes, must manage the tanks to avoid leaks, ruptures, spills and corrosion. This includes using freeboard or a containment structure (e.g., dike or trench) to prevent and contain escaping wastes, and having a shutoff or bypass system installed to stop liquid from flowing into a leaking tank.

The following paragraphs are a summary of regulations that apply to USTs under Subtitle I.

Scope of Application

An underground storage tank is defined as any tank with at least 10 percent of its volume buried below ground, including any pipes attached to the tank. Any owner or operator, including federal entities, who stores petroleum products or a substance defined as hazardous under Superfund in an underground tank must meet the new UST requirements. The UST program does not apply to:

- Tanks holding a hazardous waste regulated under the RCRA hazardous waste program (Subtitle C).
- Farm and residential tanks with a holding capacity of less than 1,100 gallons of motor fuel.
- On-site tanks storing heating oil.
- Septic tanks.
- Pipeline regulated under other laws.
- Surface impoundments.
- Systems for collecting storm water and wastewater.
- Flow-through process tanks.
- Liquid traps or associated gathering lines related to operations in the oil and natural gas industry.

The Program

The UST program outlined in the act has five parts: a ban on unprotected new tanks, a notification program, a regulatory program, state authorization, and inspections and enforcement. These parts are described below.

Ban on Unprotected New Tanks. A provision banning underground installation of unprotected new tanks went into effect on May 7, 1985. Currently, no person may install an underground storage tank unless:

- It will prevent releases of the stored substances due to corrosion or structural failure for the life of the tank.
- It is protected against corrosion, constructed of noncorrosive material or steel clad with noncorrosive material, or designed to prevent the release or the stored substances.
- The material used in the construction or lining of the tank is compatible with the substance to be stored.

The maximum penalty is $10,000 per tank for each day this provision is violated.

Notification. Subtitle I calls for a notification program that may affect several million tank owners. This program requires, in part, that owners of existing or newly installed underground storage tanks notify the state or local agency of each tank's age, size, type, location, and use.

Regulatory Program. EPA has developed regulations that specify performance standards for new underground storage tanks as well as regulations covering leak detection, leak prevention, and corrective action for both new and existing tanks. These regulations classify tanks into those containing petroleum and those holding hazardous substances.

The law specifies that the leak detection/prevention and corrective action regulations must require owners/operators of underground storage tanks to:

- Have methods for detecting releases.
- Keep records of the methods.
- Take corrective action when leaks occur.
- Report leaks and corrective actions taken.
- Provide for proper closure of tanks.
- Provide evidence, as EPA deems necessary, of financial responsibility for taking corrective action and compensating third parties for injury or damages from sudden or nonsudden releases. (States may finance corrective action and compensating programs by a fee on tank owners and operators.)

State Authorization. Several states already have or are developing regulatory programs for underground storage tanks. The new law is designed to avoid interference with those state programs and to encourage other states to press ahead with control programs. However, states still must apply to EPA for authorization to operate a UST program. The law allows states to choose either a program that will cover petroleum-containing tanks or one for hazardous substance tanks, or both. State programs must include all the regulatory elements of the federal program and must provide for adequate enforcement. After a one-to three-year grace period, state requirements must be no less stringent than federal requirements.

Inspections and Enforcement. The law provides authority for federal and state personnel to:

- Request pertinent information from tank owners.
- Inspect and sample tanks.
- Monitor and test tanks and surrounding soils, air surface water, and ground water.

Federal enforcement also is included in the new law. EPA may issue compliance orders for any violation of the UST statue or regulations. A violator who fails to comply with the order may be subject to a civil penalty of up to $25,000 per day of noncompliance. In addition, any owner who knowingly fails to notify or submits false information, or any owner or operator who fails to comply with any regulatory requirement under Subtitle I, may be subject to civil penalties of up to $10,000. Criminal penalties are not authorized under Subtitle I.

Exercises

1. Outline the major goals of RCRA's Subtitle D, Subtitle C, and Subtitle I.
2. What are the major objectives of the Comprehensive Environmental Response, Compensation, and Liability Act (CERCLA)?
3. How can the public express its opinion in regard to developing RCRA regulations?
4. Under subtitle D of RCRA, what are the minimum regulatory requirements for a state solid waste management plan?
5. What is an open dump, and how does RCRA deal with open dumps?
6. Under RCRA, what is the definition of a hazardous waste?
7. Under provisions of RCRA, who is the responsible party for identifying if a solid waste is a hazardous waste?
8. How does RCRA define an ignitable solid waste?
9. Under RCRA, what is the definition of a corrosive solid waste?
10. The EP toxicity test on an industrial waste stream gives the following information:

Component	*Concentration, mg/l*
lead	3.0
selenium	0.5
silver	10.0
chromium	3.5

    Is this stream considered to be a hazardous waste?
11. Under RCRA, what is the definition of a small quantity generator?
12. What are the responsibility of a small quantity generator in regard to hazardous waste management?
13. What are the regulatory requirements for hazardous waste generators who are not considered small quantity generators?
14. Outline the steps that must be undertaken in hazardous waste manifest routing.
15. What are the recordkeeping and reporting requirements for generators of hazardous waste?
16. What are the categories of regulations that apply to a hazardous waste TSD facility?
17. What are the technical requirements for interim status TSD facilities?
18. Under an EPA interim permit, what are the responsibilities of a TSD facility for groundwater monitoring?
19. What are the financial requirements for an interim status TSD facility?
20. What are the areas of coverage of general standards for a permit TSD facility?
21. What is the definition of an underground storage tank (UST)?
22. What requirements must be met when a UST is installed?

References

1. The Bureau of National Affairs, *The Job Safety and Health Act of 1970*, 1st Ed., Bureau of National Affairs, Washington, D. C. (1971).
2. McRae, R., *Occupational Safety and Health Compliance Manual*, 1st Ed., The Center for Compliance Information, ASPEN Systems Corporation, Germantown, Md. (1978).
3. Petersen, D., *The OSHA Compliance Manual*, 1st Ed., McGraw-Hill Book Co., New York (1975).
4. Nothstein, G. Z., *The Law of Occupational Safety and Health*, 1st Ed., The Free Press, New York (1981).
5. The Office of the Federal Register, *Code of Federal Regulations Title 29 Parts 1900–1910*, Office of Federal Register, Washington, D.C. (1985).
6. Watts, J. H., and W. I. Summers, *NFPA Handbook of the National Electric Code*, 4th Ed., McGraw-Hill Book Co., New York (1975).

7. BLT, Various safety pamphlets and business and legal reports, Bureau of Law and Business, Madison, Conn. (1988).
8. Commerce Clearing House Editorial Staff, *Occupational Safety and Health Hazard Communication Federal/State Right to Know Laws*, 1st Ed., Commerce Clearing House Chicago (1985).
9. Ingram, J. W., and J. T. Dufour, *Hazard Communication Handbook*, 1st Ed., California Chamber of Commerce, Sacramento, Calif. (1986).
10. Lowry, G. G., and R. C. Lowry, *Handbook of Hazard Communication and OSHA Requirements*, 1st Ed., Lewis Publishers, Chelsea, Mich. (1985).
11. Bland, W. F., and R. L. Davidson, *Petroleum Processing Handbook*, 1st Ed., McGraw-Hill Book Co., New York, (1967).
12. Felder, R. M., and R. W. Rousseau, *Elementary Principles of Chemical Processes*, 1st Ed., John Wiley & Sons, New York (1978).
13. Office of the Federal Register, 29 CFR 1910–120, Office of the Federal Register, Washington, D.C. (1987).
14. U.S. Environmental Protection Agency, Title III Fact Sheet, Emergency Planning and Community Right-to-Know, Washington, D.C.
15. National Response Team, *Hazardous Materials Emergency Planning Guide*, Washington, D.C. (Nov. 1986).
16. Harris, C., and D. A. Berger, *SARA Title III A Guide to Emergency Preparedness and Community Right to Know*, Executive Enterprises Publications Co., Inc., New York (1988).
17. Ninety-ninth Congress of the United States of America, *Superfund Amendment and Reauthorization Act of 1986*, H.R. 2005, Washington, D.C.
18. PARS Publishing, *SARA Title III Community Right to Know Compliance Guide*, Hamilton Square, N.J. (1987).
19. Dezelic, et al., *Summary of Emergency Planning and Community Right to Know Act of 1986*, Chicago (1986).
20. Environmental Resources Management Inc., *Community Right-To-Know Compliance Manual*, West Chester, Pa. (1986).
21. Swain, F., *Comments on Section 313 of Title III*, Small Business Administration, Washington, D.C. (1987).
22. Misco, G., *Comments on EPA's Recondsideration of Section 311 and 312 Rules for Hazardous Chemical Reporting*, Chemical Specialties Manufacturer's Association, Washington, D.C. (1987).
23. EPA *Federal Register*, Washington, D.C. (Apr. 22, 1987).
24. Environmental Protection Agency, *RCRA Orientation Manual*, EPA, Washington, D.C. (1986).
25. Wentz, C. A. *Hazardous Waste Management*, 1st Ed., McGraw-Hill Book Co., New York (1989).
26. EPA, *Regulating Hazardous Waste Facilities*, Washington, D.C. (1989).
27. EPA, *Understanding Land Disposal*, EPA, Washington, D.C. (1989).
28. EPA, *Getting Involved in the Permitting Process*, EPA, Washington, D.C. (1989).
29. EPA, *The State and Federal Partnership*, EPA, Washington, D.C. (1989).
30. EPA, *Understanding the Small Quantity Generator Hazardous Waste Rules*, EPA, Washington, D.C. (1986).

Chapter

3

Industrial Toxicology

Industrial toxicology is a science that deals with the potential harmful effects of materials, products, and wastes on health and the environment. Bringing together knowledge from many scientific fields, including chemistry, biology, pharmacology, physiology, and pathology, it has been given a name meaning "the science of poisons."[2,4,6,13]

During the twentieth century there has been an increase in concern for the health of workers exposed to toxic chemicals. Employers and workers alike are emphasizing safe work environments by implementing the principles of industrial toxicology. Since the 1960s these concerns have been particularly apparent as the field has grown rapidly. Ideally all work environments should deal only with chemicals that are considered safe and will have no harmful effects on the health of humans. However, no chemical can be considered completely safe because any chemical in a large enough amount can do some harm. Similarly, there is no precise level of chemical exposure that is safe for all people because individual human responses may vary significantly.[12,20,21,24]

The federal government has passed laws and promulgated safe exposure levels for numerous chemicals. However, the basis for these levels is a somewhat subjective interpretation of less than complete scientific information on human exposure. Thus the health hazards that tend to result from long-term or chronic exposure are more difficult to detect than the effects of acute exposure, which produces immediate health hazards.[3,9,14]

All matter and substances are composed of chemical elements and chemical compounds, so it is difficult to understand why the word "chemical" is held in such low regard in our society. The news media, environmental groups, and lawmakers have portrayed chemicals as a particularly harmful part of our health concerns. This view of chemicals as "bad" has eroded the public's trust in many industries, and could jeopardize the ability of those industries to compete effectively. It is essential to recognize at the outset of this discussion that chemical toxicity is not a function of whether or not a chemical is synthetic or natural, is an element or a complex compound, is produced by a large company or a small research laboratory, is biodegradable or not, is polar or nonpolar, or is a profitable product in the marketplace.[15,18,19,23]

Risk Management

The workplace is made up of males and females generally 18 to 70 years of age. These workers perform jobs that may be associated with toxic material exposure, typically over an entire workday. There are numerous occupations whose tasks involve contact with pollutants, which may vary widely in their toxicity and in their effects upon entering the human body.

To determine the toxicity of a substance, large quantities of the toxic material are administered to laboratory animals, to study its effect on those animals. Because the goal of animal studies is to produce an observable effect, excessive levels of the chemical usually are required in the test protocol. Also, the test results themselves are subject to interpretation; so it is difficult to assess standards for toxicity uniformly for animals, much less to translate them to human exposure.[11,22]

Chemical substances may damage living things in many ways. They may be flammable or explosive, causing burns or bodily harm. Skin irritants can be highly discomforting, and corrosive chemicals can destroy tissues, such as eyes, throat and lungs. Other chemicals produce an allergic response by sensitization, to cause dermatosis, sneezing, a runny nose, or other symptoms.[8]

Toxicology studies the adverse systemic effects of chemicals and their ability to harm an organ system or to disturb or disrupt biochemical processes within the body. A chemical may affect several different functions within the body, but with varying degrees of sensitivity among individuals. For example, the liver and the kidneys both may be affected, but the chemical effects may appear in the liver of one individual and the kidneys of another.[1,5]

Besides these basic difficulties in assessing long-term human exposure levels, it is difficult to relate toxicity assessments done under laboratory conditions to real-life human exposures. Laboratory animals normally receive constant dosage levels of the same chemical every day of their lifetime. Human exposure is vastly different, even in the work environment. Workplace exposure is not uniform because of days off, business travel, and occasional changes in job positions. Rarely do people live in the same house or community during their entire lifetime. Hence toxic exposure risk assessment is a difficult science at best. There is no infallible standardized approach for setting precise threshold levels for the toxic effects of chemicals. Instead, statistical methodology is applied to predict exposure levels that could adversely affect human health.

Once the risk has been assessed and quantified, it must be managed. For example, government regulatory agencies have determined that for every million in population, less than one excess cancer case is an acceptable risk. It should be noted that this scenario does not apply to the Delaney amendment for food additives, which does not allow any increase in cancer risk. In certain nonfood cases, zero exposure has been determined to be acceptable, thereby banning the manufacture or use of the chemical in question. Polychlorinated biphenyls (PCBs), DDT, and 2,4,5-T fall into this category.

As we cannot ban our way out of all chemical exposure, risk reduction becomes the key to risk management. Another aspect of risk management concerns the public perception of risk, which is usually vastly different from the science of toxicology. All synthetic chemicals are likely to be viewed by the general public with some suspicion, particularly if the chemical is new or is unfamiliar to the exposed individual. Long-term or chronic exposure risk often is viewed with more concern than short-term occasional or acute exposures. The continuing environmental focus on dioxin, which is a toxic chemical, constitutes an exaggerated risk operation compared to the lack of attention to known human carcinogens.

Chemicals cannot be simply classified as toxic or nontoxic because of variances in the dose level, the site of human contact with the chemical, the occupational hygiene of the worker and the general work environment, and other influences such as lifestyle, diet, heredity, weather, age, sex, and resistance to disease. In combination, these variables may be synergistic, causing greater damage in their combined effect; or their effects may be antagonistic, so that they are less damaging together than separately.

The exposure to toxic substances usually involves mixtures; so the limits set by the government for single substance exposure should not be considered completely safe, but rather as tolerable levels. As a practical matter, it is desirable to achieve the lowest workplace concentration of a toxic material that is economically and technically feasible.

Chemical, Physical, and Physiological Properties

Substances may be generally classified according to either chemical and physical or physiological properties that will influence their toxicity. Chemical properties that may be important in defining the industrial toxicology of a material include the chemical composition and the pH.[7,16]

It is impossible to perform an accurate hazard evaluation without definitive information on chemical composition. General descriptions, such as petroleum distillate, mineral spirits, or inert ingredients, are insufficient for an accurate assessment of the toxicity of a chemical mixture. Some constituents of such mixtures may be hazardous, whereas others may be inert. Specific composition information usually can be obtained from the supplier, even for proprietary products, on a confidential, need-to-know basis. Alternatives to obtaining the information from the supplier include having the substance analyzed oneself, which may be costly and time-consuming, or changing suppliers.

The pH of a substance may greatly affect its toxicity because of the corrosive nature of the chemical. The pH is easily determined with standard indicator paper or a pH meter. If the pH of a liquid is either extremely high (basic) or low (acidic), it should be considered hazardous and be managed accordingly.

Physical characteristics may cover a broad range of properties, including the following:

- Physical state (solid, liquid, or gas)
- Specific gravity
- Vapor density
- Solid melting point
- Liquid boiling point
- Vapor pressure

Each of these physical properties combines with other properties to determine the hazard degree of the substance.

It is difficult to assess the hazards of mixtures of substances because most toxicity data exist for chemicals in their pure state. The toxicity of a chemical may increase or decrease in a mixture, depending upon the circumstance. Mixtures containing hazardous toxic components should be treated as toxic unless proven otherwise.

The nature of many industrial jobs requires the performance of work that may result in some exposure to toxic substances; and most industrial operations generate more than a single toxic substance, so that workers generally are exposed to mixtures of chemicals. Toxic substances adversely affect humans either by coming in contact with external tissues, such as the eyes or skin, or by entering the body itself through the routes of inhalation, ingestion, or absorption through the skin. In the work environment the most common route for a chemical to enter the body is by inhalation, followed by skin absorption and then ingestion.

Transport of Toxic Substances in the Body

Inhalation

The inhalation of toxic substances into the air passages of the lung and on into the circulatory system is a rapid and direct means of entry into the bloodstream. Contaminants present in the air from industrial operations can gain direct entry into the respiratory system through inhalation and thereby create a workplace hazard.

There are two types of airborne toxic substances. These toxics may exist in the form of gases and vapors that are uniformly present in the atmosphere, or they may be found as solid or liquid particles that are suspended in the air. Usually a chemical reaction or a physical absorption process may be used to remove gases or vapors from air, whereas solid or liquid particles may be removed by filtration, mechanical separation, or precipitation. Toxic gases or vapors usually dissolve or liquefy on the watery surface of the pulmonary tract; otherwise they flow throughout the tract. Many factors determine the degree of entry into the circulatory system, including the toxic chemical's concentration and its exposure duration, the solubility and reactivity of the substance in the blood and body tissues, and the respiration rate of the chemical from the body.

When aerosols are inhaled, the larger particles precipitate in the upper respiratory tract, while finer particles continue deep into the respiratory tract.

The human breathing process is involuntary, being initiated by the brain, which senses oxygen, carbon dioxide, and acidity levels in the blood. Air enters the lungs, and when it is expelled, the oxygen level is reduced while the carbon dioxide level is increased. The respiratory tract and the lungs provide two main functions: an air intake and exhaust system, which is extensively subdivided in the lung; and the alveoli, which are small pockets at the end of the fine subdivisions in the lung. The alveoli diffuse oxygen into the blood and in return withdraw carbon dioxide from it. There is a thin barrier with a very large surface area between the air and the blood, allowing rapid exchange of the gases to occur. The gas–liquid transport phenomena that take place may be predicted on the basis of several factors:

- The pressure gradient between the gases in the alveoli and gases in the blood.
- The total membrane surface area for diffusion.
- The condition of the diffusion membrane.
- The volume of incoming air entering the respiratory tract per unit time.
- The volume of incoming air that actually reaches the alveoli per unit time.

The exchange of toxic gases or vapors will depend upon these alveolar–capillary membrane factors.

Asphyxiant gases reduce the partial pressure of oxygen in the alveoli, thus decreasing the oxygen flow through the membrane and entering the blood. Carbon dioxide, methane, and nitrogen are examples of gases participating in this phenomenon.

The transport of oxygen to the blood and tissues also can be blocked by other chemical compounds. For example, carbon monoxide blocks the transport of oxygen to hemoglobin, and hydrogen cyanide blocks the tissue utilization of oxygen.

Many toxic substances can irritate the lung, and in doing so they may cause asphyxiation. The action of water-soluble irritants such as ammonia or hydrogen chloride may damage the membrane's permeability or cause inflammation of the upper respiratory tract. Irritants that are less water-soluble such as chlorine or nitrous oxides will penetrate the respiratory tract and damage the bronchial tubes or even the lungs themselves.

The respiratory tract is lined with a mucus that traps particulate toxic substances as their incoming velocity decreases because of the increasing cross-sectional area. This mucus carries the particles toward the nasal cavity, eventually to be spit out or swallowed. The alveoli contain phagocytic cells that engulf smaller particles that pass through the upper respiratory tract. These cells, which will migrate up the tract, tend to increase with the inhalation of dust in air. Tobacco smoke contains components that inhibit this cell activity; so instead of leaving the alveoli, the cells accumulate.

Toxic substances that deposit in the lung can cause the irreversible development of fibrous tissue, reducing the lung capacity. Asbestos and mining dust are examples of such substances.

Skin Absorption

Human skin is an important barrier that protects body tissues from harm and retards the loss of water, an essential ingredient of life. The skin contains nerve endings that sense information about the surrounding environment, and sweat glands that evaporate water to maintain a constant body temperature, as well as pigments that protect humans from ultraviolet radiation. Lipophilic substances are readily absorbed through the skin, whereas substances that are hydrophilic are absorbed through the skin only with difficulty.

The skin is composed of epidermis, dermis, and hypodermis layers. The relatively waterproof epidermis, or outer skin layer, prevents the diffusion of fluids and is tough enough to withstand abrasive contact. The dermis, which is thicker than the epidermis, provides tear resistance and flexibility to the skin. The even thicker hypodermis layer contains the blood vessels and nerves. The sweat glands penetrate both the dermis and the epidermis.

Toxic substances have difficulty entering the body through the skin because of the epidermis layer. Chemical diffusion through the epidermis layer depends upon the nature of the chemical substance as well as the skin thickness and its condition at the point of contact. Generally water-soluble compounds pass more easily through the skin than water-insoluble substances. Diffusion occurs more readily in skin that has been damaged by physical abuse or chemical reaction than in intact skin.

Acids and bases are capable of causing severe corrosive burns when they contact the skin. The severity of a burn from a corrosive chemical depends upon the following criteria:

- Chemical type and concentration.
- Time length of contact.
- Whether the skin was covered.
- Response of first aid provider.

Ingestion

The mouth, esophagus, stomach, intestines, liver, and gall bladder are an integral part of the body during the digestion process. The gastrointestinal tract (GI tract) processes the food people eat by breaking down the complex proteins and polysaccharides in the diet into amino acids and monosaccharides.

The stomach allows alcohol, some drugs, sugar, and water to enter the bloodstream by absorption. However, much of this absorption takes place in the small intestine, while the large intestine allows the absorption of large quantities of water.

Blood Circulatory System

The average human has about 5.5 liters of blood, which is contained in a closed system—the heart, spleen, and blood vessels of the vascular system. The blood supplies the tissues with the life support chemical substances of oxygen, nutrients, and hormones. Carbon dioxide and urea waste products are transported in the blood to body organs for elimination.

Body Elimination of Toxic Substances

The removal of toxic substances from humans depends upon the route of entry and the body location. If the toxic gases or vapors entered by inhalation, they may be expelled by exhalation. Toxic substances may be removed from the skin by wiping it with a towel or washing it with soap and water. Toxic substances that are ingested may pass through the GI tract to be eliminated with the feces.

If the toxic substance enters the blood, then the elimination problem becomes more complex. The excretion of toxics in the blood is mainly a function of the kidney, as a part of the filtration process that transfers water, small molecules, and chemical ions to the primary urine stream. Nonpolar toxic substances form the kidney pass through its membranes and reenter the bloodstream, whereupon the liver changes the nonpolar substances to polar chemicals, thus facilitating their removal through the kidney. The liver also removes some toxics with bile, which is passed to the GI tract for elimination.

Carcinogens

Some toxic substances have been shown to cause malformations as a result of mutation. Mutagens, teratogens, and carcinogens all have this in common. A mutagen is a substance that can cause changes or mutations is the genetic material of cells. Teratogens cause malformed or abnormal fetuses. Carcinogens cause cancer, one of the most dreaded diseases of the human race.[17]

The public views with alarm any chemical proven to be, or suspected of being, a carcinogen. Therefore, known or suspected carcinogens have been subjected to special legislation and regulation. This public concern and the government regulatory response to cancer are expected to continue for the foreseeable future.

Relationship Between Dose and Response

In the science of industrial toxicology no chemical substance should be viewed as entirely safe or entirely harmful. An excessive concentration of any substance will produce a harmful or an undesirable effect in animals or humans. The dose–response relationship, as shown in Figure 3.1, is the most important factor in the study of industrial toxicology in humans. This relationship determines the safe and the harmful concentrations of a chemical for biological mechanisms in animals.

The dose–response relationship for any given chemical usually is determined by the administration of various dosage levels to laboratory animals to observe the resultant toxic effects. Generally, these studies define a no-observable-effect level at small dosages, some toxicity at higher dosage levels, and the death of the animal at an excessive dosage level.

The dosage–response for a chemical will vary with the species of animal used in the laboratory test program. A family of dose–response curves, as depicted in Figure 3.2 for a particular chemical, can be developed for various animals, such as mice, rats, guinea pigs, and rabbits. Some types of animals are more sensitive than others to a given chemical, and because animal sensitivity can vary greatly among species, the extrapolation of dose–response data from one test animal to another animal species should be attempted only with extreme caution. The same concern should be used in trying to apply laboratory animal toxicity data to human effects.

It also should be noted that the dose–response relationships shown for a certain chemical in Figure 3.2 may have a different effect on animal species *A*, *B*, *C*, and *D* when they are tested with other chemicals. The ordering and the shape of the animal curves in Figure 3.2 may be entirely different for a second or even an additional chemical, thereby reinforcing the prior caution regarding extrapolation to human exposure.

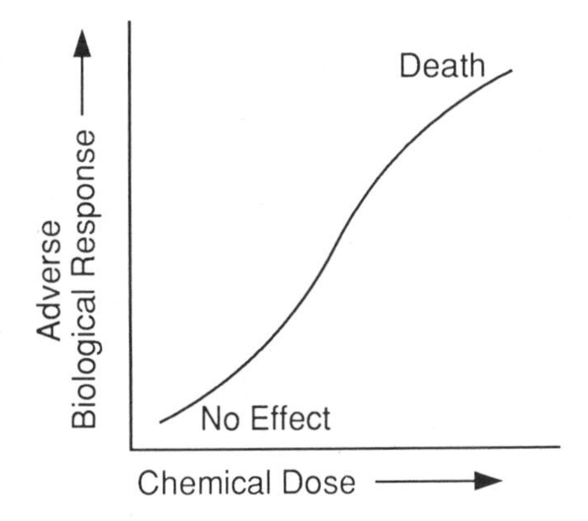

Figure 3.1. Dose–response relationship.

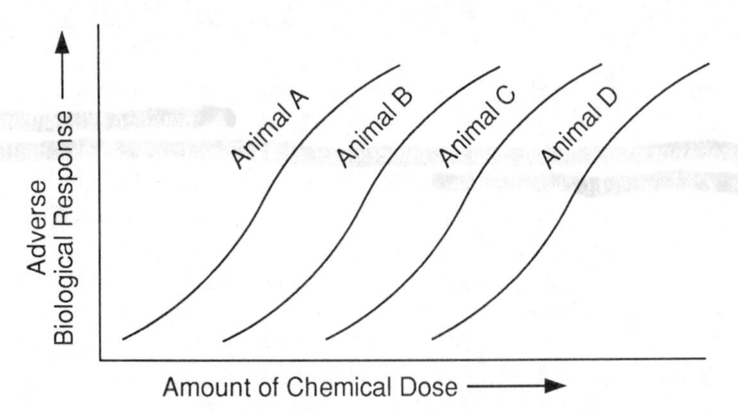

Figure 3.2. Family of dose-response curves for different species exposed to different chemicals.

Lethal Dose

The death of the test animal species is the most commonly used criterion for measuring the toxic effects of chemical exposure. The lethal dose fifty or LD_{50} (also discussed in Chapter 2, in the HCS section) describes the chemical dosage level demonstrated to be acutely lethal to 50% of the laboratory animals of a given species, who were administered the chemical under controlled laboratory test conditions. The LD_{50} units are milligrams of the chemical per kilogram of animal body weight, or mg/kg. LD_{50} values are inversely proportional to the acute toxicity of the chemical. Although the term LD_{50} values are inversely proportional to the acute toxicity of the chemical. Although the term LD_{50} is most commonly used, other lethal dose terms, such as, LD_{25} or LD_{100}, also may be reported; these terms stand for 25% and 100% animal deaths, respectively.

It is important that one know the laboratory test conditions when interpreting LD_{50} values. A reported LD_{50} value also should include the animal species tested, the path of chemical administration, the chemical purity, and the dosage level per unit of body weight. There also should be a confidence limit attached to the LD_{50}, to help quantify the error in the estimated value. For example, an LD_{50} for chemical X might be stated as follows:

chemical X (99% purity): oral–rat LD_{50} = 2,500 mg/kg (95% confidence limits)

This means when chemical X with 99% purity was administered orally to rats at 2,500 mg of chemical X/kg of rat body weight, the dosage was fatal to 50% of the rats with a 95% statistical confidence level.

If the route of chemical administration is inhalation, the airborne lethal concentration level of the doses normally will be expressed as parts per million (ppm) or milligrams per cubic meter (mg/m^3) in air per unit time. Reported LC_{50} values should include types of test condition information similar to those for the LD_{50} value. For example, an LC_{50} for chemical Y might be stated as follows:

chemical Y (95% purity): inhale–rat LC_{50} = 800 ppm/8 hr (95% confidence limits)

This means when chemical Y was inhaled by rats at an airborne concentration of 800 ppm for 8 hours, the dosage was fatal to 50% of the rats with a 95% statistical confidence level.

The LD_{50} and LC_{50} values, although not inclusive, do provide one with sufficient information to determine the relative toxicity of the chemical substance in question. The slope of the dose–

response curve provides an indication of the magnitude of the dosage range between a harmless dose and an extremely toxic level.

Generally, if the oral LD_{50} for a substance is less than 50 mg/kg or the skin LD_{50} is less than 200 mg/kg, the substance may be characterized as highly toxic. On the other hand, an oral LD_{50} > 500 mg/kg would cause the substance to be characterized as practically nontoxic.[1,3] Chemicals that cause illness or death when taken in small quantities are called poisons. In fact, a poison is legally defined as a substance that has an $LD_{50} \leq 50$ mg/kg.

Threshold Limit Values

If the dosage is small enough, no poison is lethal. On the other hand, if the dose is large enough, all substances are lethal. In between there is a rather broad dosage range that does not have a distinct dividing line to separate the lethal from the nonlethal levels. However, in order to control toxic substances adequately, some concentration must be identified below which worker exposure would be satisfactory. This "safe" exposure level, which has been called the threshold limit value or TLV, is the maximum concentration at which worker exposure could occur during an entire workday, for every day of the working life, without significant harm. The use of TLV values usually is applied to toxic substances that enter the body through inhalation into the respiratory system.

Every toxic substance has a unique TLV, which is listed by the American Conference of Governmental Industrial Hygienists (ACGIH). The TLV values are reviewed annually by the ACGIH. The establishment of TLV values and their subsequent review primarily has been based upon laboratory testing with experimental animals, coupled with industrial experience and documented human exposure. Airborne contaminates generally are classified according to three different TLV categories:

- TLV–Time Weighted Average (TWA)
- TLV–Short Term Exposure Limit (STEL)
- TLV–Ceiling (C)

The TLV–TWA is the time weighted average concentration for a normal 8-hour workday, or a 40-hour workweek, at which worker exposure may occur without ill effect. The TLV–TWA values for airborne contaminants normally are expressed as parts per million (ppm) of the chemical in air or milligrams of the chemical per cubic meter (mg/m^3) of air. These TLV values do not define specific levels between safe and harmful concentrations because of variance in test conditions and the confidence limits of the resulting data.

The TLV–STEL is the maximum concentration at which worker exposure may occur may occur for a 15-minute period without intolerable irritation, chronic or irreversible tissue change, or narcosis and increased accident proneness. The units of TLV–STEL are the same as the units for TLV–TWA, namely, ppm or mg/m^3.

The TLV–C is that concentration which never should be exceeded at any time during worker exposure, also expressed as ppm or mg/m^3. A volatile toxic liquid with a low TLV–C should be treated with extreme caution in the workplace because even a small spill could be lethal.

Time Weighted Averages

During an 8-hour workshift the concentration of a toxic substance may exceed the generally accepted TLV for that substance. It generally has been accepted that the TLV may be exceeded briefly during the normal workday if the chemical concentration also is lower than the acceptable TLV during other periods of the 8-hour shift. In order to accurately reflect this phenomenon, the TLV values typically are measured as a TLV–TWA over the 7- or 8-hour workshift. The measured concentrations of the toxic substance are used to compute the 8-hour weighted average by the following equation:

$$\text{TLV–TWA} = \sum_{i=1}^{n} X_i t_i / 8$$

where:

TLV–TWA = Threshold limit value-time weighted average, ppm
X = A time-measured concentration of the airborne toxic
t = Time period of constant toxic substance concentration, hr
n = Number of time periods studied during the 8-hour workshift

Hazardous Mixtures and Permissible Exposure Levels

The above equation applies readily to worker exposure involving only a single toxic substance. When there is a mixture of potential hazards, the accurate prediction of the harmful effect becomes more complex. For example, the OSHA-prescribed permissible exposure level (PEL) of the TLV may be somewhat above the actual measured TLV–TWA of a single substance. However, there may be several airborne contaminants present at the same time and at concentrations just under acceptable limits, which could collectively create a more serious health hazard. In other words, the combined effect of some toxic mixtures could be either far worse or conceivably less harmful than the sum of the individual effects of the substances. In order to manage this complicated situation, OSHA has adopted the principle of additive effects when considering toxic mixtures. These additive effects may be expressed by the following equation:

$$R_m = \sum_{i=1}^{n} C_i / \text{PEL}_i$$

where:

R_m = Relative ratio of the toxicity of the mixture
C_i = Concentration of a specific toxic substance, ppm
PEL_i = OSHA permissible exposure limit for that specific toxic substance, ppm

In the above calculation, if the R_m value is greater than one, the mixture concentration will exceed the PEL, although the individual PEL values are not exceeded. Appendix D contains the OSHA PEL for many chemicals encountered in the workplace.

Action and Ceiling Levels

When the TLV for a toxic substance is exceeded in the work environment, serious harm to personnel may result. In order to anticipate such a problem, a preliminary, early-on control measure called the action level (AL) has been set at one-half of the PEL. This AL provides a margin of error with an early warning signal to help assure that worker exposure levels do not exceed the safe levels of the PEL.

Ceiling levels (C) are maximum limits for single acute exposures to a toxic substance. In a somewhat similar fashion, a maximum short-term exposure limit (STEL) would be the highest concentration permitted for a brief, but specified duration. Usually the STEL is allowable for only a few minutes. Because the STEL measurement is difficult to obtain in an accurate real-time manner, there is a much greater emphasis on C level data and standards in the workplace.

Acute Toxicity

The ability of a chemical to cause systemic damage as a result of short-term exposure is called acute toxicity. Generally, acute toxicity is exhibited when the protective mechanisms of the body are overcome. Acute oral toxicity studies have been performed for many chemicals, using rats and mice as the test animals. The LD_{50} or the LC_{50} commonly is used to describe the acute toxicity effect of the chemical. These values, which are determined under controlled laboratory conditions with the experimental animals, are used to estimate the acute toxicity for humans. If a chemical has the potential for extensive human exposure, the normal testing with rats and mice is supplemented with the testing of additional animal species to help determine the effect of species variability. These additional species might include rabbits and guinea pigs. If the chemical has exhibited similar characteristics of acute toxicity for all of the animal species tested, then it probably will have similar characteristics of acute toxicity for humans. If the animal test data show wide variations, the derivation of the potential effect on humans is more complex. It usually is assumed that the effects of acute toxicity in the human are more susceptible to harmful chemicals than are the laboratory effects in the most sensitive animal species, unless there is reliable evidence to the contrary.[18]

The slopes of the dose–response curves contain useful information regarding potential chemical toxicity. These slopes may be steep or shallow, as shown in Figure 3.3, depending upon the species response to the chemical testing. If the slope is steep, there is minimal dose–response variation within the species, and probably the same type of slope will be noted in the response of humans. A steep slope also will provide greater confidence in the estimate of a safe acute dose for humans (i.e., extrapolation to the LD_0 and not the LD_{50}).

Conversely, a shallow slope on the dose–response curve translates into considerable variation of effects within the species and makes safe dosage levels more unpredictable; a shallow slope probably would be observed for human behavior, forcing the assumption that humans would be as sensitive as the most sensitive animal.

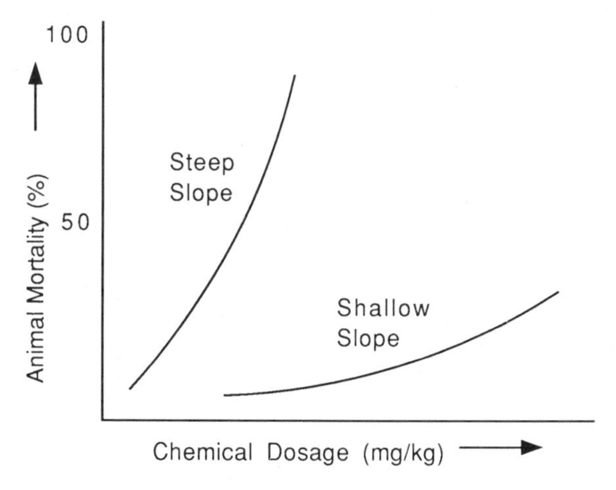

Figure 3.3. The relationship between the slope of the dose–response curve and chemical toxicity.

Chronic Toxicity

Chronic toxicity involves long-term, low-level exposure where the rate of chemical exposure is greater than the capacity of the body to eliminate the substance or render it harmless. The toxic substance will accumulate in the body under these circumstances, making the development and observation of the response much more complex and subtle.

Numerous variables are important in determining the chronic toxicity of a chemical, including the following:

- Toxicity exposure categories (inhalation, skin, oral)
- Health effects (carcinogen, mutagen, organ damage, etc.)
- Population of concern (adult, female, children, etc.)

The above factors are usually part of an experiment designed with laboratory animals. Rats and mice typically are used in the laboratory study. Other test animals generally are too costly and are of limited supply. Supplemental toxicity experiments, which study microorganisms or tissue cultures, may be useful sources of toxicity information, but they cannot replace whole test animals in the study.

Much of the chronic toxicity information in the literature is based on oral data because these data were obtained easily, and historically this type of exposure has been of concern in food consumption. It has since been recognized that inhalation and dermal exposure also are important routes of entry into the body.

Threshold Levels

Chemical effects in animals and humans have threshold levels, which describe the boundary between no-effect exposure and levels that exhibit an effect on the species studied. The threshold of a substance is the minimum quantity of chemical exposure that begins to produce an observable effect. It is a phenomenon that all substances exhibit. In the extrapolation of animal test data to humans, the threshold, as shown in Figure 3.4, is the highest chemical dosage level that does not produce a detectable health effect in a test animal species.

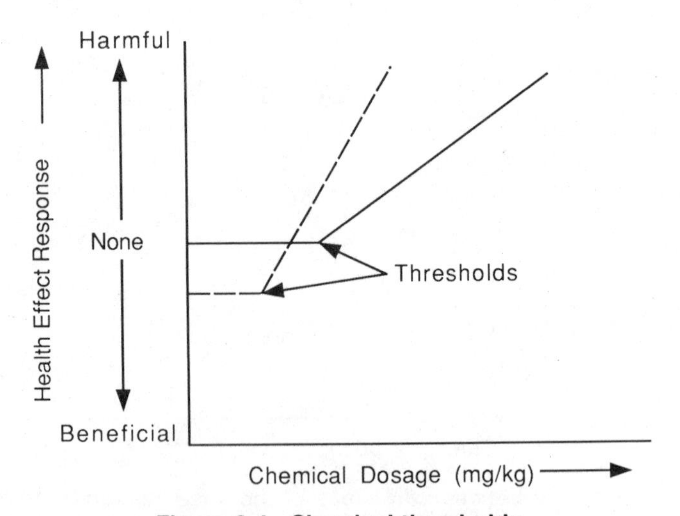

Figure 3.4. Chemical thresholds.

Effects of Toxicity

Generally, a toxic substance may have either acute or chronic health effects, or some combination in between. Acute and chronic toxicity may cause the following harmful health effects:

- Irritation
- Mutagenesis
- Teratogenesis
- Carcinogenesis
- Organ damage
- Reproductive constraints

Irritants harm the body by causing inflammation of body tissues, such as the eyes, skin, nose, and mouth, at the point of contact. The observable effects include redness, swelling, heat, and pain. Blisters also may result from the inflammation induced by toxic substances. Irritants may include liquids, solid particulates, or gases. Mineral acids, ammonia, caustic, chlorine, sulfur dioxide, nitrogen oxides, coal dust, and asbestos all are irritants when they come in contact with the body.

Mutagenesis, which is the formation of mutations or changes in hereditary material, involves either a physical change in chromosomes or a biochemical change in genes. These changes, which normally are caused by chronic toxicity, may affect present and future generations.

Teratogenesis, which is the formation of birth defects in fetuses, results from the alteration of developing cells. This typically is a chronic effect of toxicity, but in some instances it has resulted from a single exposure.

Carcinogenesis, the production of cancer, is caused by the abnormal growth or reproduction of cells. The observable health effects are tumors due to the uncontrollable cell proliferation caused by the chronic effects of the toxic substance. Sufficient scientific evidence is available to characterize known carcinogens in Appendix B and suspected carcinogens in Appendix C.

Organ damage from toxicity generally is caused by a specific toxic substance exposure. The effects of smoking on the heart and lungs, alcohol on the liver, and lead on the kidneys are examples of organ damage due to exposure to toxic substances.

The reproductive constraints of toxic substances may include reductions in fertility or live births. Tobacco and alcohol are common toxicants that can cause these types of health effects.

Toxicity of Dioxins, Dibenzofurans, and Polychlorinated Biphenyls

No other chemical substances have received as much attention from the news media as has been devoted to the chlorodibenzo-*p*-dioxins, commonly called dioxins. These compounds have a tremendous emotional impact on many people. The dioxins are a group of organic chemicals that may be formed as by-product contaminants during the manufacture of some chlorinated aromatic compounds, such as trichlorophenol, polychlorinated biphenyls, and some organic herbicides. These compounds are chlorine derivatives of the basic dibenzo-dioxin structure, shown in Figure 3.5, which contains 12 carbon and two oxygen atoms. There are essentially two benzene rings, which are bridged by the two oxygen atoms. Dibenzofurans, as shown in Figure 3.6, have a similar organic configuration that is bridged by only one oxygen atom. When these structures have been chlorinated, the various isomers of dioxins and furans result.

There are 75 possible isomers of dioxin. Studies indicate that 2,3,7,8-tetrachlorodibenzodioxin (TCDD), shown in Figure 3.7, has the most harmful effects of all the dioxin isomers. As can be seen in Table 3.1, 2,3,7,8-TCDD is among the most toxic substances ever studied with guinea pigs. As little as 0.6 $\mu g/kg$ represents an oral LD_{50} for 2,3,7,8-TCDD with guinea pigs.

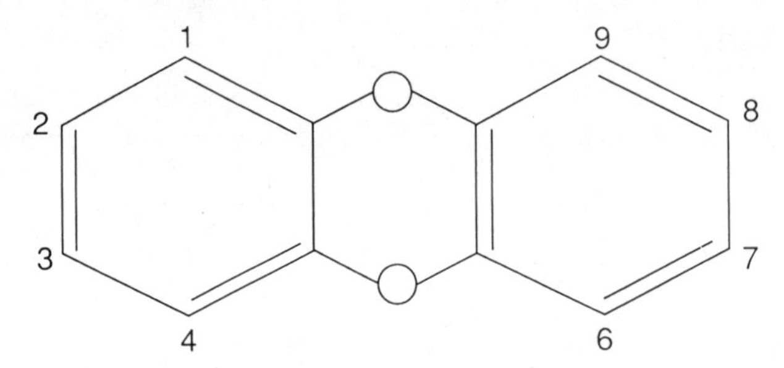

Figure 3.5. Dibenzo-*p*-dioxin molecular structure.

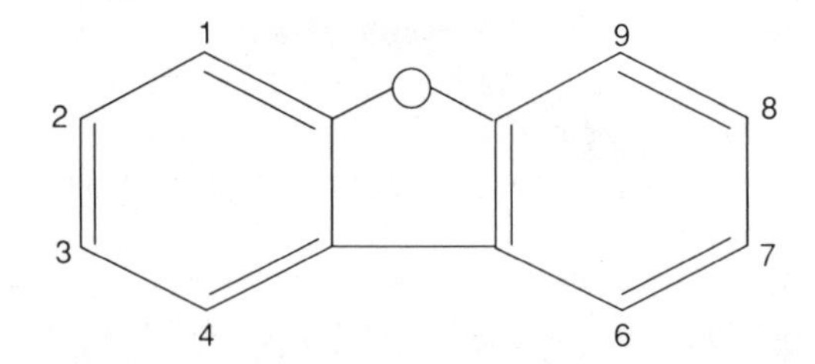

Figure 3.6. Dibenzofuran molecular structure.

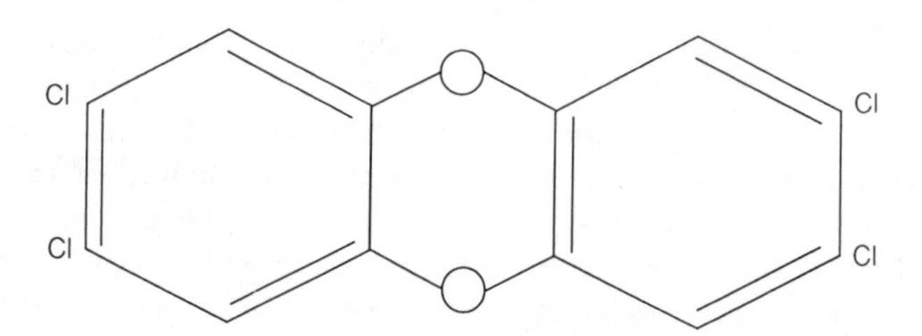

Figure 3.7. 2,3,7,8 Tetrachlorodibenzo-*p*-dioxin (TCDD) molecular structure.

Table 3.1. Minimum lethal doses for extremely toxic substances.

Substance	Test Animal	Minimum Lethal Dose (Moles/kg of body weight)
Botulinum toxin A	Mouse	3.3×10^{-17}
Tetanus toxin	Mouse	1.0×10^{-15}
Diphtheria toxin	Mouse	4.2×10^{-12}
2, 3, 7, 8-TCDD	Guinea pig	3.1×10^{-9}
Butotoxin	Cat	5.2×10^{-7}
Curare	Mouse	7.2×10^{-7}
Strychnine	Mouse	1.5×10^{-6}
Muscarin	Cat	5.2×10^{-6}
Diisopropylfluoro phosphate	Mouse	1.6×10^{-5}
Sodium cyanide	Mouse	2.0×10^{-4}

Source: U.S. Environmental Protection Agency.

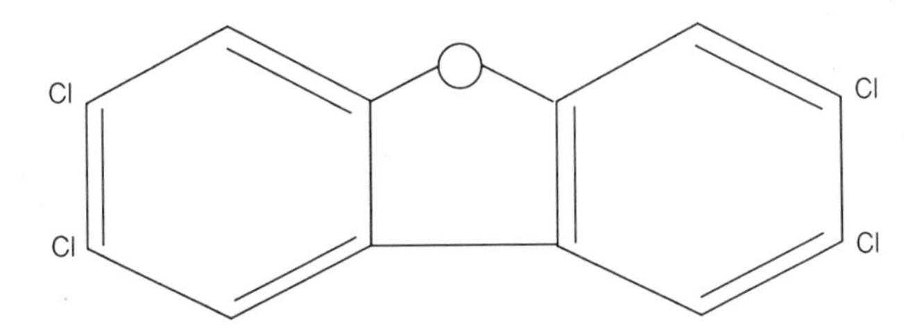

Figure 3.8. 2,3,7,8-Tetrachlorodibenzofuran (TCDF) molecular structure.

2,3,7,8-TCDD in stable in acids and bases, has a low water solubility, and decomposes rapidly at temperatures above 800°C. The isomer has a half-life in soil that ranges from 230 days in Florida to 320 days in Utah. In Severso, Italy, which has different soil microbes and conditions, the half-life may be two to three years, or possibly longer.

The uncontrolled burning of polychlorinated biphenyls (PCBs) can produce dioxins and furans, including the compound 2,3,7,8-TCDF, shown in Figure 3.8. For this reason EPA has highly regulated the incineration of PCBs.

PCBs, dioxins, and furans may cause chloracne following dermal contact. Chloracne is a skin rash, causing blackheads, that may persist for as long as several years. Dioxin causes cancer in rats and mice, but has not been shown to cause it in humans.[10]

References

1. Andelman, J. B., and D. W. Underhill, *Health Effects from Hazardous Waste Sites*, Lewis Publishers, Chelsea, Mich. (1987).
2. Anderson, K., and R. Scott, *Fundamentals of Industrial Toxicology*, Ann Arbor Science, Ann Arbor, Mich. (1981).
3. Ariëns, E. J., A. M. Simonis, and J. Offermeier, *Introduction to General Toxicology*, Academic Press, New York (1976).
4. Ballantyne, Bryan, *Current Approaches in Toxicology*, John Wright & Sons, Ltd., Bristol, U.K. (1977).
5. Bretherick, L., *Hazards in the Chemical Laboratory*, The Royal Society of Chemistry, London (1981).
6. Cooper, P., *Poisoning by Drugs and Chemicals*, 3rd Ed., Year Book Medical Publishers, Inc., Chicago (1974).
7. Deichmann, W. B., and H. W. Gerarde, *Toxicology of Drugs and Chemicals*, Academic Press, New York (1969).
8. Dreisbach, R. H., and W. O. Robertson, *Handbook of Poisoning: Prevention, Diagnosis & Treatment*, 12th Ed., Appleton & Lange, Norwalk, Conn. (1987).
9. Ellenhorn, M. J., and D. G. Barceloux, *Medical Toxicology: Diagnosis and Treatment of Human Poisoning*, Elsevier Science Publishing Co., New York (1988).
10. Fawcett, H. H., *Hazardous and Toxic Materials: Safe Handling and Disposal*, John Wiley & Sons, New York (1984).
11. Gosselin, R. E., et al., *Clinical Toxicology of Commercial Products*, 5th Ed., Williams & Wilkins, Baltimore (1984).
12. Kamrin, M. A., *Toxicology*, Lewis Publishers, Chelsea, Mich. (1988).
13. Klaassen, C. D., et al., *Casarett and Doull's Toxicology: The Basic Science of Poisons*, 3rd Ed., Macmillan Publishing Company, New York, (1986).
14. Li, A. P., *Toxicity Testing*, Raven Press, New York (1985).
15. Loomis, T. A., *Essentials of Toxicology*, Lea & Febiger, Philadelphia (1974).
16. National Research Council, *Prudent Practices for Handling Hazardous Chemicals in Laboratories*, National Academy Press, Washington, D.C. (1981).
17. Office of Technology Assessment Task Force, *Identifying and Regulating Carcinogens*, Lewis Publishers, Chelsea, Mich. (1988).

18. Ottoboni, M. A., *The Dose Makes the Poison*, Vincente Books, Berkely, Calif. (1984).
19. Proctor, N. H., and J. P. Hughes, *Chemical Hazards of the Workplace*, J. B. Lippincott Co., Philadelphia (1978).
20. Reeves, A. L., *Toxicology: Principles and Practice*, John Wiley & Sons, New York (1981).
21. Richardson, M., *Toxic Hazard Assessment of Chemicals*, The Royal Society of Chemistry, London (1986).
22. Stewart, C. P., and A. Stolman, *Toxicology Mechanisms and Analytical Methods*, Academic Press, New York (1961).
23. Williams, P. L., and J. L. Burson, *Industrial Toxicology: Safety and Health Applications in the Workplace*, Van Nostrand Reinhold, New York (1985).
24. Young, J. A., *Improving Safety in the Chemical Laboratory: A Practical Guide*, John Wiley & Sons, New York (1987).

Exercises

1. Define industrial toxicology.
2. Discuss why the term "chemical" is looked upon unfavorably by society.
3. Discuss how animal studies are used to study the toxic effects of substances.
4. Discuss the public perception of risk to chemical exposure.
5. Name several physical and chemical characteristics of hazardous substances, and discuss how these properties might interrelate in the toxicity of mixtures of hazardous wastes.
6. Discuss how toxic gases and vapors could enter the bloodstream through the process of inhalation.
7. Discuss how toxic liquids could enter the bloodstream through the process of skin absorption.
8. How does the blood circulatory system function to reduce the blood levels of hazardous wastes?
9. What are carcinogens?
10. Describe the dose–response relationship for a toxic substance in animals.
11. How may dose–response relationships vary for different animals?
12. Define the LD_{50} for a toxic substance based upon testing with rats.
13. What is the threshold limit value for a toxic substance?
14. Define TWA, STEL, and ceiling values.
15. How are TWA values determined?
16. Discuss the OSHA principle of additive effects for toxic mixtures.
17. What are action and ceiling levels for toxics?
18. Define acute toxicity.
19. Define chronic toxicity.
20. Define and contrast mutagenesis, teratogenesis, and carcinogenesis.
21. Why do dioxins have an emotional impact on many people?

Chapter

4

The Management of Personal Protective Equipment

Personal protective equipment (PPE) is needed for physical or health hazards that cannot be eliminated through engineering or administrative control techniques. Engineering control is the elimination of hazards by mechanical means or through process design. Installing a confinement guard for a machine that produces flying objects is an example of engineering control. Designing piping systems and tanks to confine hazardous chemicals and designing ventilation systems to reduce the concentration of hazardous substances in air to below the permissible exposure limit (PEL) are also examples of engineering controls. On the other hand, administrative controls are management techniques, such as changing work practices, that can result in keeping worker exposures below the permissible exposure limits. The rotation of job assignments, which can reduce worker exposure to hazardous noise or harmful air contaminants to below PEL values, is another example of administrative control. Although it is desirable to engineer or administer processes so that the need for PPE is eliminated, there is often a limit to this type of solution. When hazards cannot be eliminated by engineering or administrative controls, the use of appropriate PPE that matches the specific hazard is mandatory.[1,2]

In some obvious cases, the worker can easily realize the need to wear PPE. Examples are the use of eye protection by welders and hard hats by construction workers. It is far more difficult, however, to convince workers to wear their PPE when the probability of an accident is low but its consequences are severe.

The process of using PPE can be summarized as follows: (1) the manager identifies the hazard, (2) matches the PPE to the particular hazard, and (3) convinces workers to use the PPE, reminding them of the chances they would be taking if they worked in the presence of a hazard without appropriate protection.

Selection of the degree of protection and the type of clothing or equipment needed requires detailed examination of how the hazard might materialize and appropriate testing of the PPE to ensure that it can protect a worker against the particular hazard.

Comfort and appearance are critical factors in ensuring that equipment will be used. Where possible, workers should be allowed to make their own selection from a range of equipment that provides the required level of protection.[3]

In the selection and use of PPE, it is essential to obtain equipment that meets the performance specification standards set by such institutions as the American National Standard Institute (ANSI) and the National Institute for Occupational Safety and Health (NIOSH).

Level of Protection

The components of clothing and equipment must be assembled into a full protective ensemble that not only protects the worker from site-specific hazards but also minimizes the hazards and drawbacks of the PPE ensemble itself.

Table 4.1 lists ensemble components based on the widely used EPA protection levels A, B, C, and D. This list can be used as a starting point for ensemble creation; however, each ensemble must be tailored to the specific situation in order to provide the most appropriate level of protection.[4]

Because the type of equipment and level of protection depend on specific hazards, they should be reevaluated from time to time and the ensemble adjusted as the situation dictates. However, any changes in level of protection or equipment must be made with the approval of a safety professional.

The reasons for upgrading PPE are:

- The known or potential presence of additional hazards.
- A change in the work task that will increase exposure or potential exposure to hazardous materials.
- A request by the individual performing the task.

The reasons for downgrading PPE are:

- New information that the hazards are not so severe as originally thought.
- A change in job assignment that reduces the exposure to hazardous materials.

OSHA Requirements for PPE

OSHA's general requirements for PPE have been outlined in 29 CFR 1910.132.[5] The standard contains the following overall requirements:

1. Application: Protective equipment, including protective clothing, respiratory devices, and protective shields and barriers, must be provided, used, and maintained in a sanitary and reliable condition wherever hazards exist. The hazard may be caused by the process, operation, or environment. The PPE must protect the worker from injury through absorption, inhalation, or physical contact.
2. Employee-owned equipment: Where employees provide their own PPE, the employer is responsible for ensuring the PPE's adequacy, including proper maintenance and sanitation.
3. Design: All PPE must be of safe design and construction for the work to be performed.

The following sections describe different types of PPE for the protection of specific parts of the worker's body.

Eye and Face Protection

It is difficult to estimate the exact cost of eye accidents in industry because attention has been focused only on those accidents that have resulted in lost time. Eye injuries can be prevented by common sense, planning, and the use of the right kind of training and equipment. The cost of eye protection equipment is far smaller than the cost of eye injuries.

Although nature has provided it some protection, the eye is extremely vulnerable to injuries in the man-made industrial atmosphere. The bone structure acts as a shield against large objects. The muscles around the eye act as shock absorbers against blows. The eyebrows prevent moisture from running directly into the eyes. The eyelids and lashes act as safety curtains. If something should penetrate the eyes' defenses, the tear ducts do their best to wash away the foreign body.

Industrial operations expose the eye to a variety of hazards. It is important to identify the hazards and set up a sensible program to rectify them. Typical hazards that can cause eye and face injury are:

Table 4.1. PPE ensemble components based on EPA protection levels.

Level of Protection	Equipment	Protection Provided	Should be Used When	Limiting Criteria
A	Recommended: • Pressure-demand, full-facepiece SCBA or pressure-demand supplied-air respirator with escape SCBA. • Fully encapsulating, chemical-resistant suit. • Inner chemical-resistant gloves. • Chemical-resistant safety boots/shoes. • Two-way radio communications. Optional: • Cooling unit. • Coveralls. • Long cotton underwear. • Hard hat. • Disposable gloves and boot covers.	The highest available level of respiratory, skin, and eye protection.	• The chemical substance has been identified and requires the highest level of protection for skin, eyes, and the respiratory system based on either: —measured (or potential for) high concentration of atmospheric vapors, gases, or particulates or —site operations and work functions involving a high potential for splash, immersion, or exposure to unexpected vapors, gases, or particulates of materials that are harmful to skin or capable of being absorbed through the intact skin. • Substances with a high degree of hazard to the skin are known or suspected to be present, and skin contact is possible. • Operations must be conducted in confined, poorly ventilated areas until the absence of conditions requiring Level A protection is determined.	• Fully encapsulating suit material must be compatible with the substances involved.

Table 4.1. (*Continued*)

Level of Protection	Equipment	Protection Provided	Should be Used When	Limiting Criteria
B	Recommended: • Pressure-demand, full-facepiece SCBA or pressure-demand supplied-air respirator with escape SCBA. • Chemical-resistant clothing (overalls and long-sleeved jacket; hooded, one- or two-piece chemical splash suit; disposable chemical-resistant one-piece suit). • Inner and outer chemical-resistant gloves. • Chemical-resistant safety boots/shoes. • Hard hat. • Two-way radio communications. Optional: • Coveralls. • Disposable boot covers. • Face shield. • Long cotton underwear.	The same level of respiratory protection but less skin protection than Level A. It is the minimum level recommended for initial site entries until the hazards have been further identified.	• The type and atmospheric concentration of substances have been identified and require a high level of respiratory protection, but less skin protection. This involves —with IDLH concentrations of specific substances that do not represent a severe skin hazard; or —that do not meet the criteria for use of air-purifying respirators. • Atmosphere contains less than 19.5 percent oxygen. • Presence of incompletely identified vapors or gases is indicated by direct-reading organic vapor detection instrument, but vapors and gases are not suspected of containing high levels of chemicals harmful to skin or capable of being absorbed through the intact skin.	• Use only when the vapor or gases present are not suspected of containing high concentrations of chemicals that are harmful to skin or capable of being absorbed through the intact skin. • Use only when it is highly unlikely that the work being done will generate either high concentrations of vapors, gases, or particulates or splashes of material that will affect exposed skin.
C	Recommended: • Full-facepiece, air-purifying, canister-equipped respirator.	The same level of skin protection as Level B, but a lower level of respiratory protection.	• The atmospheric contaminants, liquid splashes, or other direct contact will not adversely affect	• Atmospheric concentration of chemicals must not exceed IDLH levels.

	• Chemical-resistant clothing (overalls and long-sleeved jacket; hooded, one- or two-piece chemical splash suit; disposable chemical-resistant one-piece suit). • Inner and outer chemical-resistant gloves. • Chemical-resistant safety boots/shoes. • Hard hat. • Two-way radio communications. Optional: • Coveralls. • Disposable boot covers. • Face shield. • Escape mask. • Long cotton underwear.		any exposed skin. • The types of air contaminants have been identified, concentrations measured, and a canister is available that can remove the contaminant. • All criteria for the use of air-purifying respirators are met.	• The atmosphere must contain at least 19.5 percent oxygen.
D	Recommended: • Coveralls. • Safety boots/shoes. • Safety glasses or chemical splash goggles. • Hard hat. Optional: • Gloves. • Escape mask. • Face shield.	No respiratory protection. Minimal skin protection.	• The atmosphere contains no known hazard. • Work functions preclude splashes, immersion, or the potential for unexpected inhalation of or contact with hazardous levels of any chemicals.	• This level should not be worn in the exclusion zone. • The atmosphere must contain at least 19.5 percent oxygen.

- Splashes of toxic or corrosive chemicals, hot liquids, and molten metals.
- Flying objects, such as chips of wood, metal, and stone dust.
- Fumes and gases, and mists of toxic or corrosive chemicals.
- Radiation.

Eye injuries generally can be classified as either burns, lacerations, or bruises. Bruises can result from blows, which can also displace the retina or lens. Lacerations usually are caused by sharp materials. Burns can be either thermal or chemical. Acids or alkaline materials can start the process of destruction of eye tissue in a progressive manner. Radiation also can be very dangerous to eye tissues and optical nerves.[6]

The correct protection will minimize the hazards. Although different types of protective equipment protect the eyes and face from different job hazards, all equipment should be designed so that it is:

- Strong, durable, and lightweight.
- Resistant to impact, penetration, and heat.
- Easy to clean.

Regular glasses, even with shatter-proof lenses, cannot provide full eye protection on the job, especially against flying objects. Also, the frames used in ordinary glasses are not strong enough to keep an impact from pushing the lens through the frame and into the eye. OSHA standards on eye protection require that safety glasses and goggles meet ANSI standards (287.1).[7]

Eye and face coverings should be snug enough to keep hazards out and comfortable enough that one can see and move around easily. Too many injuries occur because eye protection was required and issued, but not used.

OSHA Requirements for Eye Protection

OSHA requirements for eye and face protection are described in 29 CFR 1910.133.[5] The regulations require that eye and face protection be provided, by the employer, where there is a reasonable likelihood that injuries could be prevented by use of the protective equipment. The employer also is responsible for determining the suitability of equipment for protection against the specific hazards. Employees are responsible for using the provide equipment according to the employer's instructions.

Employees must be provide with adequate eye protection for operations that involve flying objects, glare, liquids, and radiation. The eye protector must:

- Be adequate for the particular hazards.
- Be comfortable when worn under the intended conditions.
- Fit snugly and not unduly interfere with the movements of the wearer.
- Be durable, be easily disinfected and cleaned, and be kept clean and in good repair.

Workers who wear corrective glasses and have job assignments that require eye protection, must be provided with one of the following:

- Prescription safety glasses.
- Safety goggles designed to fit over prescription glasses.
- Safety goggles with prescription lenses mounted behind the protective lenses.

Table 4.2. Industrial operations requiring eye and face protection.

Abrasive blasting, spray finishing, and open-surface tanks
Helicopters
Forging machines
Guarding of portable powered tools
Welding, cutting, and brazing
Pulp, paper, and paperbound mills
Textiles
Saw mills
Pulpwood logging
Telecommunications
Abrasive wheels and tools
Welding, cutting, and heating
Battery rooms and battery charging
Toxic cleaning solvents
Chemical paint and preservative removers
Mechanical paint removers
Painting

Source: Reference 8.

OSHA also requires that when the manufacturer of an eye or face protective device has indicated certain limitations for the equipment, this information be properly transmitted to the users of the equipment.

The design, construction, testing, and use of eye and face protective devices must be conducted according to the standards set forth by ANSI in the manual entitled *Occupational and Educational Eye and Face Protection*, Z87.1-1968.

Table 4.2 lists the industrial operations and processes for which OSHA's general industry standards have mandated eye and face protection.[8]

Eye and Face Protection Equipment

Eye and face protection equipment is available in many designs and styles. Employers and supervisors should be familiar with the various forms of eye and face protective devices and ensure that the type of equipment selected matches the hazard. Any equipment chosen also must meet respective design and performance requirements.

Eye protection can be furnished in heat- or chemically treated glass, plastic, wire screen, or light filtering glass. Safety glasses have extra-sturdy frames and impact-resistant lenses. Such glasses protect the eyes from flying objects from the front (and some are equipped with shields to provide protection from other directions). Safety goggles protect the eyes from dust, splashes, flying objects, and sparks from any direction.

Face shields protect the face and neck from splashes, heat, glare, and flying objects. Helmets protect the eye from sparks, splashes, and intense light. They also protect the head and may have extensions to protect the neck.

Safety Glasses and Goggles

Safety glasses with clear hardened-glass, hard plastic, or wire mesh lenses are designed for protecting eyes from moderate impacts encountered in scaling, grinding, spot welding, woodworking, and machine operations. The frames may be made of fiber, wire, plastic, or plastic-coated wire. Safety glasses with metal frames should not be worn if there is a possibility of electrical contact or in explosive atmospheres where nonsparking tools are required. Some safety glasses are equipped with sideshields to provide protection against objects coming from the side, and some are equipped with eye cup shields to provide general protection against flying objects. Most safety glasses have such features as universal nose, bridge, and adjustable temples that would allow for the stocking of a few sizes. If there is a potential chemical splash hazard, a face shield must be worn over the safety glasses.[5,8-10]

Safety glasses with Polaroid and automatically darkening lenses are recommended for eye protection in work areas where glare from the sun can create an additional hazard. Video display terminal (VDT) safety glasses are available for reducing glare and ultraviolet radiation from VDT screens.

Safety glasses and goggles with plastic frames and lenses made of lead provide eye protection from X-rays. It is extremely important that such glasses and goggles be used according to the instructions of the manufacturer. Special radar safety glasses can protect the eyes from burns by radar waves. These glasses dissipate most of the thermal energy received.

The infrared rays in laser beams have the potential of causing severe eye burns, and the ultraviolet rays generated in flash tubes flash during the pumping of lasers can be harmful to the eyes. In order to protect the eyes, workers should use special absorptive lenses while working with lasers. The lenses must be selected according to the specific wavelength of the laser beams encountered. These lenses are available for argon gas, carbon dioxide, helium, neon nitrogen, neodymium, and gallium arsenide lasers.

Cover goggles normally are worn over safety glasses to provide additional protection to the wearer's eyes or prescription lenses. These goggles also protect the safety glasses underneath them, against pitting and breaking. Cover goggles include the cup type, made of heat-treated lenses, and the wide vision type, made of plastic lenses. Goggles are recommended in such operations as grinding, machining, chipping, riveting, and work with chemicals and hot objects.[2]

Safety goggles can protect the eyes against hazards coming from any direction. Some goggles have a cup over each eye, whereas others have a frame that extends over both eyes. Although most goggles are directly ventilated to allow for air circulation around the eyes, indirectly ventilated goggles sometimes are required to keep chemical splashes and dust particles out.[8]

Although goggles come in a variety of designs and forms, they can be classified as follows:[2]

- Chemical goggles: This type is designed to protect the eyes from chemical splashes or fine dust particles. The frames usually are made of soft vinyl or rubber, and the lenses are made of either heat-treated glass or acid-resistant plastic.
- Leather mask dust goggles: These are designed to protect the eyes against noncorrosive dusts in operations such as cement manufacturing and flour mills. Wire-screen ventilators are used around the eye cup to provide for air circulation.
- Miner's goggles: Used for underground work or other high humidity areas where fogging is a serious problem, these goggles are made of a corrosion-resistant wire screen, coated with a dull black to reduce reflection.
- Melter's goggles: In this type, the lenses are coated with cobalt blue glass, and the frames are made of leather or plastic to provide protection against radiant heat.
- Welder's goggles: With lenses made of a suitable filtering material, these goggles are recommended for such operations as oxyacetylene welding, cutting, lead burning, and brazing. 29 CFR 1910.252 provides guidelines for the selection of appropriate shade numbers.
- Chipper's goggles: Used where maximum protection from flying objects is required, these goggles come in two styles, one for individuals who do not wear prescription lenses and one to fit over prescription glasses.

In situations where a chemical splash hazard exists, goggles must be corrosion-resistant and fit snugly against the face so that no splash can penetrate to the eyes. In areas where the humidity is high or rapid temperature changes take place, fogging can become a serious problem. Several methods are available to make the lenses less susceptible to fogging. These methods include but are not limited to:

- Placement of ventilating slots to maintain air flow across the inner surface of the lens.
- Use of a double lens with a hermetically sealed air space in between the lenses.
- Use of a chemical coating permanently bonded to the lens surface.

Workers who wear contact lenses must be especially careful. Contacts can absorb and trap particles and gases that could injure the eyes, and they should not be worn by persons who work in an area where they could be exposed to excessive heat, dust, corrosives, fumes, vapors, or splashes.[5,8]

Face Shields and Helmets

Face shields surround the face area and protect it against flying particles, splashes, sparks, and glare. They normally have detachable windows made of plastic, wire mesh, or both (see Figure 4.1). Nonconductive face shields are available for use by electrical workers. In potentially explosive atmospheres, nonsparking shields must be used. Goggles must be worn underneath these shields when the eyes require added protection (i.e., when the face shield is raised). Also, face shields can be attached to a hard hat. Face shields must be at least 6 inches high and 0.04 inch thick. They may be tinted for light spot welding and protection against reflected glare. They are also available with infrared absorbers to provide protection against harmful infrared and ultraviolet rays.[8]

Helmets protect the head, face, and eyes from hazards such as those encountered in welding operations (e.g., flying sparks and hot metal particles). As in the selection of any other kind of PPE, helmets must be chosen with special care to ensure compatibility with the hazard. For example, aluminized helmets protect against heat hazards, whereas others are specially designed to be chemical-resistant. Some helmets have filter lenses of various shades to protect the eyes from radiation and glare. Also available are helmets that have an enlarged bulge below the lens

Figure 4.1. Face protection. (Source: National Safety Council, Chicago)

to accommodate the wearing of a respirator. Safety glasses must be worn under helmets to provide protection when the helmet is raised.

The space helmet design combines hard-hat and face-shield impact protection and respiratory protection. Both compressed air and powered air-purifying systems are available.

The air purifying helmet and the lift-up face shield assembly provides respiratory protection combined with eye, face, and head protection. The air stream passes through coarse filters before it is supplied to the wearer. The unit which is battery-operated, is recommended for such operations as mining, quarrying, tunneling, and processing of minerals.

Eye and face protection will work better, last longer, and keep workers safer with careful cleaning and maintenance. Check the equipment before each wearing and *replace* it if:[7]

- Frames are bent or broken.
- Lenses are scratched or pitted (which can reduce impact resistance).
- The headband is loose, twisted, knotted, or sweat-soaked.

Head Protection

Most industrial head injuries are caused by heavy falling objects. These injuries, which can be serious, constitute about 10% of all occupational injuries. Head injuries on the average result in a lost time of about three weeks. Although heavy falling objects are their prime cause, head injuries can be caused by a splash of chemicals or molten materials, as well as by contact with live electrical conductors.[10]

OSHA Requirements for Head Protection

These requirements are contained in 29 CFR 1910.135. It is the responsibility of the employer to determine if any aspect of worker's job presents a risk for head injuries. Under such circumstances, the employer must ensure that appropriate protection is provided and worn. The protective helmet must provide protection against impact and penetration of falling and flying objects, and protection from limited electric shock and burn.[5,8,10] Any protective helmet provided by the employer must meet the requirements and specifications established in ANSI, Z89.1-1969, which contains the following criteria for helmets:[11]

- They have a dome-shaped shell of one-piece construction. Type I helmets (hats) have a continuous brim at least 1.25 inches wide, and Type II helmets (caps) are brimless with a peak extending forward from the crown.
- They have a crown strap of plastic, closely woven webbing, or similar material that forms a cradle to support the helmet and provide a space between the head and the shell for ventilation.
- They have headbands that are disposable and adjustable.
- They meet the physical requirements for Class A, B, or C given in Table 4.3 (applicable to both Type I and Type II helmets).

It is important that employers and supervisors conduct an on-site review of conditions at the workplace and determine the need for head protection and the type required. The employer may have to defend his or her decision after an accident or during a compliance inspection.[10]

Types of Head Protection Equipment

ANSI Type II, Class A helmets are designed for head protection from impact. They also offer limited electric shock protection. The shells must be made of fiberglass, plastic, or vulcanized fiber. They must have seamless construction and be water-resistant, slow-burning, and nonirritating to the wearer's skin.

Table 4.3. ANSI safety requirements for industrial head protection.

Category	Reference	Requirement		
		Class A	Class B	Class C
Insulation resistance	Section 7.1	2,220 V AC (60 Hz for 3 min with current leakage not exceeding 9 mA)	20,000 V AC (60 Hz for 3 min with current leakage not exceeding 9 mA)	None
Impact resistance	Section 7.2		Transmits an average force of 850 lb	
Penetration resistance	Section 7.3	Cannot be pierced more than 0.375 in. (9.5 mm)	Cannot be pierced more than 0.375 in. (9.5 mm)	Cannot be pierced more than 0.4375 in. (11.1 mm)
Flammability	Section 7.4	Burn rate of 3 in./min (7.62 cm/min)	Burn rate of 3 in./min (7.62 cm/min)	Burn rate of 3 in./min (7.62 cm/min)
Water absorption	Section 7.5	5.0%	0.5% maximum	

Source: Reference 11.

ANSI Type I, Class B helmets provide protection from electric shock and falling and flying objects. Shells must be made of fiberglass, plastic, or vulcanized fiber; have seamless construction; and be water-resistant, slow-burning, and nonirritating to the wearer's skin.

ANSI Type II, Class B helmets provide head protection from falling and flying objects and electric shock and burn. Shells must be made of aluminum, fiberglass, and vulcanized fiber; have seamless construction; and be water-resistant, slow-burning, and nonirritating to the wearer's skin.

ANSI Type I, Class C helmets protect the heads of workers from the impact of falling and flying objects where there is no exposure to electric shock or burn. Shells must be made of aluminum, fiberglass, plastic, or vulcanized fiber; have seamless construction; and be water-resistant and nonirritating to the wearer's skin.[8,11]

Hearing Protection

Noise can create physical and psychological stress and contribute to accidents by making it impossible to hear warning signals. An estimated 14 million U.S. workers are exposed to hazardous noise.

Excessive noise can destroy the ear's ability to hear; it also can put stress on other parts of the body, including the heart. Although research on the effects of noise is ongoing, it appears that noise can cause a quickened pulse rate, an elevation of the blood pressure, and a narrowing of the blood vessels. Over a long period of time, these effects may place an added burden on the heart. Workers exposed to noise sometimes complain of nervousness, sleeplessness, and fatigue. Excessive noise exposure also can reduce job performance. Because there is no cure for most effects of noise, the prevention of excessive noise exposure is the only way to avoid health damage.[12,13]

OSHA Requirements for Ear Protection

Under the Occupational Safety and Health Act (29 CFR 1910.95), every employer is legally responsible for providing a workplace free of excessive noise.[5] OSHA regulations mandate that each employer administer a continuing, effective hearing conservation program whenever employee noise exposures equal or exceed an 8-hour time weighted average (TWA) of 85 decibels,

Table 4.4. Permissible noise exposure levels.

Duration Per Day, Hours	Sound Level dBa, Slow Response
8	90
6	92
4	95
3	97
2	100
1.5	102
1	105
0.5	110
0.25 or less	115

Source: 5

referred to as the action level. The OSHA regulations also require that when employee exposure equals or exceeds those levels shown in Table 4.4, the employer must exercise feasible engineering or administrative controls. However, if such controls fail to reduce the noise level, adequate personal protective equipment must be provided.

Types of Hearing Protection Devices

Two general types of ear protection are used widely by industry: the cup muff type and the plug insert type.

Ear plugs are inner-ear protection devices designed to occlude the ear canal. They are made from soft materials, particularly rubber, neoprene, wax, cotton, fiberglass, or plastic. Ear plugs can be molded to each individual's ear; because individuals' ear canals differ in shape and size, these plugs become the property of the person to whom they are fitted.[2,9]

Ear muff protectors, such as those shown in Figure 4.2, give over-the-ear protection against

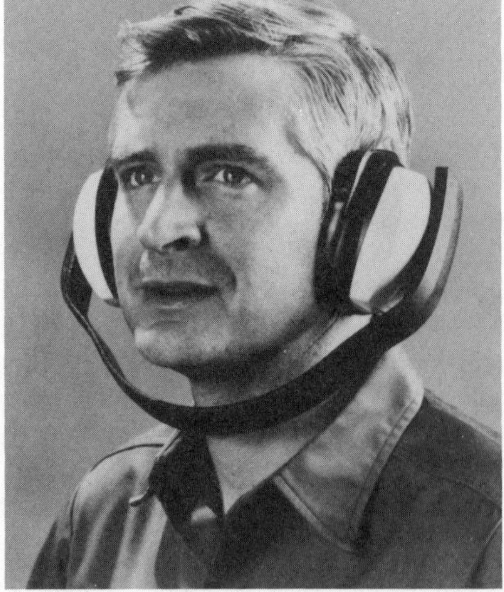

Figure 4.2. Muff type hearing protection. (Source: Mine Safety Appliances Company, Pittsburgh, PA.)

dangerous high-frequency noises. They usually allow the wearer to hear low-frequency sounds such as voice and warning signals. These muffs are made in a universal type.

Respiratory Protection

Many industrial operations generate airborne contaminants that pollute the atmosphere. Typical operations that generate airborne hazards include metal casting, grinding, welding, polishing, spray finishing, and fabric sewing. Also some industrial operations, such as the petroleum and petrochemical industry, have the potential of releasing large quantities of toxic substances into the atmosphere in case of an accident. The contaminated air can have serious adverse health effects on employees. The main objective of this section is to inform persons who have the responsibility for protecting their employees from respiratory hazards about the protective measures and protective equipment that are available. Every effort must be made to eliminate respiratory hazards through engineering and administrative control methods. However, when such controls cannot eliminate a hazard or reduce its danger to an acceptable level, the employer must provide adequate respiratory protection.[14–16]

Types of Air Contaminants

Airborne hazards can be divided into dusts, fumes, gases, vapors, mists, aerosols, and smoke:[15,17]

- Dusts: These substances consist of mechanically produced solid particles whose size can range from microscopic to visible or macroscopic. Dust particles can be inorganic, such as rock, coal, or grain, or they can be organic, as in polymeric materials. The proper evaluation of dust hazards requires information regarding particle size, chemical composition, concentration, and factors that affect dispersion.
- Fumes: These substances are formed when particles migrate and condense from the surface of a molten solid into the air. Fume particles are very fine, usually less than 1 micron in diameter, and because of their small size can be readily inhaled. In some cases, the molten solid particles can react with oxygen in air to produce a toxic oxide.
- Gases: A gas may be defined as a fluid that is above its critical point at normal temperature and pressure. Gases differ from vapors in that they can be changed to liquid only be the combined effect of pressure and temperature, whereas vapors can be liquefied by increasing the pressure alone. Gases fill the container in which they are confined. Examples are carbon monoxide, oxygen, and hydrogen cyanide.
- Vapors: A vapor may be defined as a fluid that is not in solid or liquid form and is below its critical point at normal temperature and pressure. Vapors can be condensed into the liquid phase by application of pressure alone, whereas gases require a combined effect of temperature and pressure to be condensed into liquids. Examples of vapors are steam or hydrogen sulfide in a closed container in contact with liquid.
- Mists: These substances, formed by condensation of vapors into small liquid droplets or by the fine dispersion of a liquid into small particles, are produced in such operations as foaming or atomizing. Examples are oil mists formed during metal cutting or grinding and acid mists from electroplating.
- Aerosols: These are suspensions of fine liquid or solid particles in air, which because of their small size can remain in a dispersed state for a relatively long period of time.
- Smoke: This substance consists of carbon or soot particles less than 0.1 micron in size, formed from the incomplete combustion of carbon-containing compounds.

OSHA Requirements for Respiratory Protection

OSHA's requirements for respiratory protection are contained in 29 CFR 1910.134.[5,8]

Permissible Practice

Preventing atmospheric contamination is the primary objective in attempts to control those occupational diseases caused by breathing air contaminated with harmful dusts, fogs, fumes, mists, gases, smokes, sprays, or vapors. This should be accomplished as far as is feasible by accepted engineering control measures (for example, enclosure or confinement of the operation, general and local ventilation, and substitution of less toxic materials). When effective engineering controls are not feasible, or while they are being instituted, appropriate respirators should be used.

Respirators should be provided by the employer when such equipment is necessary to protect employees' health. The respirators should be applicable to and suitable for the intended purpose. The employer is responsible for the establishment and maintenance of a respiratory protective program, which includes the requirements outlined below. The employees must use the provided respiratory protection in accordance with the instructions and training they receive.

Requirements for a Minimal Acceptable Program

The employer should provide written standard operating procedures (SOPs) for the selection and use of respirators, which should be matched to the hazards to which the worker is exposed. The employer is responsible for surveying work area conditions and the degree of employee exposure. Regular inspection of equipment and evaluation of conditions should follow to determine the effectiveness of the program. Selection criteria are those given by ANSI Standard Z88.2-1969, Respiratory Protection. The worker should be instructed and trained in the proper use of respirators and their limitations.

Respirators should be cleaned and disinfected on a regular basis. Those used by more than one worker should be cleaned and disinfected after each use. Respirators should be stored in a convenient, clean, and sanitary location. Respiratory for emergency use, such as self-contained devices, should be thoroughly inspected at least once a month and after each use.

Only persons who are physically able should be assigned to jobs requiring a respirator. This determination should be made by a local physician or health professional.

Selection of Respirators

The selection of respirators should be made in accordance with standards set forth by ANSI in Z88.2-1969.

Air Quality

Compressed air, compressed oxygen, liquid air, and liquid oxygen used for respiration should be of high purity. Compressed oxygen should not be used in supplied-air respirators or in an open-circuit, self-contained breathing apparatus that has previously used compressed air. Oxygen must never be used with air line respirators.

Any cylinder or compressor supplying air to respirators must meet both of the following requirements:

- Cylinders are tested and maintained as prescribed in DOT shipping container specification regulations (49 CFR Part 178).
- The compressor used for supplying air is equipped with necessary safety and standby devices.

The couplings on the air line should be incompatible with other gases to prevent the inadvertent flow of nonbreathable gases. Breathing gas containers should be marked in accordance with the ANSI standard method of making portable compressed gas containers (Z48.1-1954).

Procedures and Training for Respirator Use

The employer is responsible for the development of procedures for the selection, use, and care of respirators. The procedures should specify the correct respirator for each job.

If normal operation or an emergency situation creates a toxic atmosphere, the employer must prepare a written SOP for the safe use of respirators in such atmospheres. Workers must be familiar with the procedures developed by the employer. In toxic atmospheres where there is immediate danger to the life or health of the respirator wearer, there should be at least one other person present to conduct any rescue operation that might be needed. The employer must develop procedures to ensure that the respirator wearer can receive help if it is needed. (See also later section on "Use of Respirators.")

Respirators should be checked on a random basis to ensure their proper selection, use, cleaning, and maintenance. If respirators are to be handled safely, potential users of such devices must be properly trained. Every employee who may use a respirator must be properly trained on fitting procedures. Respirators should not be used if any directions prevent their proper fit, including beard growth or other facial features. If corrective lenses are required, they should be worn in such a way that they do not interfere with respirator fit. (See below, sections on "Use of Respirators" and "Respirator Fit Test.")

Identification of Gas Mask Canisters

Gas mask canisters must be properly labeled and colored, although the color code should be regarded as a secondary means of canister identification. Table 4.5 shows the distinctive colors or combinations of colors used to code gas mask canisters.

Each canister must carry, in bold letter, either of the following labels: **Canister for _______** (name for atmospheric contaminant) or **Type N Gas Mask Canister.** In addition, essentially the following wording must appear beneath the appropriate phrase on the canister label: "For respiratory protection in atmospheres containing not more than ________percent by volume of ____ (name of atmospheric contaminant)."

As gas mask canisters either remove or neutralize contaminants from air, each canister should have a proper warning label stating that it can be used only in atmospheres containing sufficient oxygen (at least 19.5% by volume).

Types of Respiratory Equipment

Respiratory protective devices can be categorized in three major types:

- Air purifier respirator: This type of equipment is designed to remove air contaminants by either mechanical or chemical action. It can be used only in atmospheres that have a sufficient amount of oxygen to support life. Because this type of equipment can remove or neutralize certain contaminants from air, it is imperative that the type and concentration of contaminants be known before any attempt is made to use an air purifier respirator.
- Atmosphere (air)-supplied respirators: This type of respirator supplies air or oxygen to the wearer from a fixed source. It is independent of the atmosphere in which it is used.
- Self-contained breathing apparatus (SCBA): This type of equipment provides oxygen or air from a cylinder carried by the user. Like supplied air equipment, it is independent of the atmosphere in which it is used. Compared to an air supplies respirator, an SCBA provides high mobility for the user.

Air Purifiers

Air purifier respirators[2,14,17] remove certain contaminants from air by either mechanical or chemical means (Figure 4.3). Prior to inhalation ambient air is passed through a filter, cartridge, or

Table 4.5. Color codes for gas mask canisters.

Atmospheric Contaminants for which Protection is Needed	Colors Assigned[a]
Acid gases	White
Hydrocyanic acid gas	White with a 1/2-inch green stripe completely around the canister near the bottom
Chlorine gas	White with a 1/2-inch yellow stripe completely around the canister near the bottom
Organic vapors	Black
Ammonia gas	Green
Acid gases and ammonia gas	Green with a 1/2-inch white stripe completely around the canister near the bottom
Carbon monoxide	Blue
Acid gases and organic vapors	Yellow
Hydrocyanic acid gas and chloropicrin vapor	Yellow with a 1/2-inch blue strip completely around the canister near the bottom
Acid gases, organic vapors, and ammonia gases	Brown
Radioactive materials, except tritium and noble gases	Purple (magenta)
Particulates (dusts, fumes, mists, fogs, or smokes) in combination with any of the above gases or vapors	Canister color for containment, as designated above, with a 1/2-inch gray stripe completely around the canister near the top
All of the above atmospheric contaminants	Red with a 1/2-inch gray stripe completely around the canister near the top

[a]Gray must not be assigned as the main color for a canister designed to remove acids or vapors.
Note: Orange must be used as a complete-body or stripe color to represent gases not included in this table. The user will need to refer to the canister label to determine the degree of protection the canister will afford.
Source: Reference 5.

Figure 4.3. Air purifier respirator. (Source: Argonne National Laboratory)

canister packed with the appropriate materials to remove or neutralize the contaminants. These respirators can have an external source of power, or they can operate through the breathing effort of the wearer.

Mechanical Filters

Mechanical filter respirators offer protection against airborne hazards such as dusts, mists, and metal fumes. These respirators are equipped with a face piece (either full face or half mask, and directly attached to the face piece is a mechanical filter, made of an appropriate fibrous material, that physically traps the airborne particles and delivers purified air to the user. The filter medium must be efficient enough to remove fine solid particles from the incoming air. Note that mechanical filters do not have the capability of removing hazardous gases or vapors. The specific type of mechanical filter that matches the airborne hazard must be selected, as there are different types of mechanical filters, each capable of removing a certain type and size of particulate matter from the air. The filters may be replaceable or may function as a permanent part of the respirator.

Chemical Canister and Cartridge Respirators

These respirators are capable of removing low concentrations of hazardous vapors and gases from breathing air. Cartridges usually attach directly to the respirator face piece. The larger-volume canisters attach to the chin of the face piece or are carried with a harness and attached to the face piece by a breathing tube.[4]

The removal of air contaminants is accomplished either physically by adsorption, or chemically by neutralization of the particular contaminants. Adsorption may be defined as the process of removing a vapor or gaseous component of a gas mixture by a solid. The type of contaminants removed and the removing efficiency of the bed are a strong function of the adsorbing medium, temperature, partial pressure of components, and gas velocity. Any adsorbing medium can hold a maximum amount of a given contaminant, and when this maximum is reached, the bed is saturated and must be either regenerated or replaced. This is an important consideration in the use of chemical cartridges and canisters. Every effort must be made to develop procedures and/or instruments to ensure that users of this type of respirator are warned before saturation of the adsorbing medium occurs.

Typical adsorbing materials used in chemical cartridges and canisters include activated charcoal, zeolites (also called molecular sieves), and silica gel. It is extremely important to realize that each adsorbing medium is capable of removing specific contaminants from the air stream. For example, activated charcoal is an excellent adsorbent for organic vapors although its adsorption efficiency in the removal of some hydrocarbon gases and carbon monoxide is quite limited. Consequently, a respirator that employs activated charcoal as the adsorbing medium should not be used for the removal of carbon monoxide.

The removal of air contaminants by chemical cartridges and canisters also can be accomplished by neutralization of the particular contaminants (neutralization being the elimination of the hazardous property by chemical reaction). Contaminated air is passed through a bed of a suitable material that acts as either a catalyst or a reactant.

Catalysts are not consumed in a chemical reaction. Their major role is to create intermediate steps for the chemical reaction with lower activation energies than that of the principal reaction. On kinetic grounds, we can say that catalysts enhance the rate of chemical reaction at a given temperature and pressure. An example of a catalyst used in chemical cartridge respirators is Hapcalite, which enhances the reaction between oxygen and carbon monoxide to produce carbon dioxide:

$$\mathrm{CO} + \frac{1}{2}\mathrm{O_2} \xrightarrow{\text{Hapcalite}} \mathrm{CO_2} + \text{heat}$$

As carbon dioxide is not a hazardous gas, the reaction has eliminated the hazardous properties of carbon monoxide. It should be noted that heat effects are associated with most chemical reactions. A reaction can generate heat (exothermic reaction), or it can consume heat (endothermic). These heat effects must be taken into account in the selection of respirators to ensure that the breathing air is not heated to intolerable levels. Some chemical cartridges contain compounds that participate directly in a chemical reaction. An example is soda lime, which reacts with carbon dioxide to produce carbonates and water.

Since air-purifying respirators can remove only certain contaminants from air, their use is limited. Specifically, they should not be used in certain conditions:[4]

- In atmospheres that do not have a sufficient amount of oxygen (19.5% by volume) to support life.
- In atmospheres containing hazardous substances in concentrations that can be immediately dangerous to life or health (IDLH).
- For entry into unventilated or confined areas with unknown concentrations of contaminants.
- In the presence or potential presence of unidentified contaminants.
- In atmospheres containing more than 2% by volume of a known contaminant.
- When the sorbent service life is unknown, or the respirator has no end-of-service-life indicator.
- In conditions that may adversely affect sorbent performance, such as high humidity.
- For contaminants that cannot be detected by odor or irritation, or do not have good warning properties.
- For contaminants that are highly irritating to the eyes.

Most chemical sorbent canisters are imprinted with an expiration date, and may be used until that date as long as they have not been opened previously. Once opened, they begin to sorb humidity and air contaminants, which will cause their efficiency and service life to decrease. Cartridges should be discarded after use but should not be used for longer than one shift or when breakthrough occurs, whichever comes first. Some respirators combine a mechanical filter with a chemical cartridge for dual or multiple exposure. It is also important to use respirators that have been approved by the Mine Safety and Health Administration (MSHA) and the National Institute for Occupational Safety and Health (NIOSH).

Supplied Air Respirators

Supplied air respirators supply breathing air to a face piece via a hose from a fixed source.[2,4,17] This type of respirator is useful against all atmospheric contaminants that are not immediately dangerous to life or health (IDLH). This limitation is imposed on this type of respirator in consideration of the fact that the air supply is not carried with the user. There is always a possibility, however remote, that the air supply equipment might malfunction and conditions of IDLH make the escape from the contaminated atmosphere impossible.

The air-line respirator is quite versatile and the most comfortable type to wear. There usually is little or no resistance to inhalation, and the flow of air produces a cool refreshing effect. In spite of these advantages, the air line reduces the wearer's mobility, and there is always danger of entanglement of the hose and the possibility of hose puncture by rough or sharp surfaces.

Supplied air respirators are available in both half and full mask types. When atmospheric contaminants are irritating to the eyes a full mask respirator should be used. The most important consideration in the use of air-line respirators is to ensure that a clean supply of air is available and the equipment that will supply the air is in proper working condition.

Supplied air respirators can be divided into air-line respirators and hose masks. Air-line respirators supply air by means of a compressor. In some plants where lubricated reciprocating compressors are used, it is possible for oil vapors and mists to contaminate the breathing air supplied by such compressors. In such instances, an appropriate mechanical filter must be used to ensure that these contaminants are removed.

Hose masks used in air-line respirators may come with or without a blower. Hose masks with a blower supply air by means of a motor-driven or hand-operated blower. The wearer can continue to inhale through the hose if the blower fails, and up to 300 feet (91 meters) of hose length is permissible. In hose masks without a blower, the wearer must provide a motivating force to pull air through the hose; up to 75 feet (23 meters) of hose length is permissible.

The following questions should be addressed in considering supplied air respirators:

- Is the atmosphere IDLH or likely to become IDLH?
- Is enough mobility provided?
- Is there danger that the air line will become entangled or punctured?
- Can a clean supply of air be provided by the compressor?

Table 4.6 summarizes the permissible upper limit concentration for air purifier and supplied air respirators.[28]

Self-Contained Breathing Apparatus (SCBA)

The main SCBA components consist of an air or oxygen supply carried by the wearer, a hose, a regulator value, and a face piece (see Figure 4.4). SCBAs are divided into positive pressure respirators and negative pressure respirators depending on the type of air or oxygen flow supplied to the face piece.

Positive pressure respirators maintain a positive pressure in the face piece during exhalation and inhalation, and are further divided into pressure demand and continuous flow types. The pressure demand type uses an exhalation valve on the face piece and a regulator maintaining a positive pressure in the face piece. In case of a leak, the regulator sends a continuous flow of air or oxygen to prevent leakage of contaminants into the face piece. Continuous flow respirators send a continuous flow into the face piece at all times. Although the flow prevents infiltration of contaminants into the face piece, it uses the air or oxygen supply much more rapidly compared to the pressure demand type.

Negative pressure respirators draw air into the face piece as a result of the negative pressure created by the wearer's inhalation. The major disadvantage of this type of equipment is that if a leak develops, the contaminated atmosphere can flow into the face piece. In situations where even the smallest leaks cannot be tolerated, the use of a positive pressure respirator is recommended.

SCBAs can be of either open circuit or closed circuit design. Open circuit devices exhaust the unbreathed air or oxygen. In closed circuit equipment, the exhaled air passes through an adsorbent to remove carbon dioxide and water, and make up air then is added to the regenerated stream (see Figure 4.5).

Most SCBAs provide cylinders containing air or oxygen. However, some designs provide an oxygen-generating breathing apparatus. These devices use closed circuits and contain an oxygen-rich compound such as a metal peroxide or superoxide that can react with moisture in exhaled air to produce oxygen. The major disadvantage of this type of equipment is that the rate of chemical reactions, and therefore oxygen generation, is a strong function of temperature.

Although SCBAs can provide protection against most types and levels of contaminants, the duration of the air or oxygen supply is an important factor in its use. This is limited by both the

Table 4.6. Permissible upper limits of respirator.

	Organic Vapor Cartridge		Supplied Air
Contaminant	Half Mask	Full Face	Full Face
Acetaldehyde	1,000 ppm	1,000 ppm	10,000 ppm
Acetic acid	100 ppm	500 ppm	1,000 ppm
Acrylonitrile	20 ppm	100 ppm	4,000 ppm
Benzene	10 ppm	50 ppm	2,000 ppm
Butyl acetate	1,000 ppm	1,000 ppm	10,000 ppm
Butyl alcohol	500 ppm	1,000 ppm	8,000 ppm

	Acid Gas Cartridge		Supplied Air
Contaminant	Half Mask	Full Face	Full Face
Chlorine	5 ppm	10 ppm	25 ppm
Chlorine dioxide	1 ppm	5 ppm	10 ppm
Formaldehyde*	—	10 ppm	100 ppm
Hydrogen chloride (hydrochloric acid)	50 ppm	50 ppm	100 ppm
Sulfur dioxide	20 ppm	50 ppm	100 ppm

	Ammonia/Methylamine Cartridge		Supplied Air
Contaminant	Half Mask	Full Face	Full Face
Ammonia	100 ppm	300 ppm	500 ppm
Methylamine	100 ppm	100 ppm	100 ppm

	Dust and Mist Filter		Supplied Air
Contaminant	Half Mask	Full Face	Full Face
Aluminum metal dust	100 mg/m^3	500 mg/m^3	10,000 mg/m^3
Calcium oxide dust	20 mg/m^3	100 mg/m^3	250 mg/m^3
Copper dust	10 mg/m^3	50 mg/m^3	2,000 mg/m^3
Cotton dust	2 mg/m^3	10 mg/m^3	500 mg/m^3
Silica (quartz)	1 mg/m^3	5 mg/m^3	200 mg/m^3
Wood dust (hardwood)	10 mg/m^3	50 mg/m^3	2,000 mg/m^3
Wood dust (softwood)	50 mg/m^3	250 mg/m^3	10,000 mg/m^3

amount of the air or oxygen available and its rate of consumption by the user. Another disadvantage of SCBAs is that they are bulky and heavy and may cause heat stress or impair movement in confined areas. SCBAs may be approved for (1) escape only, or (2) for both entry into and escape from a hazardous area. Table 4.7 summarizes different types of SCBA and their relative advantages and disadvantages. When deciding whether an SCBA is appropriate for a given situation, one should consider the following points:

- If the atmosphere is IDLH or is likely to become IDLH, a positive pressure SCBA should be used.
- The amount of air supply should match the duration of time required to perform the tasks.
- The effect of the bulk and weight of the SCBA on worker performance should be evaluated.

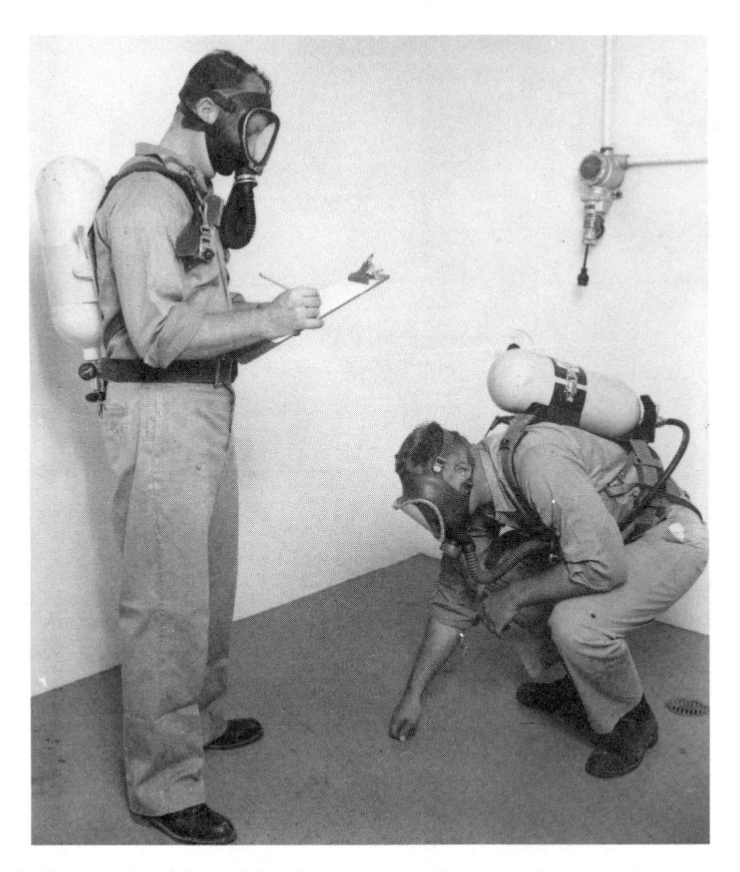

Figure 4.4. Self-contained breathing apparatus. (Source: Argonne National Laboratory)

Use of Respirators

Before using any kind of respiratory device, the employer must ensure that written standard operating procedures (SOPs), covering a complete respiratory program, have been developed. The SOPs must include information regarding the proper use of respirators, inspection, monitoring, and planning for routine, nonroutine, and emergency respirator use.[7,8,17]

The face-piece-to-face fit is extremely important in the use of respirators, as a good fit could determine the difference between life and death. Because each face piece fits a certain percentage of people, it is imperative to fit-test respirators before using them. A respirator should not be worn when conditions exist precluding a good fit. Conditions that adversely affect respirator fit include the following:

- Facial features such as long sideburns, mustache, and beard.
- Spectacles with temple bars or straps.
- A head covering that passes between the sealing surface of a respirator and the wearer's face.
- Scars, hollow temples, excessively protruding cheekbones, and deep creases in the facial skin.
- The absence of teeth or dentures, or an unusual facial configuration.

Respirator Fit Test

Respirators can be tested for good fit either qualitatively or quantitatively. In both tests a harmless agent that has an odor or irritating effect is injected into the air inside an enclosure where the test

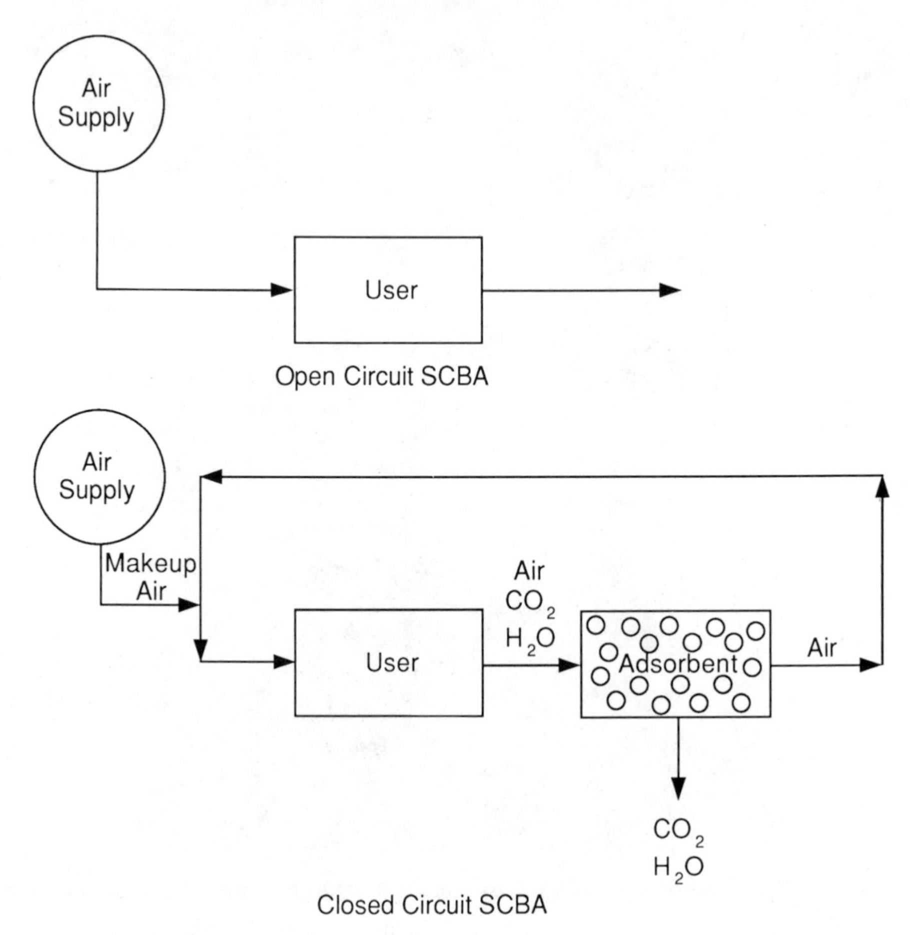

Figure 4.5. Closed and open circuit self-contained breathing apparatus.

Table 4.7. Capabilities and limitations of self contained breathing apparatus (SCBA).

Type	Operation	Advantages	Disadvantages
Entry and Escape			
Open-circuit	Air is supplied from a cylinder; unbreathed air exhausted to atmosphere.	Operates in positive pressure mode. Provides highest level of respiratory protection.	Short operating time (30–60 minutes); heavier than closed circuit equipment.
Closed-circuit	Removes CO_2 and moisture from exhaled air and recycles the unbreathed air.	• Longer operating time. • Lighter weight.	• CO_2 removal is a function of temperature. • Retains exhalation heat and heat generated as a result of CO_2 removal.
Escape Only SCBA	Air is supplied from a cylinder or an oxygen-generating source approved for escape only.	• Light weight. • Available in pressure demand and continuous flow modes.	Cannot be used for entry.

Source: Reference 4.

is conducted. In a qualitative test, the respirator wearer uses his or her senses to detect the presence of the agent inside the face-piece. In a quantitative test, special instruments are used to determine if the agent has leaked into the face piece. Two other widely used types of fit tests are the positive pressure and negative pressure tests.

In a positive pressure test, the user closes the exhalation valve and exhales into the face piece. If no air leaks into it (i.e., there is a good fit), the face piece should bulge a little.

In a negative pressure test, the user closes the inhalation valve and breathes in, holding the breath for 10 seconds. If there is a good fit, the face piece should collapse against the user's face. The manufacturer's instructions will identify which test to use.

Not everyone can use a respirator. Before making any attempt to do so, one should have a medical examination conducted by a qualified physician. Generally, individuals who have breathing problems such as asthma, heart condition, sensitivity to heat, or claustrophobia (fear of confined spaces) should not be assigned to tasks requiring the use of a respirator.

Respirators should be checked for wear and damage before use. Make sure there are no snags that could let air in, and see that all the connections are tight. Take an especially close look at rubber parts, which can deteriorate. Examine the face piece (especially the seal), headband, valves, connecting tube, fittings, and canister or cartridge.

Table 4.8 summarizes some general guidelines for selecting respirators.

Maintenance of Respirators

Regular inspection and maintenance of respirators not only is a legal requirement, but is essential to the success of any respiratory program at the workplace.

The working conditions and the hazards involved usually dictate the type of program required for the care and maintenance of respirators. However, the program should provide means for storage, repair, inspection, and cleaning. The program also must provide for inspection of all respirators before and after each use, monthly inspection for the self-contained breathing apparatus, and a record of inspection dates and findings for respirators maintained for emergency use.

The respirator maintenance program should include:[17, 18]

- Cleaning and sanitizing: Each respirator should be properly cleaned and sanitized after each use to ensure that the user is provided with a clean and sanitized device.
- Inspection for defects: Respirators must be checked before and after each use to ensure that they are free of defects. Also, after cleaning and sanitizing, the respirator must be checked

Table 4.8. Guide for selecting respirators.

Hazard	Type of Respirator to Use
Oxygen-deficient atmospheres	SCBA, air-supplied with escape SCBA
Toxic gases and vapors that are IDLH	SCBA, air-supplied with escape SCBA
Gases and vapors, not IDLH	Supplied-air, air-purifying
Particulates, IDLH	SCBA, air-purifying with full face piece, combination air-line with escape SCBA
Particulates, not IDLH	Air purifying, air-line, hose mask with blower
Combination of gases, vapors, and particulates, IDLH	SCBA, air-line with escape SCBA
Combination gases, vapors, and particulates, not IDLH	Air-line, hose mask with blower, air-purifying

Source: Reference 4.

to ensure that it is in proper working condition. All respirators intended for emergency or rescue operations must be inspected at least once a month. When inspecting a respirator, one should check the following:

- Tightness of connections
- Condition of the respiratory inlet covering
- Head harness
- Valves
- Connecting tubes
- Harness assemblies
- Filters
- Cartridges
- Canisters
- End-of-service-life indicator
- Shelf-life date
- Regulators and alarms

The inspection must also include examination of all rubber or plastic parts for signs of deterioration, and a check of the cylinders to ensure that they are fully charged. All respirators intended for emergency or rescue operations must have a record of inspection dates, findings and any remedial actions.

- Repair: Respirator problems discovered during inspection must be repaired. The following points should be taken into account:
  - Any repair should be conducted by personnel properly trained in respirator assembly.
  - Only replacement parts designed for the specific respirator should be used.
  - Any repair to reducing or admission valves, regulators, or alarms must be conducted by either manufacturer or a qualified trained technician.
  - Any instrumentation for the valve, regulator, or alarm must be approved by the manufacturer.
- Storage: Proper storage is essential in respirator maintenance. In planning for storage, the following points should be considered:
  - The storage area must protect the respirators from dust, sunlight, extreme humidity, extremes of temperature, and damaging chemicals.
  - Respirators must not be stored in such places as lockers or tool boxes.
  - Respirators intended for emergency and rescue operations must be stored in such a manner that they are readily accessible, and if they are stored in a cabinet or container, they should be clearly marked.

Summary of Respiratory Protection

Respiratory equipment provides protection from atmospheric hazards such as dust, vapor, gas, and fumes. Respirators prevent the inhalation of dangerous substances and provide an air supply in work areas that are oxygen-deficient. The potential consequences of not using respirators in toxic or oxygen-deficient atmospheres include irritation of pulmonary or mucous membranes, damage to lungs, cancer and other serious illness, and death.

The need for a respirator is determined by: the allowable exposure limits listed on the MSDS, measurement of the hazards in the work area, and the concentration of oxygen in the air in the work area.

Each organization should have standard operating procedures for selecting, maintaining, and using respirators. MSDSs are good sources of information for evaluating potential hazards and the corresponding degree of respiratory protection required. Respirators should be inspected and cleaned both at regular intervals and after each use. Also, the type of respirators used, including filter types, should be reevaluated regularly. A training program should be instituted to ensure that all potential users of respiratory equipment can use the equipment effectively and safely.[7]

The three major types of respirators are: (1) air-purifying or filtering respirators, (2) self-contained breathing apparatus, and (3) air-line breathing equipment. Air-purifying respirators are used when an atmosphere contains enough oxygen for breathing but is contaminated with smoke, dust, or vapors. If the vapors are harmful, the respirator can be fitted with a cartridge or canister that filters the air. In such cases selection of the appropriate filtering device is critical to avoid injury, and the user must be able to recognize when the filtering capacity has been depleted. Purifying respirators *must not be used* when the hazard is immediately dangerous to life or health—that is, when death or irreversible health problems can result from breathing the contaminants for a 30-minute period. SCBA or air-line breathing equipment is used when the contaminants are too hazardous to be safely removed by a purifying respirator, or the atmosphere contains less than 19.5% oxygen. SCBA is equipped with a portable air or oxygen tank, which makes it the choice for work done in hazardous atmospheres that must be confined. Air-line equipment relies on an external source of air or oxygen that is supplied by a hose; this type of equipment is especially suited for working in hazardous atmospheres for extended periods.

A respirator must fit well or it will not work. Even a tiny space between the face piece and the face could allow flow of toxic gases into the face piece.

Adherence to the following guidelines will help ensure the proper and safe use of respiratory equipment:

- Wear only the respirator you have been instructed to use.
- Check the respirator for a good fit before every wearing.
- Check the respirator for deterioration before and after use.
- Recognize indications that cartridges and canisters are at their end of service.
- Practice moving and working while wearing the respirator.
- Store respirators carefully in a protected location.

Body and Leg Protection

In recent years the availability of various new types of material for protective clothing has resulted in improved protection for workers. This section describes the process of selecting the kind of protective clothing and material needed for specific types of hazards.[19-22]

Selecting a protective garment involves analysis of the type of hazard and an evaluation of the degree of protection required. Manufacturers should be consulted about types of garments and prices and, in many cases, the type of garment material used for a specific hazardous chemical exposure. It is extremely important that the user of the protection and the manufacturer interact and exchange information during the selection process. The use should provide detailed information on the type of chemical requiring the protection, its physical form, its usage and exposure profile, and the degree of protection needed.[19]

Although all clothing is protective to some extent, it is the degree of protection that is important, particularly in the chemical industry where there is a potential for exposure to highly toxic substances.[26] Today's user of chemical protective clothing (CPC) is faced with the formidable task of selecting appropriate CPC to match specific hazards at hand. Although such factors as cost, style, availability, and exposure profile play an important role in the selection process, the most important criterion is the effectiveness of the CPC as a barrier between the user's body and the chemical(s) of interest.[21,24]

The proper selection of CPC has become increasingly important as the occupational exposure of the skin to toxic chemicals has become a national problem. According to statistics published by the Bureau of Labor Statistics, skin diseases due to exposure to chemicals account for 43% of all occupational illnesses reported by the private sector.[22]

Many industrial situations have the potential for exposing workers to toxic or dangerous chemicals, including both routine operations and emergency procedures. Some examples include:

- Handling of chemical during routine manufacturing operations
- Maintenance activities
- Acid baths in electronic industries
- Pesticide application
- Waste handling
- Emergency response operations
- Equipment failure or leaks[22]

Operations that expose workers to hazardous chemicals must be identified, and every effort must be made to eliminate the exposure risks through engineering or administrative controls. However, if these methods cannot eliminate the hazard or reduce the risk to an acceptable level, appropriate chemical protection clothing must be selected and provided.

OSHA Requirements for Body and Leg Protection

OSHA has outlined both general and specific requirements covering body and leg protection, although the types and uses of such equipment have not been specified.[8] The general requirement (1910.132) covers all equipment and conditions under which body and leg protection should be used. OSHA's specific requirements cover conditions where specific types of clothing or equipment are required. OSHA does not limit the general obligation to provide protection for all working conditions.

Types of Chemical Protective Clothing

In selecting a chemical protective clothing (CPC), the user must decide which type to purchase.[19,23] The types of CPC range from basic work clothes to fully encapsulating ensembles, with a wide variety of designs between these two extremes.

Basic items for providing splash protection for specific areas of the body, including the chest, legs, and arms, are available in materials such as PVC, natural rubber, and neoprene. Splash resistance is provided against chemicals and petroleum derivatives, depending on the material or construction. Figure 4.6 shows a simple CPC ensemble.

Chemical and acid-resistant hoods protect the worker against chemical splashes. Typical materials of construction are PVC, neoprene, and natural rubber.

An anticontamination coverall consists of a coverall, boots, and hood. A canister-type respirator can provide protection against inhalation of potentially hazardous particulates.

Reusable CPCs are used in operations such as cleanrooms and laboratories or in the manufacture of electronic devices.

When a higher degree of protection is required, CPC such as sunray suit is available, which resists penetration by alkalis, acids, hydrocarbons, and other chemicals. The material used in such a suit is made of neoprene on cotton, and the seams are vulcanized throughout. The rubberized protective suit is a two-piece neoprene-on-cotton ensemble that provides head-to-toe protection. Respiratory protection can be provided by either a supplied-air or a canister-type respirator.

When the highest degree of protection is required, a fully encapsulating suit must be used (Figure 4.7). Depending on the materials of construction, these suits can protect the wearer against a wide range of chemicals, including acids, solvents, oxidizers, toxics, carcinogens, alkalis, and so on. These ensembles are totally encapsulating, and breathing air must be supplied by a suitable respirator. In addition, body cooling may be required to minimize the effects of heat stress.

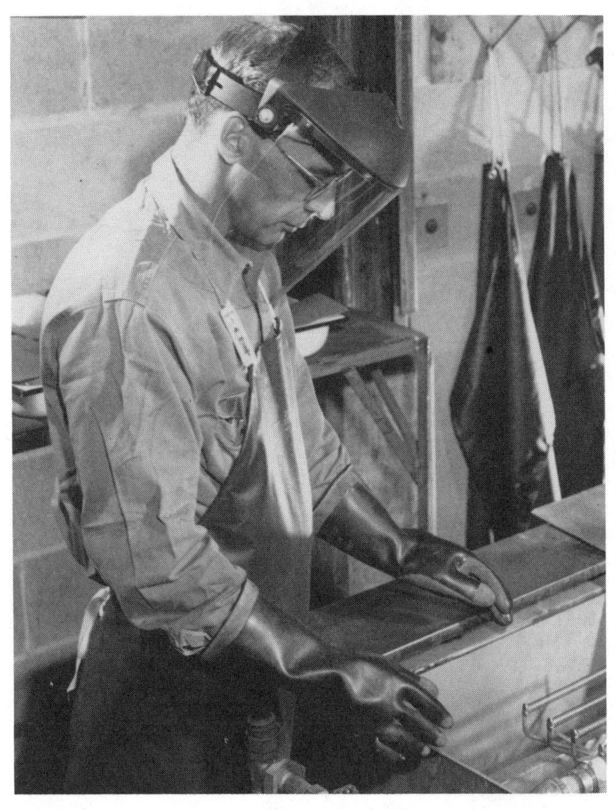

Figure 4.6. CPC Ensemble consisting of face shield and rubber gloves and apron. (Source: Argonne National Laboratory)

Protective clothing also is available in materials that can protect against flames and extreme temperatures, as shown in Figures 4.8 and 4.9.

The above examples, which represent only a few of the CPC designs available from the protective clothing industry, indicate the variety of choices available to the health and safety professional and demonstrate the complexity of selecting the proper ensemble for a specific application.

Protective Garment Materials for CPC

CPC comes in a variety of designs and materials.[6,19] Natural rubber, synthetic rubbers, and elastomers are a few of the materials available for CPC. These materials can be supported or unsupported, and in some cases combinations of different materials are used for the maximum degree of protection. However, chemicals can even penetrate plastic and rubber materials to some degree, and no single material can protect against all chemicals. A review of the garment materials available and testing of those materials against different chemicals can provide valuable insights into the process of selecting an ensemble.

Materials used for protective garments include, but are not limited to:[6]

- Butyl polymers
- Chlorinated polyethylene
- Natural rubber
- Neoprene
- Nitrile
- Polyethylene
- Polyurethane
- Polyvinyl alcohol
- PVC
- Saran
- Treated woven fabrics
- Composites

Figure 4.7. Fully encapsulated protective clothing. (Source: Argonne National Laboratory)

These materials can be supported on cotton, polyester, nylon, and other materials. Supported materials are more resistant to punctures and tears than unsupported materials.

CPC Effectiveness

The most important criterion in the selection and use of CPC is effectiveness in protecting against the chemical(s) of interest.[19,20,22] Unfortunately, there is no single universal method of rating CPC. Some manufactures have adopted descriptors such as excellent, good, fair, and so on, and some use a numerical scale for rating.

Also manufacturers' standards vary substantially; for example, one manufacturer may require a four-hour resistance for an excellent rating, while another may require a five-hour resistance for the same rating.

The rating problem is compounded further by different procedures used by different manufacturers to evaluate the effectiveness of the CPC. In the past, a degradation test was the only means of evaluating the resistance of a CPC to a given chemical. In this test, a small portion of the CPC material was dropped into a given chemical for a specified period of time, and changes in

Figure 4.8. Flame-resistant protective clothing (Source: Ryrepel Products, Newark, N.J.)

its properties, such as tear resistance, tensile strength, and so on were recorded. This test often is misleading, however, because it does not provide any information on the rate of transfer of the chemical through the CPC material.

In 1981, the American Society of Testing and Materials (ASTM) published a standard for permeation testing of protective garment materials (ASTM, F739-81). In this method the CPC material is exposed to a given chemical on one side, and the breakthrough of the chemical on the other side is monitored using instruments such as a gas chromatograph, a hydrocarbon analyzer, and so forth. The ASTM method determines two critical properties of the CPC material, breakthrough time and steady-state permeation rate.

Permeation in the diffusion of a chemical through the garment material, a phenomenon that takes place on the molecular level. Three separate steps are involved in the transfer of a chemical from the exposed side to the skin the wearer: absorption, diffusion, and desorption. Each of the three steps involved can be important at times, but the rate-determining step is the diffusion of the chemical through the CPC material. The steady-state permeation rate is defined as the rate or flux of the chemical through the CPC once breakthrough has taken place. The breakthrough time is the time required for a chemical to diffuse through a CPC material and be detected on the unexposed side.

Figure 4.9. Glove made of flame-resistant material (Source: Wolhandler Associates, Inc., New York)

The ideal CPC material has a very long breakthrough time and a very low steady-state permeation rate. Unfortunately, no material can perform ideally, and often compromises must be made. Figure 4.10 shows the breakthrough time and steady-state permeation rate for three materials. Material 1 has a long breakthrough time (an advantage); however, once breakthrough takes place, it provides a very high permeation rate (a disadvantage). Material 3 is just the opposite, offering a low breakthrough time and a low permeation rate, and Material 2 falls in between the two extremes.

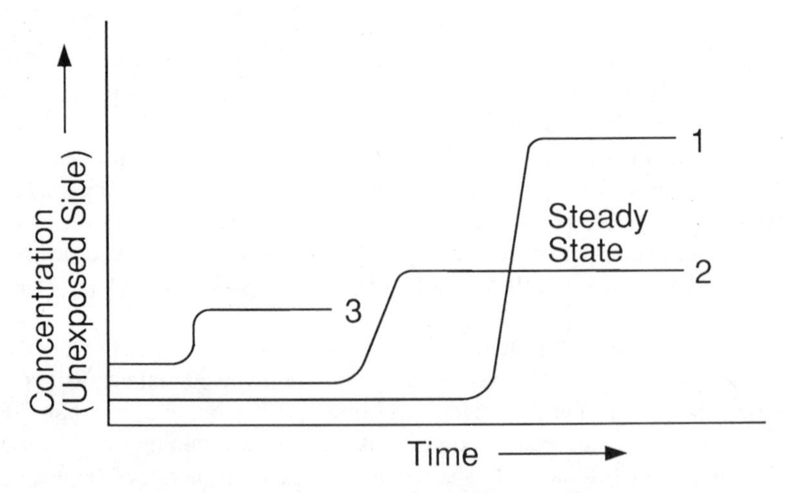

Figure 4.10. Breakthrough curves for three chemical protective clothing (CPC) materials. (Source: Reference 19)

Although ASTM permeation testing has helped the situation considerably, there are still many items of CPC material for which no permeation data are available. However, even if such data were available, there still would be a problem if one tried to compare CPC materials from different manufacturers. Differences in raw materials, processing techniques, and additives, for example, could translate themselves into different permeation and breakthrough characteristics for the same CPC material made by different manufacturers. Because it is not possible to test every CPC item against every chemical, a general method, such as use of the the solubility parameter,[20] is helpful for general comparison of CPC materials in the absence of permeation data.

The solubility parameter, long used by polymer chemists to describe the solubility of chemicals and polymers in each other, is a number assigned to a given polymer to describe its solution characteristics. The solubility parameter for liquid chemicals can be calculated using the following equation:

$$SP = \left(\frac{\Delta E_v}{V}\right)^{1/2} \qquad (1)$$

where:

SP = Solubility parameter in $MPa^{1/2}$ (MPa: megapascals)
ΔE_v = Latent molar heat of vaporization in joules/mol
V = Molar volume of liquid, cm^3/mol

The concept of solubility is an extension of the old rule that ''like dissolves like.'' In other words, the smaller the difference between solubility parameters of two chemicals, the greater the solubility of the two chemicals in each other. The solubility parameters for liquid chemicals range from a low of 15 for nonpolar compounds such as hydrocarbon liquids to a high of 47 for polar compounds such as water. The solubility parameter for materials used in CPC clothing ranges from 16 for butyl rubber (nonpolar) to 27 for polyvinyl alcohol (polar). Extending the concept of the solubility parameter to permeation resistance for CPC, one can conclude that the smaller the difference between the solubility parameter of a CPC material and a chemical, the lower the permeation resistance of the CPC material.

Another critical factor in the selection of CPC is the determination of the skin dose that the user can receive once breakthrough occurs. The dose can easily be calculated from the permeation and breakthrough data using the following equation:[22]

$$D = (t_c - t_b)(NA_S)$$

where:

D = Skin dose, mg
t_c = Time CPC has been in contact with chemical (seconds)
t_b = Breakthrough time (seconds)
N = Chemical steady-state mass flux ($mg/m^2 \cdot sec$)
A_S = The exposed area of skin, m^2

Management of CPC Selection and Use

To maximize worker protection against dangerous or toxic chemicals, the development of an effective CPC management program is essential.[4,19,23] Such a program must evaluate and balance the risks of contact with a chemical against the advantages of protection levels offered by different

CPC materials. Although the evaluation of usage and exposure profiles is critical to the selection process, the basic elements of a CPC management program should include: (1) determination of the likelihood of skin exposure, (2) consequences, (3) level of protection required, (4) making the appropriate selection and documenting the basis for the decision, and (5) training employees in proper use of CPC.

Determination of Skin Exposure

The determination of the likelihood of the skin contact by employees requires a thorough study of operations at the workplace. First, an inventory of all chemicals must be prepared, and different tasks involving chemicals must be identified. The usage profile for a particular chemical can be divided into light use, normal production, splash protection and escape, entry and maintenance, and cleanup and disposal. Worker exposure during the first two activities is minimal. In activities that require splash protection and escape, the worker normally can exit the area and use a safety shower. Entry into a hazardous area for maintenance has the potential for frequent worker exposure, and cleanup and disposal operations pose long-term exposures to unknown chemicals.

Next the degree of chemical threat in performing a given task must be identified from literature sources or MSDS and by taking into account factors such as the following:

- Chemical or chemical family—chemicals that belong to the same family pose similar hazards although the severity of such hazards may be different (e.g., alkalis, acids, aromatics, etc.).
- Physical state (solid, liquid, vapor, gas).
- Physiological effects (corrosive, toxic, irritating, target organ chemical, etc.).
- Exposure limits (TLV, PEL, skin LD_{50}, etc.).

Consequences of Skin Contact

In determining the consequences of skin contact, the primary question is whether direct skin contact can result in illness or injury. An examination of the literature or the MSDS is needed to clearly identify the physiological effects of skin contact. Some chemicals can cause irritation or minor dermatitis when they contact the skin, and some can cause sensitization and allergic reactions. Another primary question, which also deserves attention, is whether the chemical can be absorbed through the skin and enter the bloodstream, and, if so, what physiological effects can result (e.g., target organ chemicals).

Level of Protection

Users of CPC are becoming increasingly aware of the importance of laboratory test data under simulated occupational environments; materials and designs that are not supported by quantitative data will succumb to those in the marketplace that are.

Although there is no single method of determining the level of protection offered by a CPC material, permeation and breakthrough time data can be used to calculate important factors such as the steady-state skin dose (equation 2). This information, coupled with the degree of chemical threat, can help in determining the adequacy of CPC for a given task.[19,23]

The U.S. Coast Guard sponsored a study to determine the level of protection of CPC for members of emergency response teams. The toxic hazards rating scale summarized in Table 4.9 was chosen for recommending the degree of protection. This study[4] resulted in the following recommendations for protection against chemicals that irritate the skin or are absorbed by it:

- Hazard rating of 0: no protection required.
- Hazard rating of 1 or 2: nonsealed suit required.
- Hazard rating of 3: fully encapsulating suit required (with SCBA)

If carcinogens might exist in the area, a fully encapsulating suit is required.

Table 4.9. Effects of hazardous chemicals according to sax.

Hazard	Rating	Effects
None	0	No health effects under normal conditions.
Slight	1	Accute effect that disappears shortly after exposure.
Moderate	2	Can cause reversible or irreversible effects; not IDLH.
High	3	Can cause death or permanent injury; IDLH.
Unknown	4	No human data are available.

Source: Reference 19.

The levels of protection discussed above were based on protection against hazardous liquids or gases for a period of two hours. The level of protection should be adjusted appropriately for longer or shorter exposure profiles.

As foregoing discussion indicates the most important factor in proper determination of the level of protection for CPC materials is the availability of laboratory data. In the absence of such data, the determination would be at best qualitative. In reviewing laboratory data, one should remember that the rate of permeation is a function of several factors, including clothing material type and thickness, manufacturing method, the concentrations(s) of the hazardous substance(s), temperature, pressure, humidity, the solubility of the chemical in the clothing material, and the diffusion coefficient of the permeating chemical in the clothing material. Thus the permeation rate and breakthrough time may vary, depending on these conditions.[4]

CPC Material Selection

Studies of CPC materials have been conducted by the government and by private industry. At one time, the main criterion for material selection was degradation resistance, but later it was found that breakthrough time and permeation resistance are the critical factors in garment material selection—a material with high degradation resistance could have a very low permeation resistance.[19] The most appropriate clothing material will depend on the chemicals present and the task to be accomplished. Ideally, the material that is chosen resists permeation, degradation, and penetration.[4] No CPC material can provide adequate protection against all chemicals; so a balance must be established among likelihood of exposure, consequences of contact, available levels of protection, and budgetary and other business constraints.[23]

The selection of chemical-protective clothing is a complex task that should be performed by personnel with training and experience. Under all conditions, clothing is selected by evaluating its performance characteristics against the requirements and limitations of task-specific conditions. If possible, representative garments should be inspected before purchase, and their use and performance discussed with someone who has had experience with the clothing under consideration. In all cases, the employer is responsible for ensuring that the personal protective clothing (and all PPE) necessary to protect employees from injury or illness that may result from exposure to hazards at the work site is adequate and of safe design and construction for the work to be performed (see OSHA standard 29 CFR Part 1910.132–1910.137).[4]

In garment material selection, other factors, such as thickness and whether the material is supported or unsupported, affect both the breakthrough time and the permeation rate. Thus, manufacturers should be consulted regarding their specific materials' properties.

Accessory Equipment. Accessory equipment such as gloves, boots, respiratory support, and the communication system must be considered once the degree of protection and chemical threat are defined and the garment material has been selected. Cooling is important with fully encapsulating suits because heat stress can limit the time within the suit. Table 4.10 summarizes types of body protection, clothing, and accessories.[4]

Table 4.10. Protective clothing and accessories.

Body Part Protected	Type of Clothing or Accessory	Description	Type of Protection	Use Considerations
Full Body	Fully encapsulating suit	One-piece garment. Boots and gloves may be integral, attached, and replaceable, or separate	Protects against splashes, dust, gases, and vapors.	Does not allow body heat to escape. May contribute to heat stress in wearer, particularly if worn in conjunction with a closed-circuit SCBA; a cooling garment may be needed. Impairs worker mobility, vision, and communication.
	Nonencapsulating suit	Jacket, hood, pants, or bib overalls, and one-piece coveralls.	Protects against splashes, dust, and other materials but not against gases and vapors. Does not protect parts of head or neck.	Do not use where gas-tight or pervasive splashing protection is required. May contribute to heat stress in wearer. Tape-seal connections between pant cuffs and boots and between gloves and sleeves.
	Aprons, leggings, and sleeve protectors	Fully sleeved and gloved apron. Separate coverings for arms and legs. Commonly worn over nonencapsulating suit.	Provides additional splash protection of chest, forearms, and legs.	Whenever possible, should be used over a nonencapsulating suit (instead of using a fully encapsulating suit) to minimize potential for heat stress. Useful for sampling, labeling, and analysis operations. Should be used only when there is a low probability of total body contact with contaminants.
	Firefighters' protective clothing	Gloves, helmet, running or bunker coat, running or bunker pants (NFPA No. 1971, 1972, 1973), and boots.	Protects against heat, hot water, and some particles. Does not protect against gases and vapors, or chemical permeation or degradation. NFPA Standard No. 1971 specifies that a garment consist of an outer shell, an inner liner, and a vapor barrier with a minimum water penetration of 25 lb/in^2 (1.8 kg/cm^2) to prevent the passage of hot water.	Decontamination is difficult. Should not be worn in areas where protection against gases, vapors, chemical splashes, or permeation is required.

	Proximity garment (approach suit)	One- or two-piece overgarment with boot covers, gloves, and hood of aluminized nylon or cotton fabric. Normally worn over other protective clothing, such as chemical-protective clothing, firefighters' bunker gear, or flame-retardant coveralls.	Protects against brief exposure to radiant heat. Does not protect against chemical permeation or degradation. Can be custom-manufactured to protect against some chemical contaminants.	Auxiliary cooling and an SCBA should be used if the wearer may be exposed to a toxic atmosphere or needs more than 2 or 3 minutes of protection.
	Blast and fragmentation suit	Blast and fragmentation vests and clothing, bomb blankets, and bomb carriers.	Provides some protection against very small detonations. Bomb blankets and baskets can help redirect a blast.	Does not provide hearing protection.
Eyes and Face	Splash hood		Protects against splashes. Does not protect adequately against projectiles.	
	Safety glasses		Protect eyes against large particles and projectiles.	If lasers are used to survey a site, workers should wear special protective lenses.
	Goggles		Depending on their construction, can protect against vaporized chemicals, splashes, large particles, and projectiles (if constructed with impact-resistant lenses).	
	Sweat bands		Prevents sweat-induced eye irritation and vision impairment.	
Ears	Ear plugs and muffs		Protect against physiological damage and psychological disturbance.	Must comply with OSHA regulation 29 CFR Part 1910.95. Can interfere with communication. Use of ear plugs should be carefully reviewed by a health and safety professional because chemical contaminants could be introduced into the ear.
	Headphones	Radio headset with throat microphone.	Provide some hearing protection while enabling communication.	Highly desirable, particularly if emergency conditions arise.

Table 4.10. (*Continued*)

Body Part Protected	Type of Clothing or Accessory	Description	Type of Protection	Use Considerations
Hands and Arms	Gloves and sleeves	May be integral, attached, or separate from other protective clothing.	Protect hands and arms from chemical contact.	Wear jacket cuffs over glove cuffs to prevent liquid from entering the glove. Tape-seal gloves to sleeves to provide additional protection.
		Overgloves.	Provide supplemental protection to the wearer and protect more expensive undergarments from abrasions, tears, and contamination.	
		Disposable gloves.	Should be used whenever possible to reduce decontamination needs.	
Foot	Safety boots	Boots constructed of chemical-resistant material.	Protect feet from contact with chemicals.	
		Boots constructed with some steel materials (e.g., toes, shanks, insoles).	Protect feet from compression, crushing, or puncture by falling, moving, or sharp objects.	All boots must at least meet the specifications required under OSHA 29 CFR Part 1910.136 and should provide good traction.
		Boots constructed from nonconductive, spark-resistant materials or coatings.	Protect the wearer against electrical hazards and prevent ignition of combustible gases or vapors.	
	Disposable shoe or boot covers	Made of a variety of materials. Slip over the shoe.	Protect safety boots from contamination. Protect feet from contact with chemicals.	Covers may be disposed of after use, facilitating decontamination.
General	Knife		Allows a person in a fully encapsulating suit to cut his or her way out of the suit in the event of an emergency or equipment failure.	Should be carried and used with caution to avoid puncturing the suit.
	Flashlight or lantern		Enhances visibility in buildings, enclosed spaces, and the dark.	Must be intrinsically safe or explosion-proof for use in combustible atmospheres. Sealing the flashlight in a plastic bag facilitates decontamination. Only electrical equipment

			approved as intrinsically safe, or approved for the class and group of hazard as defined in Article 500 of the National Electrical Code, may be used.
Personal dosimeter		Measures worker exposure to ionizing radiation and to certain chemicals.	To estimate actual body exposure, the dosimeter should be placed inside the fully encapsulating suit.
Personal locator beacon	Operated by sound, radio, or light.	Enables emergency personnel to locate victim.	
Two-way radio		Enables field workers' communication with personnel in the support zone.	
Safety belts, harnesses, and lifeline		Enable personnel to work in elevated areas or enter confined area and prevent falls. Belts may be used to carry tools and equipment.	Must be constructed of spark-free hardware and chemical-resistant materials to provide proper protection. Must meet OSHA standards in 29 CFR Part 1926.104.

Source: Reference 4.

Employee Training

An important part of any CPC management program is the training of all employees who use the CPC.[19,23] Training should focus on the selection process that was used and should include hands-on training, especially for fully encapsulating suits. Specifically, employees should be trained on the following topics:

- Types of hazards posed as a result of chemical exposure and their consequences.
- Degree of protection provided by the CPC.
- Procedures for proper care of the CPC.
- CPC limitations.
- Procedures for donning CPC.
- Procedures for doffing CPC.
- Use of accessory items.
- Methods of sealing.
- Decontamination procedures.

Donning and doffing nonencapsulating suits is a rather easy task; however, for a fully encapsulating suit a buddy system should be used. The donning procedure is further complicated by accessory equipment such as overgloves, overboots, body cooling equipment, breathing apparatus, and communications equipment. The assistance of a coworker will ensure that all the accessory equipment is installed, and that the ensemble has been correctly donned and sealed. The buddy system will also minimize the time required for suiting up.

Other Considerations

In addition to permeation, degradation, and penetration, several other factors must be considered during clothing selection.[4] They affect not only chemical resistance, but also the worker's ability to perform the required task. The following checklist summarizes these considerations:

- *Durability:* Does the material have sufficient strength to withstand the physical stress of the task(s) at hand? Will the material resist tears, punctures, and abrasions? Will the material withstand repeated use after contamination/decontamination?
- *Flexibility:* Will the CPC interfere with the workers' ability to perform their assigned tasks?
- *Temperature effects:* Will the material maintain its protective integrity and flexibility under temperature extremes?
- *Ease of decontamination:* Are decontamination procedures available on site? Will the material pose any decontamination problems? Should disposable clothing be used?
- *Compatibility with other equipment:* Does the clothing preclude the use of another, necessary piece of protective equipment (e.g., suits that preclude hardhat use in hardhat area)?
- *Duration of use:* Can the required task be accomplished before contaminant breakthrough occurs, or before degradation of the CPC becomes significant?

Heat Stress and Other Physiological Factors

Use of some CPC can generate a considerable risk of developing heat stress, and may result in problems ranging from heat fatigue to serious illness and even death. Heat stress is caused by a number of interacting factors such as environmental conditions, type of clothing, degree of physical activity, and the general health condition of the worker. In situations where there is a potential for the worker to develop heat stress, the heat transfer characteristics of the CPC must be considered in the CPC selection process. Also, regular monitoring and other precautionary measures become vital in minimizing heat stress.[4,25,26]

The most important heat transfer characteristic of the CPC material is its thermal insulation value (CIO), which is a measure of the capability of the CPC to dissipate heat by heat transfer modes other than evaporation. The larger the value of the CIO, the lower its heat transfer capability and the greater the risk of heat stress. In hot working environments or when the work rate is high, the CPC material with the lowest CIO value should be considered, given that other properties are equivalent.

To understand the contribution of CPC to heat stress, one must consider the mechanism of body heat generation and dissipation. The main requirement for normal body function is to maintain the deep body temperature at 37°C (98.6°F) ± 1°C (1.8°F); and to maintain this thermal equilibrium, a constant exchange of heat between the body and the environment is required.[25] The amount of heat that must be transferred from the body to the environment is a function of: (1) heat generated within the body (metabolic heat); (2) the heat gained, if any, from the environment; (3) type of CPC worn; and (4) environmental conditions, such as air temperature, velocity, humidity, and so on.

Modes of Heat Transfer

Heat may be defined as energy in transit. This energy is transferred by three different mechanisms: conduction, convection, and radiation; and the driving force for heat transfer is a temperature gradient. Heat is always transferred from a high- to a low-temperature region.

Conduction is the transfer of thermal energy on a molecular level from one molecule to the next. When a rod is placed in a fireplace, heat is transferred from the point of contact to the other end by conduction. Similarly, part of the heat generated within the body is transferred through the CPC by conduction. The most important characteristic of a material with respect to conductive capability is thermal conductivity; the higher this value, the more conductive the material is.

Convection is the mechanism of interphase heat transfer between a solid and a gas or a liquid and a gas. The heat transfer between the surface of a CPC and ambient air is achieved by the convection mechanism. The property governing the ease or difficulty with which heat can be transferred is called the heat transfer coefficient and is a function of the properties of the two phases involved.

Radiation is the electromagnetic mechanism of heat transfer, involving transfer of energy from a source to a receiver. For example, a worker standing near an open flame will receive most of the thermal energy by radiation. The material property that determines the amount of heat transfer by radiation is called emissivity.

Heat Balance

The heat balance equation for a worker can be written as follows:

$$\Delta Q = (q_M - W) \pm q_C \pm q_R - q_e \qquad (3)$$

where

ΔQ = Change in body heat content
q_M = Heat generated by the total metabolism
W = External work performed by the worker
q_C = Convective heat transfer between the worker and the ambient air
q_R = Radiative heat transfer between the worker and the surroundings
q_e = Heat loss by evaporation

In order to maintain thermal equilibrium in the worker's body and avoid a deep-body temperature rise and the possibility of heat stress, ΔQ in equation (3) must be kept as close to zero as possible.

The rate of convective heat exchange between a person's skin and the ambient air immediately surrounding the skin is a function of the difference in temperature between the ambient air (t_a) and the mean weighted skin temperature ($\bar{t}_{sk}$) and the rate of air movement over the skin (V_a). This relationship for a standard worker wearing a customary one-layer work clothing ensemble can be described as follows [a standard worker being defined as one who weighs 70 kg (154 lb) and has a body surface area of 1.8 m^2 (19.4 ft^2)]:[25]

$$q_C - 0.65\ V_a^{0.6}(t_a - \bar{t}_{sk}) \tag{4}$$

where:

q_C = Convective heat exchange in Btu/hr
V_a = Air velocity in feet per minute (fpm)
t_a = Air temperature in °F
$\bar{t}_{sk}$ = Mean weighted skin temperature, usually assumed to be 95°F (35°C)

Although the rate of radiative heat transfer to or from a worker's body is a function of the fourth power of the temperature gradient and emissivity, an acceptable approximation for a standard worker wearing one-layer clothing can be described as follows:[25]

$$q_R = 15.0\ (\bar{t}_w - \bar{t}_{sk}) \tag{5}$$

where:

q_R = Radiant heat exchange in Btu/hr
$\bar{t}_w$ = Mean radiant temperature in °F
$\bar{t}_{sk}$ = Mean weighted skin temperature

The evaporation of water (sweat) from the skin surface results in a heat loss from the body. The maximum evaporative capacity (and heat loss) is a function of air motion (V_a) and the water vapor pressure difference between the ambient air (p_a) and the wetted skin at skin temperature (p_{sk}). The equation for this relationship for the customary one-layer-clothed worker is:[25]

$$q_E = 2.4 V_a\ 0.6(p_{sk} - p_a) \tag{6}$$

where:

q_E = Evaporative heat loss in Btu/hr
V_a = Air velocity in fpm

Effects of Clothing on Heat Exchange

Clothing serves as a barrier between the skin and the environment to protect against hazardous chemical, physical, and biologic agents. A clothing system also will alter the rate and amount of heat exchange between the skin and the ambient air by convection, radiation, and evaporation. When calculating heat exchange by each or all of these routes, one must apply correction factors that reflect the type, amount, and characteristics of the clothing being worn when the clothing differs substantially (i.e., more than one layer and/or greater air and vapor impermeability) from the customary one-layer work clothing.

The correction factors can be calculated from heat transfer considerations such as thermal properties of the CPC material. These correction factors have been determined for a variety of environmental and metabolic heat loads and three clothing ensembles.[25] The customary one-layer clothing ensemble was used as the basis for comparisons with the other clothing ensembles.

The information presented above should be used, to the extent possible, in selecting the type of CPC material that will minimize the hazards of heat stress.

Monitoring

Because the incidence of heat stress depends on a variety of factors, all workers, even those not wearing protective equipment or clothing, should be monitored by measuring:[4]

- Heart rate
- Oral temperature
- Body water loss

Table 4.11 summarizes signs and symptoms of heat stress.[26]

Summary of Body and Leg Protection

Although the CPC selection process is still developing, it requires consideration of both the capabilities and the limitations of materials. No single garment material can provide protection against all hazardous chemicals, but manufacturers can provide detailed information on the performance of their materials.

The following steps should be used in the selection and use of CPC:

- Define the chemical threat:
  - Identify the chemical or chemical class.
  - Identify the physical state of the chemical.
  - Determine the exposure pathway of the chemical.
  - Review the literature for information such as the TLV or PEL.

Table 4.11. Signs and symptoms of heat stress.

- Heat rash may result from continuous exposure to heat or humid air.
- Heat cramps are caused by heavy sweating with inadequate electrolyte replacement. Signs and symptoms include:
  —muscle spasms
  —pain in the hands, feet, and abdomen
- Heat exhaustion occurs from increased stress on various body organs including inadequate blood circulation due to cardiovascular insufficiency or dehydration. Signs and symptoms include:
  —pale, cool, moist skin
  —heavy sweating
  —dizziness
  —nausea
  —fainting
- Heat stroke is the most serious form of heat stress. Temperature regulation fails, and the body temperature rises to critical levels. Immediate action must be taken to cool the body before serious injury and death occur. Competent medical help must be obtained. Signs and symptoms are:
  —red, hot, usually dry skin
  —lack of reduced perspiration
  —nausea
  —dizziness and confusion
  —strong, rapid pulse
  —coma

Source: Reference 26.

- Define the exposure profile:
  - Determine whether the user will be exposed to vapor, splash, or deluge of the chemical.
  - Determine the frequency of exposure (i.e., limited, intermittent, or continuous).
  - Specify the type of exposure—for instance, escape, maintenance and repair, or entry and cleanup.
- Determine the degree of protection required, using sources such as:
  - OSHA and EPA guidelines.
  - MSDSs.
  - User experience.
- Select the garment design and material, using sources such as:
  - CPC manufacturer's data.
  - ACGIH *Guidelines for the Selection of Chemical Protective Clothing*.

Hand and Arm Protection

The incidence of occupational hand injuries is second only to back injuries. Over 30,000 disabling hand injuries occur yearly. The high incidence of work-related hand injuries dictates that hand protection should be a primary concern of employers.

Some of the most common hand injuries are partial or complete amputation; lacerations, cuts, and puncture wounds, which can be superficial or can penetrate muscles and tendons; burns, which can be caused by heat, chemicals, or electricity; fractures, which can range from dislocation to crushing by a heavy object; and dermatitis and other skin irritation, which are usually caused by contact with chemicals. In general, hazards to hands are presented by hot surfaces, objects, and other materials; chemicals; sharp or pointed objects, such as broken glass, knives, and punches; moving parts of machines and tools, such as lathes and drills; rough materials; and heavy loads or equipment, such as lift trucks.

Most occupational hand injuries are rather serious, resulting in up to two weeks of lost time. Many people have the wrong notion that any glove protects their hands against injuries; in fact, accidents and injuries to the hand happen when the wrong type of glove is used. Employers must choose proper hand protection for employees based on the type of hazard they will encounter.[8]

OSHA Requirements for Hand Protection

OSHA has both general and specific requirements for hand protection. The general requirements, contained in 29 CFR 1910.132, state that protective equipment must be provided by the employer wherever it is necessary to protect employees from hazards. The specific requirements appear in other sections and apply to specific industries or conditions.

Types of Hand and Arm Protection

Gloves

Most accidents involving hands and arms can be classified under four main hazard categories: chemicals, abrasions, cutting, and heat. There are gloves available that can protect workers from any of these individual hazards or any combination thereof. It is important, however, as in the selection of any other type of PPE, to match the glove to the particular hazard. Gloves can be classified into several types, as described below.

Cloth Gloves

These gloves are useful for light-duty materials handling for protection against cuts or abrasions. They also provide insulation for protection against moderately hot or cold conditions. Several kinds are available:

- Cotton gloves can provide good moisture absorbing capability, and are used in light-duty work.
- Polyester gloves, which are made by copolymerization of an alcohol and a carboxylic acid, are useful in protecting the hands when working with certain chemicals. These gloves should not be used under extremes of temperature, as hot conditions can easily degrade the polymeric material, and cold conditions can change the physical properties of the glove (such as strength and brittleness). The resulting damage to the glove may expose unprotected hands to temperature extremes.
- Wool gloves can provide good insulation for the hands in cold environments.
- Terry cloth is a good insulator and sometimes is used in temperature extremes, although in dealing with extremely cold or hot conditions the right kind of glove must be selected to provide protection against the particular temperature extremes (see below).

It should be emphasized that cloth gloves are intended for general light-duty work. These gloves should not be used for protection against hazardous chemicals.

Rubber Gloves

These gloves provide hand and arm protection against chemicals. Some of the more commonly used rubber and plastic gloves are made of the following materials:

- Neoprene: This compound is similar to polyisoprene (natural rubber). It can provide protection against most solvents, alkalis, greases, oils, and alcohol. Neoprene also is resistant to hydrofluoric acid.
- Natural rubber: Gloves made of this material can provide protection against most chemicals that are soluble in water. Because of its low glass transition temperature, it provides good flexibility under cold conditions. Note that the glass transition temperature, defined as the temperature below which the polymeric material becomes brittle and amorphous like glass, is an important consideration in the selection of plastic gloves that might be used under extremely cold conditions.
- Polyvinyl chloride (PVC): This material is obtained from the polymerization of vinyl chloride and gloves made of PVC are resistant to most acids and alkaline chemicals. However, gloves made of rubberized PVC do not provide adequate protection against petroleum derivatives, mainly because most rubberized (plasticized) PVC materials are filled with hydrocarbon oils.[27] The fillers accomplish two purposes: (1) they make the polymer softer and easier to process; and (2) because the oils are less expensive than the polymer, they substantially reduce costs. However, because ''like dissolves like,'' if a plasticized PVC glove were used to handle petroleum products, the fillers would dissolve in the oil, resulting in a substantial reduction in breakthrough time and a substantial increase in the steady-state permeation rate.
- Nitrile: This material is manufactured by the copolymerization of acrylonitrile and butadiene. Gloves made of this copolymer show good resistance to cuts, abrasions, and punctures, and provide good protection against oils.

Insulating Gloves

Although asbestos gloves can withstand temperatures of up to 1,000°C, these gloves have been replaced by synthetic fibers such as Nomex and Kevlar (E. I. du Pont de Nemours & Co. trade names) because of the health problems posed by asbestos.

Other Glove Materials

Other types of gloves are used against specific hazards. For example, aluminized gloves are used for protection from flames and intense heat, and metal mesh gloves protect against cuts and punctures associated with using power saws, knives, and other sharp or piercing objects. These gloves should not be used around electrical equipment.

Gloves made of thin plastic or rubber are used in clean rooms and in the food processing industry. These gloves protect against germs carried on the hands.[8]

Besides gloves, there are other types of hand protection such as mittens, finger cots, sleeves, hand pads, and barrier creams. These glove alternatives should be used where working conditions dictate their use.

Remember these safety tips for hand protection: Tools and machines must have guards that prevent the hands from contacting the point of operation, power train, or other moving parts. To protect the hands from injury due to contact with moving parts, do the following:

- Ensure that guards are always in place and used.
- Always lock out machines or tools and disconnect the power before making repairs.
- Treat a machine without a guard as *inoperative.*
- Do *not* wear gloves around moving machinery, such as drill presses, mills, lathes, and grinders.

Testing, Selecting, and Using Hand Protection

Testing Gloves

The employer is responsible for ensuring that the gloves chosen will protect employees from hazards. A significant amount of testing is required.

Gloves should be tested for penetration and permeation. Depending on the conditions under which the glove is used, these tests may last a few minutes or a few hours. Most gloves used for protection against hazardous chemicals are made of polymeric materials. The molecular structure of the polymer can be adversely affected in certain physical or chemical environments, which in turn would adversely affect the permeation rate and breakthrough time.

OSHA and ANSI standards are specific in describing requirements for electrical workers' gloves. The rubber material is categorized at different voltages. Use of the wrong glove by an electrical worker could result in death.

Most glove manufacturers provide on-site testing. Testing should be done for the following: cut, puncture, abrasion, and tear resistance; dexterity; comfort; gripping ability; and permeability of chemicals. Gloves used regularly should be inspected after three months of use and every six months thereafter.

Table 4.12 summarizes breakthrough time and steady-state permeation rate for seven different types of glove material against 49 chemicals.[28] Study of this table reveals the importance of matching the glove to the hazard. For example, butyl offers a breakthrough time of 9.6 hours and a permeation rate of 0.066 $mg/m^2/sec$ against acetaldehyde, whereas unsupported neoprene has a breakthrough time of 10 minutes and a relatively high permeation rate. Obviously, gloves made of neoprene do not provide adequate protection against acetaldehyde.

Selecting Gloves

The most critical aspect of choosing hand protection is to match the glove to the specific hazard. Not only must the hazard and the glove match, but some input from employees can facilitate glove selection, not only for protection against hazards but also with regard to comfort and dexterity.

Table 4.12. Protection offered by clothing of synthetic polymers vs the more common industrial chemicals

	VITRON® (10 mil)		BUTYL (17 mil)		SILVER SHIELD™ (3 mil)		NITRILE (22 mil)		PVA		NEOPRENE UNSUPPORTED	
Chemical	Breakthrough Time (hrs)	Permeation mg/m²/sec	Breakthrough Time (hrs.)	Permeation mg/m²/sec	Breakthrough Time (hrs)	Permeation mg/m²/sec	Breakthrough Time (hrs)	Permeation mg/m²/sec	Breakthrough Time (hrs)	Permeation mg/m²/sec	Breakthrough Time (hrs)	Permeation mg/m²/sec
Acetaldehyde	NR	—	9.6 hrs	0.066	>6 hrs	ND	NT	—	—	—	10 min.	<9000
Acetic Acid (Glacial)	NT	—	NT	—	NT	—	1.9 hrs	221	—	—	7 hrs	—
Acetic Acid (50%)	NT	—	NT	—	NT	—	>8 hrs	ND	—	—	—	—
Acetone	NR	—	>17 hrs	ND	>6 hrs	ND	NT	—	—	—	5 min.	<900
Ammonium Hydroxide (29%)	NT	—	NT	—	NT	—	>8 hrs	ND	—	—	>6 hrs	—
Aniline	NR	—	>8 hrs	ND	>8 hrs	ND	1.2 hrs	3	$1\frac{1}{2}$ hrs	<9	35 min.	<9
Benzene	6 hrs	.012	NR	—	>8 hrs	ND	27 min.	97	7 min.	<0.9	—	—
Butyl Acetate	NR	—	1.9 hrs	7.61	>6 hrs	ND	1.7 hrs	24	ND	<0.9	—	—
p-t Butyltoluene	>8 hrs	ND	1.7 hrs	8	>8 hrs	ND	NT	—	—	—	—	—
Carbon Disulfide	>16 hrs	ND	NR	—	RD	—	20 min.	86	ND	<0.9	—	—
Carbon Tetrachloride	>13 hrs	ND	NR	—	>6 hrs	ND	5.7 hrs	8	ND	<0.9	—	—
Chloroform	9.5 hrs	0.46	NR	—	NR	—	NT	—	ND	<0.9	—	—
Chloronaphthalene	>16 hrs	ND	NR	—	>8 hrs	ND	NT	—	ND	<0.9	—	—
Cyclohexane	>7 hrs	ND	1.1 hrs	20.3	>6 hrs	ND	>8 hrs	ND	—	—	—	—
Cyclohexanol	>8 hrs	ND	>11 hrs	ND	>6 hrs	ND	NT	—	6 hrs	<0.9	$2\frac{1}{2}$ hrs	<9
Cyclohexanone	NR	—	>16 hrs	ND	>6 hrs	ND	NT	—	—	—	—	—
Dibutyl Phthalate	>8 hrs	ND	>16 hrs	ND	>6 hrs	ND	NT	—	ND	<0.9	2 hrs	<0.9
1,2 Dichloroethane	6.9 hrs	.81	2.9 hrs	53	>6 hrs	ND	16 min.	292	—	—	—	—
Diisobutyl Ketone (80%)	1.2 hrs	90.6	3.3 hrs	41.2	>6 hrs	ND	NT	—	ND	<0.9	—	—
Dimethyl Formamide	NR	—	>8 hrs	ND	>8 hrs	ND	25 min.	41	—	—	10 min.	<90
Dioxane	NR	—	>20 hrs	ND	>8 hrs	ND	NT	—	—	—	—	—
Divinyl Benzene	>17 hrs	ND	2.2 hrs	238	>8 hrs	ND	NT	—	—	—	—	—
Ethyl Acetate	NR	—	7.6 hrs	3.4	>6 hrs	ND	NT	—	Nd	<0.9	15 min.	<90
Ethylamine (70% in water	NR	—	>12 hrs	ND	NR	—	NT	—	—	—	—	—
Ethyl Alcohol	NT	—	NT	—	NT	—	>8 hrs	ND	—	—	$1\frac{1}{2}$ hrs	<9
Ethyl Ether	NR	—	NR	—	>6 hrs	ND	NT	—	>6 hrs	<0.9	10 min.	<90

Table 4.12. (*Continued*)

Chemical	VITRON® (10 mil)		BUTYL (17 mil)		SILVER SHIELD™ (3 mil)		NITRILE (22 mil)		PVA		NEOPRENE UNSUPPORTED	
	Breakthrough Time (hrs)	Permeation mg/m²/sec	Breakthrough Time (hrs.)	Permeation mg/m²/sec	Breakthrough Time (hrs)	Permeation mg/m²/sec	Breakthrough Time (hrs)	Permeation mg/m²/sec	Breakthrough Time (hrs)	Permeation mg/m²/sec	Breakthrough Time (hrs)	Permeation mg/m²/sec
Formaldehyde (37% in water)	>16 hrs	ND	>16 hrs	ND	>6 hrs	ND	>8 hrs	ND	—	—	2 hrs	<0.9
Furfural	3.6 hrs	14.8	>16 hrs	ND	>8 hrs	ND	NT	—	ND	<0.9	20 min.	<90
n-hexane	>11 hrs	ND	NR	—	>6 hrs	ND	>8 hrs	ND	ND	<0.9	45 min.	<900
Hydrazine (70% in water)	NR	—	>8 hrs	ND	2.1 hrs	1.0	>8 hrs	ND	—	—	ND	—
Hydrochloric Acid (37%)	RD	—	RD	—	>6 hrs	ND	>8 hrs	ND	—	—	ND	—
Methylamine (40% in water)	>16 hrs	ND	>15 hrs	ND	1.9 hrs	2.0	NT	—	—	—	$4\frac{1}{2}$ hrs	<90
Methylene Chloride	1 hr	7.32	NR	—	1.9 hrs	0.002	NT	—	17 min.	<0.9	—	—
Morpholine	1.9 hrs	97	>16 hrs	ND	>8 hrs	ND	NT	—	3 hrs	<0.9	—	—
Nitrobenzene	>8 hrs	ND	>23 hrs	ND	>8 hrs	ND	1 hr	15	>6 hrs	<0.9	—	—
Nitropropane	NR	—	>8 hrs	ND	>8 hrs	ND	NT	—	>6 hrs	<0.9	5 min.	<900
Pentachlorophenol (1% in kerosene)	>13 hrs	ND	NR	—	>8 hrs	ND	NT	—	7 min.	<900	6 min.	<0.9
n-Pentane	>8 hrs	ND	NR	—	>6 hrs	ND	NT	—	ND	<0.9	30 min.	<900
Phenol (85% in water)	>15 hrs	ND	>20 hrs	ND	NT	—	>8 hrs	ND	30 min.	<90	3 hrs	<90
Propyl Acetate	NR	—	2.7 hrs	2.86	>6 hrs	ND	NT	—	2 hrs	<9	—	—
Sodium Hydroxide (50%)	RD	—	RD	—	>6 hrs	ND	>8 hrs	ND	—	—	ND	—
Sulfuric Acid (3 molar)	RD	—	RD	—	>6 hrs	ND	>8 hrs	ND	—	—	3 hrs	—
Tetrachloroethylene	>17 hrs	ND	NR	—	>6 hrs	ND	NT	—	—	—	—	—
Toluene	>16 hrs	ND	NR	—	>6 hrs	ND	28 min.	25	15 min.	<9	—	—
Toluene Diisocyanate	>16 hrs	ND	>8 hrs	ND	>8 hrs	ND	>8 hrs	ND	ND	<0.9	—	—
1,1,1-Trichloroethane	>15 hrs	ND	NR	—	>6 hrs	ND	2.2 hrs	44	1 hr	<0.9	—	—
Trichloroethylene	7.4 hrs	0.24	NR	—	>6 hrs	ND	9 min.	62	30 min.	<0.9	—	—
Vinyl Chloride	4.4 hrs	0.098	NR	—	>8 hrs	ND	NT	—	—	—	—	—

NR = Not Recommended; NT = Not Treated; RD = Resists Degradation not tested for permeation; ND = None Detected; < = Less Than; ≫ = Greater Than; — = Data not available.
Source: (28).

Using Hand Protection

Wearing the right size glove is important. Also, the right glove must be used for the hazard present. In addition, use safe work practices to prevent injuries caused by tools and machines.

To prevent chemical burns, follow the work practices outlined below:[7]

- Read instructions and warnings on chemical container labels and MSDSs before working with any chemical.
- Keep chemical containers closed when not in use.
- Cover minor cuts so chemicals cannot get into them.
- For each different chemical, wear the correct glove.
- Make sure gloves are clean and have no rips or holes.
- Rinse gloves carefully before removing them to avoid contamination.
- Clean gloves before putting them away.
- Store gloves in a cool, dark, dry place.
- Wash your hands thoroughly with soap and water after working with chemicals.
- Do not use strong solvents or chemicals such as turpentine, kerosene, or gasoline to clean your hands.

Hand Protection Summary

To prevent hand injuries, remember to:

- Be aware of hazards to your hands.
- Follow safe procedures when using tools, machines, and chemicals, both on the job and off.
- Keep your hands clean.
- Rotate tasks and take breaks to avoid repetitive type injuries.
- Report and treat injuries immediately.

Foot Protection

A survey of occupational foot injuries undertaken by the Bureau of Labor Statistics indicated that 75% of all industrial foot injuries were experienced by workers who were not wearing protective footwear.[8] This study also revealed that in most accidents involving safety show wearers, the seriousness of the injury was substantially reduced by the protective equipment.

OSHA Requirements

OSHA has three types of standards pertaining to foot protection. The general rules for personal protective equipment contained in 29 CFR 1910.132 require the employer to provide adequate foot protection to protect workers from hazards that have a potential for causing injury to the foot. OSHA also requires that any safety shoes provided must meet the requirements set forth by ANSI in its standard for industrial foot protection (ANSI, Z41, 1983). OSHA's specific requirements pertain to the use of various types of foot protection under occupational working conditions where it has determined that the need for foot protection exists.

Safety Shoes

Safety shoes come in a variety of materials and designs. They provide a protective toe box that is made of stiff material such as steel and designed in such a way as to protect the wearer's toes against heavy objects. Safety shoes must satisfy the requirements for compression and impact resistance set forth by ANSI. Figure 4.11 shows an example of a safety shoe.

Figure 4.11. Protective footwear. (Source: National Safety Council, Chicago)

References

1. Srachta, B., What you should know about personal protective equipment, *National Safety News*, National Safety Council, Chicago, pp. 47–51 (Mar. 1985).
2. National Safety Council, *Take the Extra for Personal Protection*, Chicago (1975).
3. King, W. R., and J. Magid, *Industrial Hazard and Safety Handbook*, 1st Ed., New Buttersworths, London, pp. 342–343 (1979).
4. NIOSH, *Occupational Safety and Health Guidance Manual for Hazardous Waste Site Activities*, National Institute for Occupational Safety and Health (1987).
5. Office of the Federal Register, *Code of Federal Regulations*, Vol. 29, Part 1910, National Archives and Records Service Administration, Washington, D.C. (1987).
6. Handley, W., *Industrial Safety Handbook*, 2nd Ed., McGraw-Hill, Berkshire, England, pp. 351–353 (1977).
7. BLR, Various safety pamphlets and business and legal reports, Bureau of Law and Business, Madison, Conn. (1988).
8. Best, *Best's Safety Directory*, 28th Ed., A. M. Best Co., Oldwick, N.J., pp. 234–517 (1988).
9. Boley, J. W., *A Guide to Effective Industrial Safety*, 1st Ed., Gulf Publishing Co., Houston, Tex., pp. 37–39 (1977).
10. Parmeggiani, L., *Encyclopedia of Occupational Health and Safety*, 3rd Ed., International Labor Office, Geneva, Switzerland, p. 1009 (1983).
11. ANSI, *Protective Headwear for Industrial Workers Requirements*, Z89.1-1986, American National Standard Institute, New York (1986).
12. Stepkin, R. L., and R. E. Mosley, *Noise Control: a Guide for Workers and Employers*, 1st Ed., American Society of Safety Engineers, Des Plaines, Ill., p. 1. (1984).
13. Cheremisinoff, P. N., and P. P. Cheremisinoff, *Industrial Noise Control Handbook*, 1st Ed., Ann Arbor Science, Ann Arbor, Mich., p. 5 (1978).

14. National Safety Council, Respiratory Protective Equipment, Data Sheet I-734-Rev. 87, National Safety Council, Chicago.
15. Srachta, B. J., What you should know about personal protective equipment, *National Safety News*, National Safety Council, Chicago (Mar. 1985).
16. National Safety Council, Dress right for safety respiratory protection, *National Safety News*, National Safety Council, Chicago (Mar. 1983).
17. ANSI, *Respirator Protection*, Z88.2, American National Standard Institute, New York (1980).
18. National Safety Council, Respirator maintenance so simple but most important, *National Safety News*, National Safety Council, Chicago (Apr. 1981).
19. Slote, L., *Handbook of Occupational Safety and Health*, 1st Ed., John Wiley & Sons, New York (1987).
20. Spence, M. W., A proposed basis for characterizing and comparing the permeation resistance of chemical protective clothing materials, *Proceedings* of Symposium on Performance of Protective Clothing, ASTM Committee F-23, Raleigh, N.C. (July 1984).
21. Henry, N. W., How protective is protective clothing?, *Proceedings* of Symposium on Performance of Protective Clothing, ASTM Committee F-23, Raleigh, N.C. (July 1984).
22. Mansdorf, S. Z., Risk assessment of chemical exposure hazards in the use of chemical protective clothing, *Proceedings* of Symposium on Performance of Protective Clothing, ASTM Committee F-23, Raleigh, N.C. (July 1984).
23. Coletta, G. C., and M. W. Spence, Managing the selection and use of chemical protective clothing, *Proceedings* of Symposium on Performance of Protective Clothing, ASTM Committee F-23, Raleigh, N.C. (July 1984).
24. Parmeggiani, L., *Encyclopedia of Occupational Health and Safety*, 3rd Ed., International Labor Office, Geneva, Switzerland (1983).
25. NIOSH, *Occupational Exposure to Hot Environments*, National Institute for Occupational Safety and Health, Cincinnati, Ohio (1986).
26. U.S. EPA, *Standard Operating Safety Guides*, Office of Emergency and Remedial Response, Hazardous Response Support Division, Edison, N.J. (Nov. 1984).
27. Rosen, S. L., *Fundamental Principles of Polymeric Materials*, 1st Ed., John Wiley & Sons, New York, p. 314 (1982).
28. Lab Safety, General Catalog, Lab Safety Supply, Janesville, Wis. (1988).

Exercises

1. Describe engineering and administrative measures for hazard elimination and control.
2. Summarize the process of using PPE.
3. What are the reasons for upgrading or downgrading the level of personal protection?
4. Under the OSHA standard for PPE, what is the employer's responsibility when employees provide their own PPE?
5. What are the requirements that an eye protector must meet under the OSHA standard in 29 CFR 1910.133?
6. What are the choices for an employee who needs prescription eye protection?
7. What type of protective goggles would you recommend for maximum eye protection?
8. What are the methods used for making eye protectors resistant to fog?
9. What are the requirements for ANSI Type I, Class B head protective helmets?
10. Under what circumstances should hearing protective devices by provided by the employer?
11. What are the seven categories of airborne hazards? What is the difference between an airborne hazard in the gaseous form and one in the vapor form?
12. What are the OSHA requirements for a cylinder or compressor that supplies air to respirators?
13. What are the OSHA requirements for the maintenance and care of respirators?
14. List the major categories of respiratory protective devices.
15. Describe the processes by which a chemical canister or cartridge respirator removes contaminants from air.
16. List at least five conditions under which air-purifying respirators should not be used.
17. List the points which must be considered prior to use of an air-line respirator.

18. In [illegible] s indicated that the conditions are immedi[illegible] rator leaks cannot be tolerated. What type of [illegible] y?
19. W[illegible] ntained breathing apparatus (SCBA)?
20. B[illegible] tten standard operating procedure (SOP). W[illegible] m SOP?
21. L[illegible]
22. D[illegible] and describe how they affect the selection o[illegible]
23. D[illegible] ed in the proper selection of CPC material.
24. A [illegible] ime of three hours has been in contact with a [illegible] te is 5 $mg/m^2/sec$ and the area of skin in contact with the glove is 225 cm^2, calculate the amount of chemical on the skin of the wearer after four hours of contact with the chemical.
25. What are the basic elements of a CPC management program?
26. In a CPC management program, what are the steps which must be undertaken in determination of skin exposure?
27. List at least five topics that should be used in the training of employees who may use CPC.
28. Describe the heat effects which can contribute to a worker's heat stress while wearing CPC.
29. What are the bodily factors that should be measured in monitoring for worker heat stress?
30. What are the main hazard categories that involve hand injuries?
31. Why do plasticized PVC gloves not provide adequate protection against petroleum derivatives?
32. List different tests that would help in the proper selection of gloves.

Chapter

5

Management of Fire Hazards

Energy, an essential component of any industrial process, usually is obtained by the combustion of a fuel. Combustion, a heat-generating process, is the chemical reaction between oxygen and a fuel, which requires three main ingredients: fuel, oxygen, and an ignition source, or a sufficiently high temperature. The ignition temperature is a function of the properties of the fuel, which can be in the solid, liquid, or gaseous state. In this chapter, we are concerned with managing the hazards of combustion in the workplace.

Although solid materials such as wood and natural or synthetic polymers are involved in most fires (or instances of accidental combustion), the mishandling of flammable liquids and gases is one of the major causes of industrial fires.

Flammable liquids and gases cannot cause fire by themselves; however, in the presence of an ignition source, they have contributed to many major fires. The misuse of flammable liquids alone is responsible for 15 to 18% of industrial fire losses. The improper handling of flammable gases is responsible for a smaller percentage of industrial losses, but it is still a major factor in the overall loss picture. To reduce the losses now being incurred, it is important to have a basic understanding of the elements of combustion and the properties of flammable liquids, gases, and solids.[1,2]

Combustion

Combustion can be defined as the rapid chemical reaction of oxygen with a fuel. When a fuel is burned, the carbon in the fuel reacts with oxygen to form either carbon dioxide or carbon monoxide:

$$C + O_2 \rightarrow CO_2 + \text{heat}$$
$$C + 1/2\ O_2 \rightarrow CO + \text{heat}$$

Carbon monoxide is produced by the incomplete combustion of a carbon-containing fuel.

Most fuels contain hydrogen as well as carbon, and the hydrogen in the fuel reacts with oxygen to produce water. For example, the combustion of propane (C_3H_8) produces water, CO_2, and CO:

$$C_3\ H_8 + 5O_2 \rightarrow 3CO_2 + 4H_2O + \text{heat}$$
$$C_3\ H_8 + 7/2O_2 \rightarrow 3CO + 4H_2O + \text{heat}$$

Some fuels contain other elements such as sulfur and nitrogen. These elements are converted to their respective oxides in a combustion reaction.

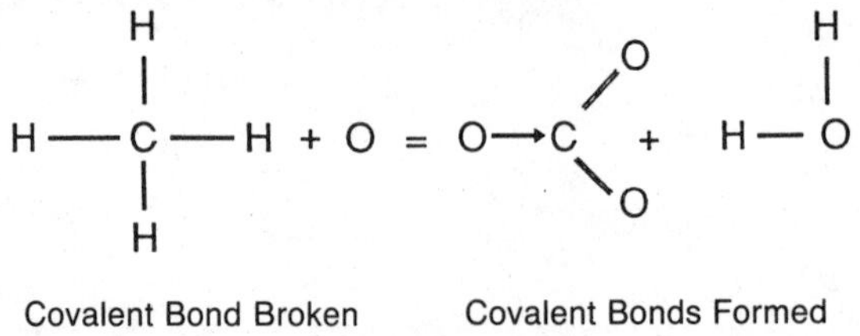

Covalent Bond Broken   Covalent Bonds Formed

Figure 5.1. Cleavage and formation of bonds when methane is converted to carbon dioxide and water.

Heat of Combustion

All chemical reactions involve the formation and the cleavage of chemical bonds between atoms; for example, when natural gas (methane) reacts with oxygen, the covalent bond in the oxygen molecule as well as the bonds between carbon and hydrogen atoms in the methane molecule must be broken. On the other hand, assuming that the combustion reaction produces carbon dioxide and water, covalent bonds between carbon and oxygen atoms in the CO_2 molecule and hydrogen and oxygen atoms in the water molecule must be formed, as shown in Figure 5.1.

On a molecular level, there is a certain amount of energy associated with each chemical bond. In a chemical reaction, the sum of the energy of the chemical bonds in the products is usually different from the bond energies in the reactants. Depending on the magnitude of the bond energies involved, a chemical reaction can produce or consume energy (some reactions show no net energy change), with the energy existing in the form of thermal energy or heat. Chemical reactions that consume heat are called endothermic, and those that produce heat, such as combustion, are called exothermic. The difference between the bond energies of products and reactants is called the heat of reaction, and, in the case of combustion, is known as the heat of combustion. This quantity usually is expressed in terms of energy per unit of mass of the fuel. Table 5.1 summarizes heat of combustion data for some typical fuels.[3]

Heat Transfer

Fires can generate large quantities of heat from the combustion of fuels. The heat thus generated is transferred to other objects, and this heat release can cause the other materials to explode, combust, or disintegrate. Also, the combustion or disintegration of materials such as synthetic polymers can produce deadly toxic gases. Thus an understanding of different modes of heat transfer is essential to a sound fire engineering and control program.[4,5]

Heat may be defined as "energy in transfer." The flow of thermal energy always takes place from an object or region at high temperature to a point at lower temperature. Heat can never flow from a low-temperature point to a high-temperature point unless external mechanical work is performed.

Heat transfer takes place via three different mechanisms: radiation, convection and conduction.

Table 5.1. Heat of Combustion of Typical Fuels.

Fuel	Heat of Combustion, Btu/lb
Flammable liquids	16,000–2,100
Flammable gases	20,000–23,000
Hydrogen	60,000
Coal	12,500

Source: Reference 3.

Radiation

An important mechanism of heat transfer from a fire is radiation, which is the electromagnetic propagation of thermal energy. Whereas other modes of heat transfer require an intervening medium, heat transfer by radiation can take place in an absolute vacuum. Indeed, this is the mechanism by which the earth receives its thermal energy from the sun. Heat transfer by radiation is accomplished by the motion of thermal waves, which travel with the speed of light (186,000 miles/sec) and can be absorbed, transmitted, or reflected by other bodies.

The amount of heat transfer by radiation from a body maintained at a temperature above absolute zero can be determined from the Stefan–Boltzmann law:

$$\frac{Q}{A} = \epsilon \, \sigma \, T^4 \tag{1}$$

where:

Q/A = Heat flux, energy/time-area
ϵ = Emissivity of radiating body
σ = Stefan–Boltzmann constant
T = Absolute temperature, °F + 460 or °C + 273
A = Area of heat transfer

The important point about equation (1) is that the rate of heat transfer is proportional to the fourth power of the temperature, which thus becomes critical in fire situations. For example, consider two fires, one at 1,000°F and the other at 2,000°F, radiating heat to two objects having the same surface area and emissivity. The ratio of heat transfer to the two objects can be calculated as follows:

$$\frac{Q_1}{Q_2} = \frac{(2{,}000 + 460)^4}{(1{,}000 + 460)^4} \approx 8.0$$

It is interesting to note that, in the example above, when the temperature doubles from one fire to another, the amount of heat transfer by radiation increases by a factor of about 8.

Convection

Heat transfer by convection takes place as a result of fluid motion.[4,5] When a cold fluid comes into contact with a hot surface, the film adjacent to the hot surface picks up thermal energy. This energy, in turn, is transmitted to the bulk of the cold fluid, either by natural mixing or as a result of external forces. When the mixing takes place as a result of difference in densities of the hot and cold fluids, the heat transfer mechanism is called natural convection. On the other hand, when mixing is accomplished by means of an external force, such as wind or a pump, the heat transfer mechanism is known as forced convection.

In a fire situation, convective heat transfer takes place as a result of the movement of air and the combustion products. This movement is a critical factor, as it determines the direction in which a fire will spread.

The rate of convective heat transfer is proportional to the temperature difference and the area of heat transfer:

$$Q \propto A \, \Delta T \tag{2}$$

where:

Q = Rate of convective heat transfer, energy/time
A = Area of transfer
ΔT = Temperature difference

Equation (2) can be transformed to an equality by multiplying it by a constant:

$$Q = H\,A\,\Delta T \tag{3}$$

The constant of proportionality in equation (3), called the heat transfer coefficient, is a function of the physical properties of the fluid. Note that the transfer area (i.e., geometry) and the temperature difference must be defined before a heat transfer coefficient can be calculated from heat transfer data.

Conduction

Conduction is the mechanism of thermal energy transfer from one molecule to another. The molecules near a hot surface absorb heat with the result that their temperature and kinetic energy increase. This energy is transmitted to the next molecule, raising its temperature and kinetic energy. Heat transfer by conduction is similar to the transfer of mechanical energy from one billiard ball to another. Consider a wall of surface area A and thickness L with one side maintained at a temperature of T_1 and the other side at a lower temperature of T_2, as shown in Figure 5.2. The rate of heat transfer can be assumed to be proportional to the temperature gradient per unit thickness of the wall and the transfer area:

$$Q \propto \frac{A\,(T_1 - T_2)}{L} \tag{4}$$

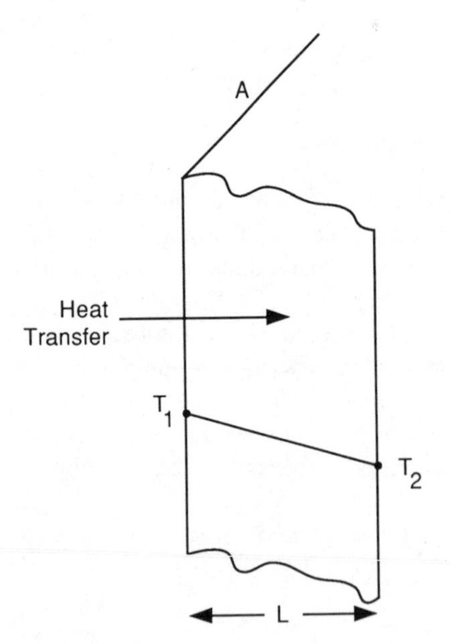

Figure 5.2. Heat flow by conduction through a wall.

Table 5.2. Thermal Conductivities of Solids, Btu/hr-ft-°F.

Material	Thermal Conductivity
Iron, pure	39.0
Iron, cast	29.0
Steel	33.0
Asphalt	0.43
Building brick	0.40
Portland cement	0.17
Glass	0.50
Plaster	0.20
Cork	0.026
Glass wool	0.02
Sawdust	0.034

Source: Reference 4.

If we multiply equation (4) with a constant, K, the following equation results:

$$Q = K\frac{A}{L}(T_1 - T_2) \tag{5}$$

The proportionality constant in equation (5) is called the thermal conductivity. Different materials have different thermal conductivities; for example, metals have high thermal conductivities and are good conductors of heat, whereas plastics generally have low thermal conductivities and are poor conductors of heat (good insulators).

Conduction heat transfer can play an important role in fire situations, especially in areas where flammables are stored. For example, in a flammable storage area heat can be transferred to the outer layer of a drum by radiation and convection from an ongoing fire. This heat transfer will raise the surface temperature of the drum, and heat then will be transferred through the drum to the flammable material by conduction, which can cause the temperature inside the drum to increase and eventually may ignite the drum's contents. In the construction of storage areas for flammables, it is important to use materials with a low thermal conductivity and a high temperature degradation resistance, such as concrete or brick. Table 5.2 summarizes the thermal conductivities of some commonly used solid materials.

The Combustion Process and Fire Development

The combustion process can be envisioned as taking place in six different regions, as shown in Figure 5.3. The three regions within the material are the degradation or vaporization region, the heating region, and the intact region. The three zones in the surroundings are the combustion products region, the flame region, and the preflame region.

- Vaporization or degradation region: In this region, heat, which is transferred from the preflame zone, causes the molecules to absorb energy and either disintegrate (in the case of solids) or vaporize (in the case of liquids) into the preflame region.
- Heating region: This is the area right below the degradation region where the fuel molecules are beginning to react to the heat generated by combustion.
- Intact region: This is the portion of the fuel where no effects of the fire can be detected. Its

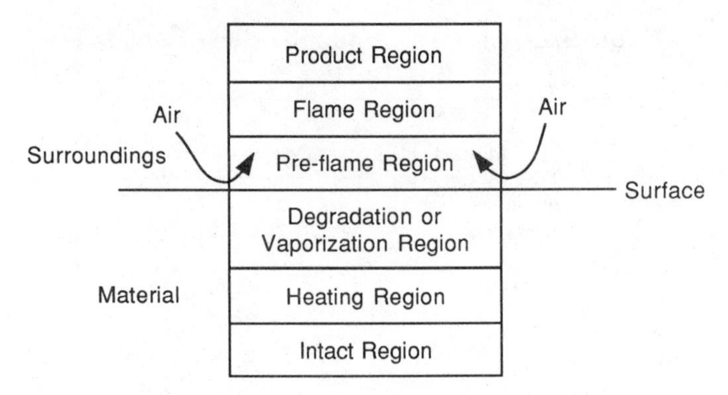

Figure 5.3. Combustion model. (Adapted from Reference 6)

location is a strong function of the thermal conductivity of the burning material and the burning time.

- Preflame region: The flammable mixture is produced in this area.
- Flame region: The flammable mixture from the preflame region is ignited here and generates large amounts of heat and light. Most of the heat generated in this area travels away from the burning material by radiation and convection. The remainder is transferred to lower regions to support degradation or vaporization of fuel into the preflame region.
- Combustion products region: Here the products of combustion begin to cool and form the smoke or toxic gases that are associated with many fires.

Once a fire is started, it can spread, diminish, or remain constant, depending on the manner in which heat produced by combustion is transferred to the fuel. The fire triangle shown in Figure 5.4 demonstrates the three essential components of a fire: fuel, oxygen, and heat. It should be pointed out that the fourth essential component of a fire is a chemical chain reaction. The removal of any of these components will prevent a fire in flammable or combustible materials and will diminish an ongoing fire. In a fire, the heat generated by the exothermic combustion reactions is transferred by radiation and convection to the surroundings and by convection and conduction to the fuel. The difference between the rate of heat generation and the rate of heat dissipation determines whether a fire will spread, diminish, or remain constant. The energy balance of an ongoing fire is represented by the following equation:

$$\frac{\Delta Q}{\Delta t} = \dot{m}\,\Delta Hc - \delta \qquad (6)$$

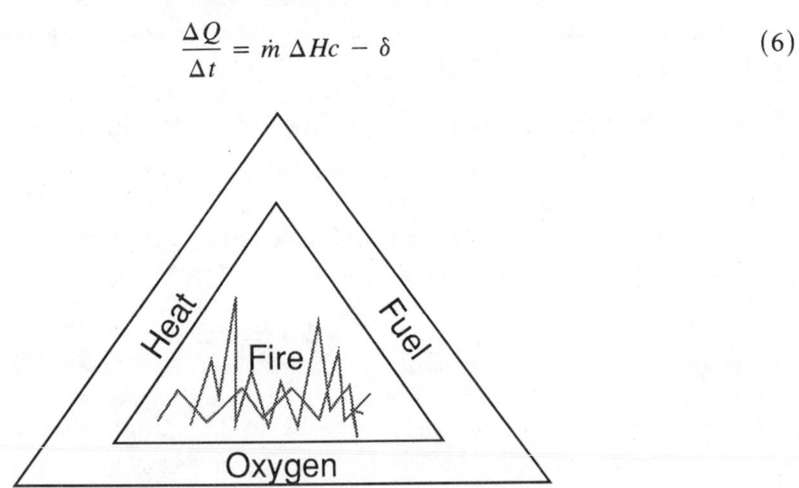

Figure 5.4. Fire triangle, showing the necessary components of a fire.

where:

$\Delta Q/\Delta t$ = Rate of heat accumulation or depletion
$\dot{m}$ = Rate of fuel consumption mass/time
ΔHc = Heat of combustion of the fuel, energy/mass
δ = Total heat transfer, energy/time
t = Time

When $\Delta Q/\Delta t$ is positive, an ongoing fire will become larger because the accumulation of heat will cause more vaporization or degradation of the fuel. On the other hand, when $\Delta Q/\Delta t$ has a negative value, an ongoing fire will diminish with time because the heat transferred to the surroundings is more than the heat generated by the combustion reactions. In controlled fire situations such as industrial furnaces and power plants, the value of $\Delta Q/\Delta t$ is zero, which means that the amount of heat produced by the combustion reactions equals the amount of heat transferred to the surroundings.

Elements of Combustion

Oxygen

As shown in Figure 5.4, oxygen is an essential component of any combustion reaction. In a fire situation, the degree of oxygen supply determines both the efficiency of the fire and the nature of the combustion products formed. For example, a fire in a poorly ventilated area would have a low rate of combustion; but for lack of an ample supply of oxygen, the carbon in the fuel would react incompletely with oxygen to form large amounts of deadly carbon monoxide gas. Although combustion can be viewed as an oxidation reaction, most oxidation reactions take place at a much slower rate than combustion reactions. For example, the rusting of iron and the darkening of an apple are examples of oxidation reactions with negligible amount of heat generation. It also should be pointed out that some fuels, such as organic peroxides and oxidizers, have oxygen built into their molecule and can support combustion.

Fuels

A fuel is a necessary component of all combustion reactions. The fuel can be present in solid, liquid, vapor, or gaseous form.

Solid Fuels

Examples of solid fuels that can support combustion are wood, coal, textiles, paper, plastics, and some metals such as magnesium and cesium.

The majority of fire fatalities in the United States are caused by fires in homes, and wood is the most commonly used construction material employed in family homes as well as many other residential dwellings. Wood or wood products can be treated with fire-retardant chemicals to reduce their flammability.[7]

In recent years, plastics have assumed a significant role in the construction of buildings. It is estimated that about 25% of the plastic industry output is used in the construction industry. There are thousands of polymeric materials now in use, with different properties and formulations.

The fire behavior of plastics is largely dependent on the chemical composition of the polymer molecule, as well as the type of additive or filler material used. Plastic materials are manufactured by the polymerization or the copolymerization of organic monomers, which contain carbon and hydrogen in their molecules; so most polymeric materials can be classified as combustible solids. The nature and the toxicity of the combustion products of plastics depend on the monomers' molecular composition, as most polymeric materials disintegrate into original monomers, which may be hazardous, and some can react with oxygen to produce hazardous gases.

Liquid Fuels

The attractive forces between a liquid fuel's molecules are much weaker than those in solid fuels. As a result, the liquid molecules can move freely, and can adapt themselves to the shape of the container in which they are placed. It is difficult to draw a line between solid, liquid, and gaseous fuels. A liquid fuel can exist in solid, vapor, or gaseous form, depending on the conditions of temperature and pressure.[1]

The improper handling and storage of liquid fuels have been the major cause of many disastrous fires. It is imperative that the physical and chemical properties of these liquids, especially those relating to fire prevention, be understood. Liquid fuels have been classified as flammable and combustible liquids (see next section).

Flammable and Combustible Liquids

Properties of Flammable and Combustible Liquids

In order to handle flammable and combustible liquids safely, it is necessary to understand some of their properties. The following discussion describes flammable liquids in terms of flash point, vapor pressure, fire point, explosive range, autoignition temperature, spontaneous heating, specific gravity, vapor density, evaporation rate, water solubility, boiling point, viscosity, and latent heat of vaporization.

Flash Point

Flash point is defined as the lowest temperature at which a liquid can generate enough vapor above its surface to ignite in the presence of a source of ignition. Two methods of measuring flash points are:

1. The Pensky-Martens closed tester, which is used for the determination of flash points of fuel oils, asphalts, and other viscous liquids.
2. The tag closed-cup tester, which is used for the measurement of flash points of liquids below 175°F.[8]

The rate of vapor generation increases with temperature, so that ventilation and handling requirements are a strong function of ambient temperature, as shown in the following examples:[9]

- A liquid with a flash point of −22°F, such as carbon disulfide, requires special handling, especially with respect to ventilation, at normal ambient temperatures.
- A liquid with a flash point of 90°F, such as styrene, poses no great danger where the temperature is below 70°F. Special care is required, however, when temperature exceeds 90°F.
- A liquid with a flash point of 232°F, such as ethylene glycol, is dangerous only in areas where temperatures of 232°F or higher are encountered.

Vapor Pressure

In a closed container, some of the liquid molecules escape into the space above the surface of the liquid. Some of these molecules can condense back into the liquid phase, and some will stay in the vapor phase. This vaporization and condensation of molecules will continue until the rates of vaporization and condensation become equal. When this condition is achieved, no further change in the amount or composition of the two phases take place, and they are said to be at equilibrium. The pressure exerted by a vapor on its liquid at equilibrium is called the vapor pressure of the liquid at the temperature of the system.[10] The equilibrium and the vapor pressure are strong functions of temperature. When the temperature of a system at equilibrium is changed, a new condition of equilibrium will be reached, with a new vapor pressure.[10]

A qualitative plot of vapor pressure at different temperatures for a single-component system is shown in Figure 5.5. This is also called the phase diagram for the system. Vapor–liquid equilibrium exists under those conditions of pressure and temperature falling on the line separating the liquid and vapor phases. At any given temperature, the pressure corresponding to the vapor–liquid line is called the vapor pressure, and at a given pressure, the temperature corresponding to the vapor–liquid line is called the boiling point of the liquid. As can be noted from Figure 5.5, a liquid has an infinite number of vapor pressures, depending on the temperature and an infinite number of boiling points depending on the pressure.

For example, in Figure 5.5, P_1 is the vapor pressure at T_1, and T_1 is the boiling point of the liquid at P_1. By the same token, P_2 is the vapor pressure at T_2, and T_2 is the boiling point at P_2. It is extremely important to indicate the temperature in reporting vapor pressure data, as the latter can change substantially with temperature. This is important when one is dealing with closed containers of flammable liquids, as vapor pressure is also a measure of the amount of force that the vapor exerts on the walls of the container. For example, if the vapor pressure data for a hydrocarbon liquid are available at 70°F, one must realize that on a hot afternoon when the temperature is 100°F, the vapor pressure and the force exerted on the container walls will be substantially higher than at 70°F. This is a critical consideration in selecting a proper storage area and containers for flammable liquids. The amount of force on the container walls can be calculated from the following expression:

$$F = P \cdot A$$

where:

P = Vapor pressure at the system temperature
A = Area of the container not occupied by liquid
F = Force on the container wall

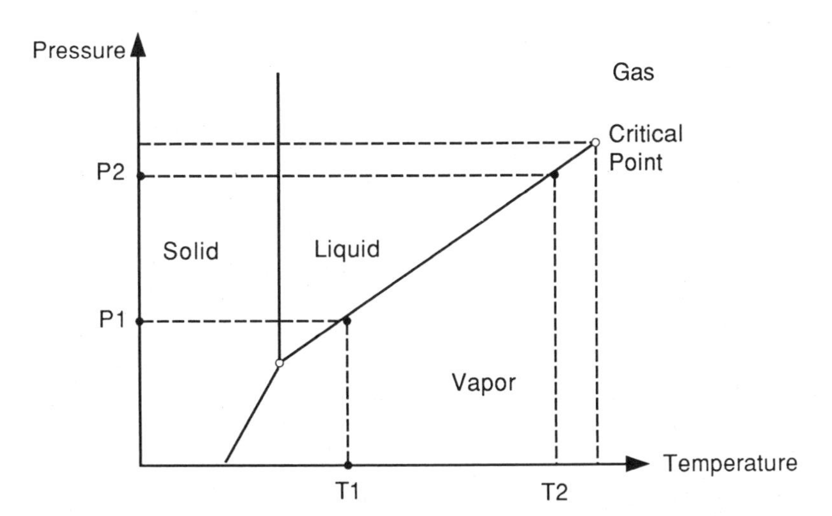

Figure 5.5. Phase diagram for a single-component system.

Table 5.3. Vapor Pressure of Some Common Hydrocarbons.

Compound	Vapor Pressure, psia	Temperature, °F
Cyclohexane	0.918	50
n-Heptane	0.398	50
Toluene	0.240	50
n-Octane	0.537	100
Benzene	0.88	50
Xylene	0.326	100

Source: Reference 8.

Table 5.3 summarizes vapor pressure data for some of the more commonly used liquid hydrocarbons.[8]

The phase diagram shown in Figure 5.5 also demonstrates the difference between a gas and a vapor. As can be noted, a gas is a vapor that is above its critical point. A vapor, on the other hand, is a fluid that is below its critical point and is not in the liquid phase. Vapors can be condensed into the liquid phase by the application of pressure alone, whereas gases can be condensed into the liquid phase only by a combination of both pressure and temperature effects. An understanding of the difference between the properties of a vapor and a gas is important in the safe handling of flammable liquids and gases, as will be explained later.

Definition of Flammable and Combustible Liquids. The National Fire Protection Association (NFPA) defines a flammable liquid as any liquid that has a flash point below 100°F and a vapor pressure not exceeding 40 psia at 100°F. A combustible liquid, on the other hand, is any liquid that has a flash point between 100°F and 200°F.[1] Flammable liquids are further classified into the groups summarized in Table 5.4, depending on their flash points and boiling points.

Combustible liquids are also classified into three groups, as summarized in Table 5.5.

Table 5.4. Definition of Classes of Flammable Liquids.

Class	Definition
IA	Flash point below 73°F; boiling point below 100°F.
IB	Flash point below 73°F; boiling point at or above 100°F.
IC	Flash point at or above 73°F, but below 100°F.

Source: Reference 1.

Table 5.5. Definition of Classes of Combustible Liquids.

Class	Definition
II	Liquids have flash points at or above 100°F and below 140°F.
IIIA	Liquids have flash points at or above 140°F and below 200°F.
IIIB	Liquids have flash points at or above 200°F.

Source: Reference 1.

Fire Point

Fire point is the lowest temperature at which a flammable liquid in an open container gives off enough vapors to continue to burn once it has been ignited. The fire point usually is slightly higher than the flash point. It is important to understand the difference between fire point and flash point; fire point data should not be used when a decision should be based on the flash point.

Explosive or Flammable Range

The explosive or flammable range encompasses those concentrations of flammable vapor or gas in air that can ignite in the presence of a source of ignition.[1,9] The term lower flammability limit (LFL) is used to describe the minimum concentration of vapor in air below which a flame will not propagate through the mixture. The term upper flammability limit (UFL) is used to describe the maximum concentration of flammable vapor in air above which a flame will not propagate. Any mixture of flammable vapor and air that is below its LFL is said to be too lean to burn, and any mixture above its UFL is said to be too rich to burn. Percentages vary widely, depending on the substance in question. Table 5.6 summarizes the UFL and LFL for some of the more commonly used flammable liquids.

Lower flammability limits for flammable liquids can be calculated from the vapor pressure data at the flash point temperature, using the following expression:

$$\text{LFL} \times 100 = \frac{P^*}{14.7}$$

where P^* is the vapor pressure of the liquid in psia at its flash point temperature.

The LFL and the UFL for a material are key factors in determining the amount of safety ventilation needed and the extent to which the venting of explosive pressures may be necessary.

Autoignition Temperature

Autoignition temperature is the lowest temperature that will produce combustion in the absence of an ignition source.[1,11] In areas where relatively high temperatures are used (above 200°F), special attention must be paid to ensure that liquids are not exposed to temperatures above their autoignition temperature. The standard testing method for autoignition temperatures of petroleum products is described in ASTM D 2155.

Table 5.6. Flammability Range for Some Common Flammable Liquids.

Substance	LFL, Vol. % in Air	UFL, Vol. % in Air
Gasoline	1.3	7.6
Acetone	2.6	12.8
Carbon disulfide	1.3	44.0
Ethanol	4.3	19.0
Methanol	5.5	36.5
Jet fuel	0.8	6.2
Toluene	1.3	7

Sources: References 11 and 12.

Spontaneous Heating

Some flammable liquids generate heat when mixed with oxygen as a result of slow oxidation (i.e., spontaneous heating).[11] If the heat generated is not removed, it causes the temperature to rise, and eventually ignition will take place. Oil- or paint-soaked rags in closed containers have caused many such fires, in which the oxidation rate depends on the material. The rate of oxidation of a material should be considered in decision-making about the storage and disposal of waste materials in the workplace.

Specific Gravity

The specific gravity (sp. gr.) describes the density of a liquid relative to the density of water. Liquids with a specific gravity less than one are lighter than water. Most flammable liquids are lighter than and insoluble in water, and will float on its surface. A few are heavier and insoluble in water and will sink in it. Specific gravity information is vital because it sets the requirements for fire extinguishing agents, storage, overflow drains, and other safety precautions. For example, fires in carbon disulfide (CS_2) (sp. gr. 1.263) can be extinguished with either CO_2, dry chemicals, or water. Fires in gasoline (sp. gr. 0.7) may be extinguished with CO_2, dry chemicals, or foam. Foam is useless on CS_2 because the flammable vapors penetrate through the foam blanket and reignite.

Specific gravity also affects storage requirements. For example, ethylene dichloride is heavier and insoluble in water, so it is feasible to store it in an open container, provided that a layer of water is placed over it.

Table 5.7 summarizes the specific gravity of some commonly used liquids.[8]

Vapor Density

Vapor density is a measure of the relative densities of vapors and gases compared to air under the same conditions of temperature and pressure. A vapor density of 2 indicates that the vapor or gas is twice as heavy as air and would collect in low areas. Vapor density data normally are reported for standard temperature and pressure conditions. Because gases and vapors are compressible, and temperature has a large effect on pressure and volume, changes in these conditions would greatly affect the vapor density.

Table 5.7. Specific Gravity of Some Common Liquids.

Compound	Specific Gravity
Acetaldehyde	0.783
Acetone	0.797
Ammonia	0.597
Benzene	0.879
Carbon disulfide	1.263
Carbon tetrachloride	1.595
Ethyl alcohol	0.790
Methyl alcohol	0.792
Methyl ethyl ketone (MEK)	0.805
Phenol	1.071
Propyl alcohol	0.804
Styrene	0.903
Toluene	0.866

Source: Reference 8.

Vapors generated by most flammable liquids are heavier than air (vapor density greater than one), so the ventilation for flammable liquids should be at or near floor level. Most flammable gases, on the other hand, have a vapor density of less than one and, thus are lighter than air. Ventilation systems for these flammable gases should be placed above the floor, depending on the point of generation.

When the rate of vapor production is high, positive-displacement safety ventilation is required. Fresh air intakes should be located so that they will assist the safety ventilation process.[11]

Evaporation Rate

The evaporation rate is the rate at which a liquid is converted to vapor at a given temperature and pressure. The evaporation rate of most liquids is reported relative to butylacetate (evaporation rate of one). A liquid with an evaporation rate less than one evaporates more slowly than butylacetate.

The evaporation rate is a strong function of pressure and temperature. As the temperature increases, the molecules in the liquid phase gain energy and can break the intermolecular forces in the liquid and migrate into the vapor phase. Therefore, the evaporation rate increases with an increase in temperature. Pressure has an opposite effect on the evaporation rate; as pressure increases, the molecules in the vapor phase are driven back into the liquid, and those molecules in the liquid trying to go to the vapor phase must overcome the pressure energy. Therefore, as the pressure goes up, the evaporation rate decreases.

Water Solubility

Many flammable liquids, such as alcohols, ethers, and ketones, are completely or partially soluble in water, and mixing these liquids with water reduces their flammability and eliminates static hazards. Some flammable liquids, such as alcohols, become completely nonflammable when sufficiently diluted with water. Fires in liquids that are soluble in water can be extinguished by dilution with water.

Boiling Point

The boiling point of a liquid is defined as the temperature at which the vapor pressure of the liquid becomes equal to the external pressure on the surface of the liquid. For example, water boils at 212°F when the external pressure is one atmosphere; in other words, the vapor pressure of water at 212°F is equal to one atmosphere. As vapor pressure is a function of temperature, when the external pressure on the surface of a liquid changes, so does its boiling point. For example, a camper trying to boil water at a high elevation would notice that water boils well below 212°F there, because the external pressure on the surface of water has decreased, and a lower temperature is required to create a vapor pressure equal to the external pressure. By the same token, when the pressure on the surface of a liquid is increased, the boiling point of the liquid goes up. Theoretically, it is possible to boil water at any temperature (even at freezing—0°C, 32°F) by adjusting the pressure.

An understanding of the relationship between pressure and boiling point is important in the safe handling of flammable liquids. The boiling point reported for most materials is measured at one atmosphere (normal boiling point), but, as with water, the reported boiling point can change when external pressure changes. For example, a flammable liquid with a normal boiling point of 100°F has a much lower boiling point in Denver, Colorado, where the elevation is approximately one mile above sea level.

Viscosity

Viscosity is a measure of the resistance of a fluid to flow as a result of internal friction between fluid layers. The higher the viscosity, the more difficult it is to make the fluid flow. Because one of the major contributing factors to the production of static electricity in flammable liquids is friction, special care should be exercised in the transport and handling of highly viscous flammable or combustible liquids.

Latent Heat of Vaporization

When a pure liquid is at its boiling point temperature, the liquid molecules migrate into the vapor phase. Under these conditions, the liquid temperature remains constant in spite of the fact that heat must be supplied to the liquid to maintain boiling. This heat does not contribute to an increase in temperature, so it is called the latent heat. By definition, the amount of heat required to vaporize a unit mass of liquid at normal boiling point is called its latent heat of vaporization. This heat effect should be considered when flammable liquids are exposed to temperatures at or near their boiling points.

Basic Safeguards for Flammable Liquids

The following is a discussion of basic safeguards for fire safety when dealing with flammable liquids.

Hazard Isolation

A fire in a drum of flammable liquid located outdoors in an isolated area can result only in the loss of the liquid. If the drum were in a small, detached storage shed, the loss would at most include the storage shed and its contents. If the same drum were to burn in a heavily populated building, the results could be disastrous.

The following is a list of recommendations for the safe storage of flammable liquids:[11]

1. Storage in buried tanks (such as gasoline tanks).
2. Storage in aboveground tanks that are diked, have proper drainage, and are located away from important buildings.
3. Storage in sheds that are made of fire-resistant materials and are located away from important buildings.
4. Storage in isolated rooms on the first floor; if upper stories are used for the storage of flammables, adequate drainage must be provided, and the floors must be leak-proof.

Substitution

Flammable liquids sometimes may be substituted by relatively safe materials. These materials should be stable and nontoxic, and should either be nonflammable or have a high flash point.[1]

Confinement of Liquids

Procedures should be developed to confine flammable liquids inside equipment, piping, or storage containers. The escape of materials should be minimized, and arrangements must be made for the safe removal of material that might escape.

Explosion Venting

Explosion venting should be provided in areas where there is a potential for a flammable liquid vapor to form a mixture with air within its explosive range. Examples include areas where flammable liquids are stored or used.[1]

Ventilation of Storage Areas

Ventilation helps to prevent flammable liquid fires and vapor–air explosions. Ventilation systems are designed to confine, dilute, or remove the maximum normal amount of vapor release.[11]

The ventilation system must be designed in such a way to maintain an air–vapor mixture that is at least 25% below the LFL. Positive displacement ventilation devices are recommended for flammable liquids, as well as for combustible liquids if there is a potential that they might be heated to temperatures near their flash points.

Elimination of Ignition Sources

All nonessential ignition sources must be eliminated where flammable liquids are used or stored.[11] The following is a list of some of the more common potential ignition sources:

- Open flames, such as cutting and welding torches, furnaces, matches, and heaters: These sources should be kept away from flammable liquids operations. Cutting or welding on flammable liquids equipment should not be performed unless the equipment has been properly emptied and purged with a neutral gas such as nitrogen.
- Chemical sources of ignition: Some flammable liquids generate heat as a result of slow oxidation reactions with air; this heat can raise the temperature to above the ignition temperature.
- Electrical sources of ignition such as d.c. motors, switches, and circuit breakers: These sources should be eliminated where flammable liquids are handled or stored. Only approved explosion-proof devices should be used in these areas. (See Chapter 7.)
- Mechanical sparks: These sparks can be produced as a result of friction. Only nonsparking tools should be used in areas where flammable liquids are stored or handled.
- Static sparks: These sparks can be generated as a result of electron transfer between two contacting surfaces. The electrons can discharge in a small volume, raising the temperature to above the ignition temperature. Every effort should be made to eliminate the possibility of static sparks. Also proper bonding and grounding procedures must be followed when flammable liquids are transferred or transported.
- Smoking: As smoking provides an open flame, it must be prohibited in areas where flammables are stored or used.

Removal of Incompatibles and Other Safety Precautions

Materials that can contribute to a flammable liquid fire should not be stored with flammable liquids. Examples are oxidizers and organic peroxides, which, on decomposition, can generate large amounts of oxygen. Also, these steps may be taken:

Provide fire protection. Fires occur despite the efforts that are put into preventing them. Fire extinguishers and other firefighting equipment must be installed and must be appropriate for the nature of the flammables and their use.

Provide an inert atmosphere. If a liquid flammable container cannot be fully emptied or purged, it is advisable to keep a layer of an inert gas such as nitrogen over the surface of the liquid, a practice called inerting.

Educate and train. Employees who handle flammable liquids must be properly trained in handling methods and actions to take in an emergency.

Storage Facilities

Flammable liquids normally are stored in drums or tanks. Either type may be located inside buildings or outside. If the storage is located inside the building, the limits on the quantity of flammables set by codes must be followed. The following paragraphs list basic considerations for storage facilities.[11]

Storage locations for indoor drum storage, in order of preference, are: (1) an isolated building made of noncombustible materials, or (2) an attached, cut-off building or room. In either case, the enclosure must be vented to prevent the accumulation of an explosive atmosphere. Also, adequate drainage must be provided to remove any spilled or burning flammable liquids. Trapped floor drains leading to a safe location are the most common form of drainage.

Outdoor drum storage areas should not slope toward buildings or sewer systems. Such areas should be kept free of weeds and divided into wide lanes for fire department access. Fire hydrants should be strategically located. Remember that an unpurged empty drum or a partially filled drum is far more dangerous than a full one.

Regulations and codes for the location and spacing of aboveground and underground storage tanks should be consulted to ensure that the tanks meet the requirements. The following checklist should be helpful in the inspection of storage tanks:

- What is contained in the tank?
- Is the tank well maintained (visual inspection)?
- Is a drainage system provided and operable?
- Is the electrical equipment suitable for the flammable liquid?
- Has adequate firefighting equipment been provided?
- Is an inerting system required? Is it operable?
- Are diked areas free of weeds?
- Are standard operating and emergency procedures posted and enforced? (This requirement is particularly important for pumping flammable liquids in and out of storage tanks.)
- Are floating roofs functioning so that potential vapor spaces can be controlled?
- Are tanks purged before reconditioning?
- Are appropriate safety and warning signs posted?

Transfer Methods

The safest method for the transfer of flammable liquids uses positive displacement pumps, which normally are driven by either electricity or steam. Every attempt should be made to eliminate all sources of ignition where flammable liquids are being transferred. Also, static charges that can develop within a flammable liquid must be discharged by proper bonding and grounding procedures.

Basic fire safety rules for two industrial operations, mixing and paint-spraying, are summarized below. These can be used as a guideline in developing safety measures for other industrial operations that handle flammable liquids.

Mixing Operations

Mixing operations are common in industries such as the manufacture of paints, lacquers, varnishes, or petroleum products. Mixing can generate heat due to friction and resistance to mixing. Both fire and explosion hazards may be present, depending on the flammable liquids involved, the method of handling, the degree of confinement, the provisions for ventilation, and the temperatures of the area and the liquid.[11]

Mixing operations involving flammables present all the fire hazards common to flammable liquids. Also, explosion hazard potentials exist at ambient temperatures if (1) the flammable liquids being used have flash points considerably lower than ambient temperature (25°F or less), or (2) the operation involves heating flammable liquids up to or slightly higher than their flash point.

As with any other industrial operation, mixing operations should entail very careful planning to ensure that the operation is conducted safely. The following guidelines can be helpful in planning for a mixing operation:

- Buildings or rooms should be located and selected appropriately.
- Drainage and venting should be provided for buildings.
- Equipment that poses an explosion hazard should be vented or inerted.
- The area should be ventilated to prevent a dangerous concentration of flammable vapors.
- Ignition sources must be eliminated.
- Heating systems must be spark-free and should operate at temperatures below the autoignition temperatures of the liquids.
- Adequate fire protection equipment must be provided.
- Equipment should be used to monitor operating conditions, such as temperatures, pressures, liquid levels, agitation, and other conditions.
- Electrical equipment must be ignition-free or moved to an appropriate location.

Paint Spraying

Although there are many types of paint-spraying operations, all types have the following safety requirements in common:

- Never locate a paint operation in such a way that an operator can be trapped by a fire (e.g., an operator with his or her back to a wall).
- Clean booths and ducts regularly (usually after four hours of operation).
- Remove paint-soaked rags and waste promptly, and transfer it to a covered metal container for temporary storage. These rags can generate heat when in contact with air and eventually catch fire.
- Use grease or paper to protect sprinkler heads.
- Use nonsparking tools and machinery.
- Make sure that ventilation is turned on before paint spraying is started.
- Make sure that electrical equipment fits the hazard.
- Provide watertight floors and adequate drainage.

Flammable Gases

Overview

All substances can exist in solid, liquid, vapor, or gaseous form, depending on conditions of temperature and pressure. The intermolecular forces in solid materials are relatively strong, so that solids have both their own shape and volume. The intermolecular forces in liquids are weaker, compared to solids, and liquid molecules have some freedom of motion. Although liquids do not have a shape of their own and assume the shape of the container in which they are placed, they do have their own volume, which is independent of the container volume. Gaseous molecules have enough kinetic energy associated with them to completely overcome their weak intermolecular forces, and thus have neither a shape nor a volume of their own; they assume the shape and occupy the entire volume of the container in which they are placed.

Definition of Flammable Gas

A gas is considered to be flammable if it satisfies either of the following criteria:

- At ambient temperature and pressure it has a UFL of 13% or less by volume.
- At ambient temperature and pressure it has a flammability range that is wider than 12%.

Properties of Flammable Gases

Generally, flammable gases pose the same type of fire hazards as flammable liquids and their vapors. Some differences in properties, however, warrant discussion.

Flammable gases have flash points only when they are liquefied at cryogenic temperatures. These temperatures are extremely low and normally are not used in most industrial operations. Therefore, the use of the flash point or the fire point for flammable gases would not serve any purpose in fire prevention.

It is important to distinguish between a flammable vapor and a flammable gas. As was mentioned before, vapors exist below their critical point and can be liquefied by the application of pressure alone. Gases, on the other hand, exist above their critical point and can be liquefied by the combined effects of temperature and pressure.

Flammable gases usually have a wider explosive range, as compared to flammable vapors. This should be taken into account when designing the ventilation system and electrical equipment in a flammable gas atmosphere. Although all flammable vapors are heavier than air, most flammable gases are lighter than air. This property would dictate the nature of the design for ventilation systems.

All flammable gases are lighter than water, and many flammable gases are soluble in water to some extent. Spontaneous heating is not a problem with flammable gases.

Flammable gases can be compressed, whereas flammable liquids are only slightly compressible. This difference in compressibility affects fire safety for flammable gases. In order fully to understand the effects, it is necessary to review the fundamental gas laws and their applications (see below).

The Ideal Gas Law

The ideal gas law can be derived from kinetic theory by assuming that: (a) the volume occupied by gas molecules is negligible, (b) the molecules do not exert any intermolecular forces on each other, and (c) the molecules collide elastically with the walls of their container. The ideal gas law can be expressed as:

$$PV = NRT$$

where:

P = Absolute pressure of the gas.
V = Volume or volumetric flow rate of the gas.
N = Number of moles or molar flow rate of the gas.
R = The universal gas constant, whose value depends on units of P, V, N, and T.
T = Absolute temperature of the gas in degrees Rankine or Kelvin.

The ideal gas law indicates that the volume occupied by a gas is directly proportional to its temperature and inversely proportional to its pressure—a relationship that is important in dealing with flammable gases. When a flammable liquid container breaks, the liquid volume does not change appreciably, but when a container of gas under pressure breaks, the volume of the gas can increase substantially.

From the gas law, we can derive the following relationships:

$P_1V_1 = P_2V_2$ Isothermal conditions; only gas pressure and volume change.)

$\frac{T_1}{T_2} = \frac{P_1V_1}{P_2V_2}$ (Gas temperature, pressure, and volume change.)

$V_1T_1 = V_2T_2$ (Isobaric change; pressure is constant, and only temperature and volume change.)

The subscripts 1 and 2 in the above equations refer to two different states of the gas.

We will apply the above expressions to a 500 ft^3 container of flammable liquid and a 500 ft^3 container of a flammable gas under 1,000 psi pressure:

- The 500 ft^3 container of flammable liquid breaks, and the volume of flammable liquid does not increase appreciably.
- The 500 ft^3 container of a flammable gas breaks, relieving pressure from 1,000 psi to atmospheric (14.7 psi); the volume of the gas does change:

$$V_2 = \frac{P_1 V_1}{P_2} = \frac{(1{,}000 + 14.7)\ 500}{14.7} = 34{,}514\ \text{ft}^3$$

As the above calculation shows, with the change in gas pressure, the gas volume increases from 500 ft^3 to 34,514 ft^3. This large volume of gas will mix with air, and, depending on its explosive range, can pose severe fire and explosion hazards for the entire volume it occupies.

In calculating the dangers of flammable gases, the room volume is important. In a small room, even if its pressure is declining, enough gas might remain in the room to destroy a wall. The same would not be true of a released liquid.

Some examples of flammable gases, their flammability ranges, and specific gravities are summarized in Table 5.8.

Basic Safeguards for Flammable Gases

Many of the safeguards for flammable liquids also apply to flammable gases (i.e., isolation, confinement, ventilation, etc.) In dealing with compressed or liquefied flammable gases, other properties such as toxicity, reactivity, and corrosivity also must be taken into account. Also, a gas that is flammable could produce toxic combustion products.

Products of Combustion and Their Effect on Safety

The products of combustion can be classified into four major groups: (a) fire gases, (b) flame, (c) heat, and (d) smoke. Depending on the chemical structure of the fuels involved and the availability of oxygen, these products can be produced in varying amounts. A combination of inhaled toxic gases and obscured vision can contribute to physical incapacitation, loss of coordination, faulty judgment, disorientation, restricted vision, and panic. Delay or prevention of escape leads to injury or death from further inhalation of toxic gases or the infliction of thermal burns.[1,12,13]

Table 5.8. Combustion Properties of Some Flammable Gases.

Gas	LFL	UFL	Specific Gravity (air = 1)
Ethylene	2.7	34.0	0.97
Natural gas	4.5	14.0	0.70
Butane	1.9	8.5	2.0
Carbon monoxide	12.5	74.0	0.97
Acetylene	2.5	81.0	0.90
Anhydrous ammonia	16.0	25.0	0.60
Vinyl chloride	4.0	22.0	2.15
Hydrogen	4.1	74.2	0.069

Sources: References 1 and 11.

Fire gases

Fire gases are those products of combustion that are in the gaseous state at normal temperature.

The nature and the toxicity of fire gases strongly depend on the chemical structure of the burning material, the oxygen supply, and the temperature. Most fuels contain carbon, which can be converted to either carbon dioxide or carbon monoxide. Carbon dioxide is formed when the supply of oxygen is well in excess of the stoichiometric requirements, and the air and fuel are premixed to obtain a uniform distribution of oxygen. Carbon monoxide, on the other hand, is formed as a result of the incomplete combustion of carbon-containing materials. In most fire situations, the amount of oxygen available is not sufficient for complete combustion, and a considerable amount of deadly carbon monoxide gas is formed. Also, depending on the molecular composition of the burning material, a number of other gases, such as sulfur oxides, hydrogen cyanide, and nitrogen oxides, can be formed.

Although heat and flame contribute to death or injury in many fires, the major cause of death in most fires is the inhalation of toxic gases and smoke. Also many gases that are nontoxic, such as carbon dioxide, decrease the relative concentration of oxygen in air and can cause fatalities due to oxygen deficiency.

The concentration of the gases, the time of exposure, and the physical condition of the individual determine the toxic effects of the gaseous products of combustion. Specific hazard situations are discussed in the following paragraphs.

Carbon Monoxide. As a result of the incomplete combustion of carbon-containing compounds, CO is produced in large amounts during a fire. Although CO is not the most toxic of fire gases, it is 210 times more reactive with blood than is oxygen. Thus, CO poisons humans by a form of asphyxiation. Carbon monoxide reduces the oxygen-carrying capacity of blood and interferes with necessary gaseous exchange, and it can be dangerous even at low concentrations. Inhalation of air containing 0.4% CO can cause death in less than one hour. Inhalation of air containing 1% carbon monoxide is fatal in less than one minute.[1]

Carbon Dioxide. Fires produce large amounts of carbon dioxide (CO_2). Carbon dioxide itself is not a toxic gas; its major contribution to death and injury results from its reducing the relative concentration of oxygen in air. As a result, an affected person's breathing rate increases to compensate for the oxygen deficiency. The rate of breathing can double with the presence of as little as 3% carbon dioxide in air.[1] This condition contributes to the overall hazard of a fire gas environment by causing accelerated inhalation of toxicants and irritants.

Oxygen Deficiency. During a fire, the concentration of oxygen in air is reduced by two mechanisms. Oxygen is consumed in combustion reactions, and its relative concentration in air drops with the production of combustion products. For example, consider a room that contains 20 moles of air. The number of moles of oxygen in the room is 4.2, and the number of moles of nitrogen is 15.8 (air contains approximately 21% oxygen and 79% nitrogen). Assume that 2 moles of carbon are burned in this room, and that all the carbon is converted to carbon dioxide.

$$2C + 2O_2 \rightarrow 2CO_2$$

Each mole of carbon consumes one mole of oxygen to form one mole of carbon dioxide. After completion of the reaction, the number of moles of each component in the room is as follows:

Number of moles of O_2 consumed $= 2$
Number of moles of CO_2 produced $= 2$
Number of moles of O_2 remaining $= 4.2 - 2 = 2.2$
Number of moles of N_2 remaining $= 15.8$

Total number of moles in the room = $2(CO_2) + 2.2(O_2) + 15.8(N_2) = 20$. The new concentration of O_2 in the room is:

$$\frac{\text{Number of moles of } O_2}{\text{Total number of moles}} = \frac{2.2}{20} = 0.11$$

As can be noted, as a result of CO_2 production the concentration of oxygen in the air has dropped from 21% to 11%. Notice, too, that both the consumption of O_2 by the chemical reaction and the production of CO_2 contribute to the creation of an oxygen-deficient atmosphere. Table 5.9 summarizes the effects of oxygen deficiency on a person's safety.[1]

Hydrogen Cyanide. Hydrogen cyanide, a deadly gas, can be produced during fires from the combustion of materials such as wool, silk, acrylonitrile, acrylates, agricultural chemical, rodenticides, and polyurethane. Some of the hazards of hydrogen cyanide are summarized in Table 5.10.

Sulfur Dioxide.[13] This gas can be produced during fires from the combustion of sulfur-containing materials. A strong irritant, it is intolerable well below lethal concentrations. Sulfur dioxide is a highly toxic gas with a TLV of 5 ppm, and can cause death if inhaled for 10 minutes at a concentration of 500 ppm.

Ammonia. During fires, ammonia can be generated from the combustion of materials such as wool, silk, fertilizers, explosives, acrylonitrile, and nylon. Ammonia has a TLV of 50 ppm, and although, it is not classified as a highly toxic or toxic gas, its prolonged inhalation at concentrations above the TLV can cause edema of the respiratory tract and suffocation. It is also highly irritating and erosive to the skin and mucous membranes. Contact of the gas with the eyes can cause severe visual disorder.[13]

Table 5.9. Effects of Oxygen Deficiency on a Person's Safety.

Oxygen Concentration in Air, Vol. %	Effects
17	Loss of coordination.
10–14	Person may exercise faulty judgment, become quickly fatigued.
6–10	Loss of consciousness, person must be revived with fresh air to prevent death.

Source: Reference 1.

Table 5.10. Hazardous Properties of Hydrogen Cyanide.

Flammability:	Flash Point: −18°C Ignition temperature: 538°C LFL–UFL: 6–41%
Toxicity:	TLV = 10 ppm LC_{50} inhalation—rats (1 hr): 544 ppm LD_{50} oral—rats: 3,700 micrograms/kg

Source: Reference 13.

Hydrogen Chloride.[1,13] This gas can be generated by the combustion of materials such as polyvinyl chloride (PVC), dyes, perfumes, agricultural chemicals, and some fire-retardant materials. Hydrogen chloride has a TLV of 5 ppm and can cause severe irritation of the skin, bronchitis, and rapid throbbing of the heart.

Hydrogen Sulfide.[13] During fires, hydrogen sulfide can be generated as a result of the incomplete combustion of sulfur-containing compounds such as wool and rubber. This gas, which smells like rotten eggs, has a TLV of 20 ppm and an inhalation LC_{50} of 750 ppm (rats, 1 hour). Exposure to hydrogen sulfide at concentrations above 0.05% can be fatal; at lower concentrations it is capable of causing skin burns, respiratory problems such as bronchitis, and systematic complications such as headache and dizziness.

Nitrogen Dioxide.[1,13] This gas can be generated from the combustion of nitrogen-containing compounds such as fabrics, cellulose nitrate, celluloid, catalysts, and polymerization inhibitors. The compound is toxic and has a TLV of 5 ppm and an inhalation LC_{50} of 67 ppm (rats, 1 hour). Severe skin burns, cyanosis, muscle spasms, and convulsions are among the symptoms of overexposure to hydrogen fluoride. Exposure to concentrations above 400 ppm for 10 minutes can be fatal.

Hydrogen Bromide. This gas, which can be produced during fires, has a TLV of 3 ppm and an inhalation LC_{50} (rats, 1 hour) of 2,858 ppm. It is a strong irritant to eyes, skin, mucous membranes, and respiratory organs. Exposure to concentrations of 500 ppm for 10 minutes can cause death.

Acrolein.[1,13] This compound can be generated during fires as a result of the pyrolysis of polyolefins and cellulosic materials. It is highly toxic with a TLV of 0.1 ppm and an oral LD_{50} in rats of 50 mg/kg. Its vapors can cause inflammation of the eyes, bronchitis, nausea, vomiting, diarrhea, and unconsciousness. Exposure to concentrations as low as 100 ppm for 10 minutes can be fatal.

Flame

Combustion reactions are exothermic in nature, producing large quantities of heat, and this thermal energy possesses both convective and radiative components. The radiative component is seen as flame in the flame zone. Much of the heat produced in the flame zone is transferred by convection and radiation to the surroundings and can cause severe thermal burns in individuals, depending on their distance from the flame zone.[1]

Heat

Heat is the combustion product that spreads fire in buildings. Heat creates a physical danger to humans through exposure to hot gases and thermal radiation. Heat can cause a series of events ranging from minor injury to death.

Smoke

Incomplete combustion of most organic materials generates finely divided particulate matter and liquid droplets known as aerosols, which appears as smoke.[1,12] Because the average size of smoke particles is the same as the wavelength of visible light, smoke can scatter the light and adversely affect the vision. It has been estimated that the inhalation of smoke and fire gases is responsible for 50 to 75% of the fatalities in fires.[12]

Fire Hazard Identification (NFPA 704 System)

The increasing use of a wide variety of hazardous chemicals in industry has created the need for a simple fire hazard identification system, with the objective of preserving the life and safety of individuals who could be exposed to hazardous substances during a fire.

NFPA has developed a standard system (NFPA 704) with easily recognizable color markings that, at a glance, can show the major health and physical hazards during a fire. This system can provide quick, on-the-spot information to protect the lives of firefighting personnel and identify any threats to the public safety as well.[14]

The NFPA 704 system divides the fire hazards of a material into four major categories: health, flammability, reactivity, and special. The system uses a number rating system, ranging from 0 to 4, to identify the severity of hazard in each category, with 0 indicating no hazard and 4 indicating the most severe hazard.

Although this system is easy to use, the ratings in each hazard category must be done by individuals who are familiar with the chemical behavior of the materials in question under fire conditions. It is possible for a material that is highly toxic to produce nontoxic combustion products, and for a material such as PVC that is not hazardous to produce extremely toxic fumes such as vinyl chloride and hydrogen chloride during a fire.

The hazard information in NFPA 704 system is provided on a diamond with special background colors assigned to each type of hazard, as shown in Figure 5.6.

Degrees of Hazards

Health (Blue Background)

The health hazard of a material during a fire is defined by a single exposure that can last anywhere from a few seconds to an hour. Health hazards can be the result of vaporization or the chemical reaction of a material during a fire. For example, sulfur-containing compounds can generate toxic

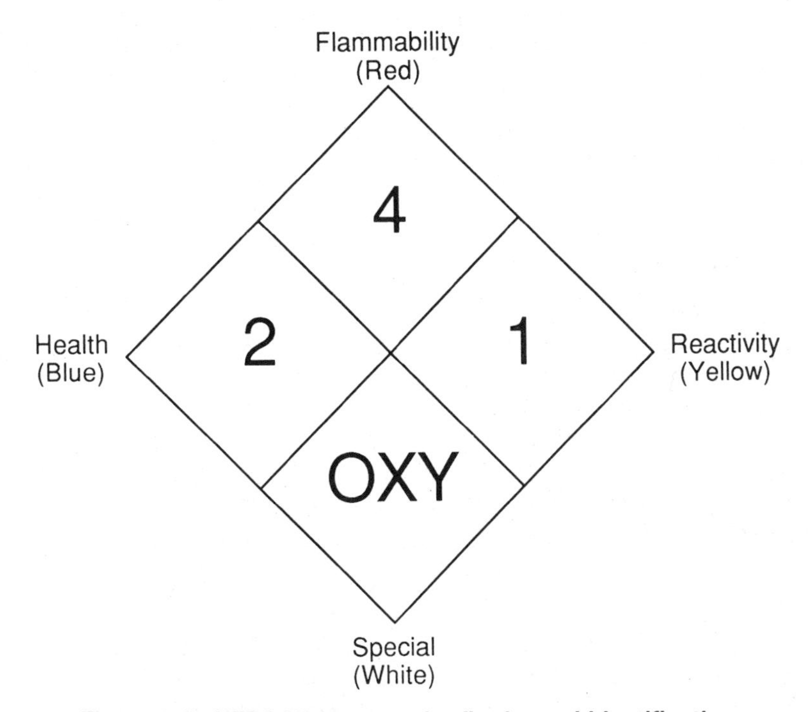

Figure 5.6. NFPA 704 system for fire hazard identification.

sulfur dioxide gas, and materials such as silk are capable of producing deadly hydrogen cyanide when exposed to high temperatures. The following list explains the degrees of health hazard:

4 Materials are too dangerous to health during fires. Inhalation or absorption by skin of combustion products can cause death. A fully encapsulating suit to protect against inhalation and skin absorption is required.
3 Materials are extremely dangerous to health. Appropriate protective clothing and equipment are required for entry into areas containing these materials.
2 Materials pose moderate health hazards. Areas may be entered with appropriate respiratory and eye protection.
1 Materials pose only slight health hazards. Use of appropriate respiratory protection is recommended.
0 Materials pose no health hazard.

Flammability (Red Background)

The flammability of materials is rated as follows:

4 Extremely flammable gases or flammable liquids with very low flash points.
3 Extreme fire hazard; gases or liquids that can be ignited under normal temperature conditions.
2 Moderate hazard; materials that have flash points above normal temperature and must be slightly heated before they can be ignited.
1 Slight hazard; materials that have flash points considerably above room temperature. These materials must be preheated before ignition can take place.
0 No hazard; materials that are stable during fire and do not react with water.

Reactivity (Yellow Background)

Degrees of reactivity, or stability, are assigned on the basis of how stable the material is, either alone or in the presence of water:

4 Materials that in themselves are capable of hazardous polymerization, decomposition, explosion, or any other hazardous chemical reaction. If a chemical is rated in this category, the area must be evacuated during a fire.
3 Materials that can detonate or explode in the presence of an initiating source. This category includes materials that are sensitive to shock at high temperatures and pressures, or that can react violently with water.
2 Materials that are capable of undergoing violent chemical reaction but do not detonate. This category includes materials that can undergo chemical reaction with rapid release of energy, such as polymerization. Materials that can react violently with water or form explosive mixtures with water also fall in this category.
1 Materials that are stable under normal conditions of temperature and pressure but may become unstable at elevated temperatures, or materials that are capable of reacting with water, resulting in the slight release of heat.
0 Materials that are stable under fire conditions and do not react with water.

Special (White Background)

The fourth space in the NFPA 704 diamond is reserved for any special hazards that the material might pose in a fire. For example, materials that react violently with water should carry the symbol W̶ (do not use water) in this space.

Fire Control

Oxygen, fuel, and a sufficiently high temperature are the three essential components of any fire (see Figure 5.4). Removal of any of these components would prevent a fire from starting and would extinguish an ongoing fire. Because the fuel component usually cannot be removed during a fire, most fire control techniques are aimed at the elimination of oxygen or temperature reduction by cooling. Flame inhibition by chemical means also is used widely in the control of ongoing fires.

Water is the most commonly used extinguishing agent for the control of fires by cooling. It is extremely effective on ordinary combustible fuel fires such as wood, cloth, and paper. The effectiveness of water on flammable and combustible liquid fires depends on the flash point, physical and combustion characteristics, temperature, area of the burning surface, and quantity of liquid involved.[1]

Water usually is effective in extinguishing fires in flammable or combustible liquids with any of the following characteristics:

- Liquids having a flash point of 200°F or higher: The basic action of water is to cool the surface of the liquid to the point where vaporization of the liquid into the preflame zone is minimized.
- Liquids with a specific gravity greater than 1: A few flammable and combustible liquids such as carbon disulfide are heavier and insoluble in water, and water can extinguish their fires by separation of air and the vapor layers. The fire would stop for lack of the oxygen needed for the combustion reactions.
- Water-soluble liquids: Certain flammable liquids such as alcohols are soluble in water. The principal action of water on fires involving these liquids is dilution so that the flash point of the mixture is above the solution temperature.

It should be noted that fires in flammable liquids with a low flash point that are not water-soluble and are lighter than water cannot be extinguished with water.

Oxygen Removal

Many fires are extinguished by the elimination of oxygen from the flammable vapors. For example, flooding the surface of a burning fire with carbon dioxide would constitute placing a barrier between the oxygen in the air and the flammable vapors above the surface of the liquid. Also, fire in many materials that can react violently with water is extinguished by the elimination of oxygen.

Flame Retardants

Fires can be extinguished by the use of flame-retardant agents such as halogenated hydrocarbons, commonly known as halons. Although the exact extinguishing mechanism is not well understood, it is believed that the halogen radical absorbs the chain carrier radicals, and thus stops the chain reactions necessary for the combustion reactions to proceed.

Fire Extinguishers and Extinguishing Methods

Almost all fires are small at the outset. The use of a fire extinguisher that matches the class of fire, by a person who is well trained, can save both lives and property. Portable fire extinguishers must be installed in workplaces regardless of other firefighting measures. The successful performance of a fire extinguisher in a fire situation largely depends on its proper selection, inspection, maintenance, and distribution. Remember that notification of the fire department should not be delayed in the hope that a fire extinguisher will put out the fire.

Materials Used in Fire Extinguishers

Carbon Dioxide. Used extensively to extinguish fires in most flammable liquids, ordinary combustibles, and electrical fires, carbon dioxide can put out fires by forming a barrier between the oxygen and the flammable vapors. Carbon dioxide extinguishers were used extensively during World War I and World War II. By 1950, however, dry chemicals had replaced them for flammable-liquid fires.

Dry Chemicals. The history of dry chemicals as fire extinguishing agents goes back to the early 1800s; in 1928, a cartridge-operated dry chemical extinguisher was developed; and in 1943, an improved, finely granulated agent was produced, which was further improved in 1947. Borax and sodium bicarbonate were among the first dry chemical agents so used. Dry chemicals can be classified into the following groups:

- Regular and ordinary chemicals: These agents are effective on flammable liquid and electrical fires.
- Multipurpose dry chemicals: These chemicals are effective on flammable liquid fires as well as electrical fires.

Dry chemicals can extinguish a fire by a combination of dilution, cooling, radiation shielding, and flame-retardant actions. The term ''dry chemical'' should not be confused with ''dry powder.'' The latter refers to an extinguishing agent suitable for combustible metal fires.

Foam. This fire extinguishing agent usually is used on flammable liquid fires. The amount of foam required for a flammable liquid fire varies widely with conditions of the fire. For small fires, enough foam should be used to cover the surface of the fire a few inches. Foam can be produced by either mechanical or chemical means. Mechanical foam-producing devices inject air into a water-concentrate solution of a foam-producing liquid, and are replacing chemical foam-producing devices for safety reasons. Foam can extinguish a fire by forming a cooling blanket that prevents the transfer of flammable vapors from the surface of the liquid.

Halogenated Agents. Halogenated fire extinguishing agents are hydrocarbons in which one or more atoms of hydrogen have been replaced by halogen atoms. This substitution not only eliminates the flammability characteristics of the hydrocarbon molecule but also imparts a flame-retardant capability to the halogenated compound.

Halogenated compounds used in fire extinguishers are commonly known as halons. There are different types of halons, depending on the type of the hydrocarbon, the number of hydrogen atoms substituted, and the type of the halogen atoms. Halons normally are characterized by a four- or five digit number, such as halon 1301 or halon 10001. The first digit indicates the number of carbon atoms in the parent hydrocarbon molecule; the second digit indicates the number of atoms of fluorine; and the third, fourth, and fifth digits indicate the number of atoms of chlorine, bromine, and iodine respectively. For example, halon 1301 is formed from methane by substitution of three atoms of hydrogen with fluorine and one atom of hydrogen with bromine, and there are no chlorine or iodine atoms in halon 1301. As was mentioned before, halogenated compounds stop combustion reactions by interfering with the progress and development of combustion-intermediate free radicals.

Although halons can be effective on ordinary combustible, flammable-liquid, and electrical fires, the toxicity of their decomposition products must be considered in their selection. For example, when halon 1301 is exposed to temperatures around 900°F, in the presence of hydrogen, it can form hydrogen fluoride and hydrogen bromide gases.

Dry Powder. The term ''dry powder'' is used to describe extinguishing agents suitable for use on combustible metal fires. This type of extinguishing agent should have the capability of removing the heat of combustion without reacting with the metal. Because each combustible metal fire has its own characteristics, there is no single dry powder that can be used on all metals. Use of the wrong kind of dry powder on a metal fire can lead to the release of highly toxic or flammable gases.

Dry powders come in a variety of forms and compositions, each specifically designed for certain metal fires. For example, ''Met–L–X Powder,'' which is composed of sodium chloride with additives such as tricalcium phosphate, metal stearates, and a thermoplastic, is suitable for sodium, potassium, and magnesium fires. This extinguishing agent also has been used successfully against zirconium, uranium, titanium, and aluminum fires.[1]

Classification of Fires and Extinguishers

NFPA has divided fires into four different classes, and different extinguishing agents are required for the different classes of fires.[1]

- Class A includes fires in ordinary combustible materials (such as wood, cloth, paper, rubber, and many plastics). Water can cool this type of fire by bringing the temperature below the ignition point. Dry chemicals can retard combustion, and halogenated agents can interrupt the combustion chain reaction.
- Class B includes fires in flammable or combustible liquids, flammable gases, greases, and similar materials. Although water can be used on these fires, depending on the properties of the fuel, the use of other extinguishing agents such as dry chemicals and halons is recommended.
- Class C includes fires in energized electrical equipment, where the extinguishing agent must be nonconductive to avoid electrical shock. Halons are widely used to extinguish electrical fires. It should be noted that a Class C fire can be extinguished using Class A or B extinguishing agents when the electrical equipment on fire is deenergized.
- Class D includes fires in combustible metals (such as magnesium, titanium, zirconium, sodium, and potassium). The extinguishing agent should absorb heat without reacting with the metal.

Some fire extinguishers can be used on just one class of fire, and some are designated for use on two or three classes. No extinguisher is suitable for all four classes of fire. Thus, extinguishers must be clearly marked as to the class(es) of fire for which they can be used.

Class A and B extinguishers have a numerical rating on their label to indicate the relative effectiveness of the extinguisher. For example, an extinguisher for Class A fires has a numeral that precedes the letter A; this numeral indicates the relative extinguishing capacity of the extinguisher. This means that an extinguisher with an 8-A rating can put out a larger fire than one with a 4-A rating, but not necessarily twice as large a fire; the ratings are only relative.

Class B extinguishers are rated based on the amount of flammable liquid in a flat pan that can be extinguished during a laboratory test. An extinguisher rated 20-B can put out a much larger fire than one rated 5-B.

Dry powder extinguishers do not have a letter or numeral rating because an extinguisher for one metal may not be effective on a different metal fire. These extinguishers carry a plate indicating the type of metal fire against which the extinguisher is effective.

Extinguishers that are effective against more than one class of fire carry multiple letters and numeral ratings to indicate the class of fire and the extinguisher's effectiveness.

Location and Marking of Extinguishers

Fire extinguishers must be installed in locations that provide easy access to the user.[1] In an emergency it is of utmost importance to locate and use fire extinguishers quickly while the fire is still small.

Extinguisher markings should be durable and visible from 3 feet away. If a pictograph marking system is used, the decal should be visible from the front as the extinguisher hangs. The locations of extinguishers should be marked by painting a red rectangle about 8 to 10 feet above them. It is also good practice for the surface on which it is mounted to be painted red.

Selection, Operation, and Distribution of Fire Extinguishers

Selection

Before an extinguisher is selected, several points must be considered: (1) the type of combustibles present, (2) who will use the extinguisher, (3) the location of the extinguisher, and (4) the type of chemicals present and their possible reaction with the extinguishing agent. Also, one must determine whether the extinguishing agent is effective on the specific hazards present, and whether the extinguisher's ease of operation and maintenance requirements are reasonable.[1]

Operation and Use

The effectiveness of an extinguisher depends on the training of the individual using it. Because many extinguishers discharge their contents in about 8 to 15 seconds, it is important that personnel be trained on the use of different types of extinguishers. Information on the use and operation of different types of fire extinguishers can be found in the NFPA handbook.[1]

Distribution of Fire Extinguishers

Extinguishers must be distributed is such a way that the amount of time needed to travel to their location and back to the fire does not allow the fire to get out of control. OSHA requires that the travel distance for Class A extinguishers not exceed 75 feet. The maximum travel distance for Class B extinguishers in 50 feet because flammable liquid fires can get out of control faster than Class A fires. If a Class B extinguisher has a low rating, the travel distance must be reduced accordingly. There is no maximum travel distance specified for Class C extinguishers, but they must be distributed on the basis of appropriate patterns for Class A and B hazards. The maximum travel distance for Class D fire extinguishers should not exceed 75 feet.[1]

Inspection and Maintenance of Fire Extinguishers

Once an extinguisher is selected, purchased, and installed, it is the responsibility of the owner to inspect, maintain, and test it to ensure that it is in proper working condition.

- Inspection: A visual inspection of fire extinguishers must be conducted monthly to ensure that the extinguisher (1) is in its designated place, (2) is visible, (3) is not blocked, (4) is filled with the extinguishing agent, (5) has not been tampered with, and (6) does not have any signs of physical damage. The maintenance tag should be checked to determine the date of the last maintenance check.
- Maintenance: Maintenance involves a complete and thorough examination of each extinguisher. A maintenance check includes disassembling the extinguisher; examining all its parts; cleaning and replacing any defective parts; and reassembling, recharging, and, where appropriate, repressurizing the extinguisher.

Table 5.11. Hydrostatic Test Intervals for Extinguishers.

Extinguisher Type	Test Interval (years)
Soda-acid	5
Cartridge-operated water and/or antifreeze	5
Stored pressure water and/or antifreeze	5
Wetting agent	5
Foam (stainless steel)	5
Aqueous film-forming foam (AFFF)	5
Dry chemical with stainless steel shell	5
Carbon dioxide	5
Dry chemical, stored pressure, with mild steel, brazed-brass, or aluminum shell	12
Dry chemical, cartridge- or cylinder-operated with mild steel shells	12
Halon 1211	12
Halon 1301	12
Dry powder, cartridge- or cylinder-operated, with mild steel shells	12

Source: Reference 15.

- Hydrostatic testing: Hydrostatic testing of portable fire extinguishers is required to detect corrosion, physical damage, improper assembly of parts, and damage to the mechanical properties of metal parts. OSHA requires that hydrostatic testing of extinguishers be conducted according to the schedule outlined in Table 5.11.

References

1. NFPA, *Fire Protection Handbook*, 15th Ed., NFPA, Quincy, Mass. (1981).
2. Marchant, E. W., *Fire and Buildings*, 1st Ed., Harper & Row Publishers, New York, pp. 18–19 (1973).
3. Meidl, J. H., *Explosive and Toxic Hazardous Materials*, Glencoe Press, Beverly Hills, Calif. (1970).
4. Schultz, N., *Fire and Flammability Handbook*, 1st Ed., Van Nostrand Reinhold Co., New York (1985).
5. Kern, D. Q., *Process Heat Transfer*, 1st Ed., McGraw-Hill Book Co., New York (1950).
6. Moore, G. R., and D. E. Kline, *Properties and Processing of Polymers for Engineers*, 1st Ed., Prentice-Hall, Englewood Cliffs, N.J.
7. Bugbee, P., *Principles of Fire Protection*, National Fire Protection Association, Quincy, Mass. (1984).
8. Bland, W. F., and R. L. Davidson, *Petroleum Processing Handbook*, 1st Ed., McGraw-Hill Book Co., New York (1967).
9. Marchant, E. W., *Fire and Buildings*, 1st Ed., Harper & Row Publishers, New York (1973).
10. Felder, R. M., and R. W. Rousseau, *Elementary Principles of Chemical Processes*, 1st Ed., John Wiley & Sons, New York (1978).
11. Whitman, L. E., 1979, *Fire Prevention*, 1st Ed., Nelson Hall, Chicago (1979).
12. Bugbee, *Principles of Fire Protecton* (see reference 7).
13. The International Technical Information Institute (ITI), *Toxic and Hazardous Industrial Chemicals Safety Manual*, 11th Ed., ITI, Tokyo (1986).
14. NFPA, *Fire Protection Guide on Hazardous Materials*, 8th Ed., NFPA, Quincy, Mass. (1984).
15. Office of the Federal Register, *Code of Federal Regulations*, 1910.157, Washington, D.C. (1985).

Exercises

1. How can heat of combustion be calculated from molecular bond energies of reactants and products of combustions?

2. In a refinery, a gasoline pump that is on fire has a temperature of 1,700°F and is radiating heat to a nearby flammable storage tank. How much must the fire temperature be lowered to reduce heat transfer by radiation by a factor of 10?
3. Define the convective mechanism of heat transfer, and distinguish between natural and forced convection. Describe how convective heat transfer contributes to the spread of fires.
4. A flammable liquid is stored in a 55-gallon drum having a surface temperature of 80°F and a thickness of 0.5 inch. The liquid has a flash point of 75°F, and the drum is made of carbon steel with a thermal conductivity of 39.0 Btu/hr-ft-°F. What is the maximum allowable heat transfer rate by conduction to keep the temperature of the liquid at 5°F below its flash point?
5. What are the six zones associated with a combustion reaction or fire?
6. A flammable liquid that has a heat of combustion of 8,000 Btu/lb is consumed by a fire at a rate of 2 lb/min. The total amount of heat transfer by radiation and convection is 10,000 Btu/min, and the heat of vaporization of the flammable liquid is 5,000 Btu/lb. What is the minimum amount of heat that must be removed by an extinguishing agent in order to stop this fire?
7. What is flash point, and how would it affect the ventilation and handling requirements for flammable liquids?
8. A flammable liquid has a vapor pressure of 181 mm Hg at its flash point temperature. Calculate the lower flammability limit for the vapor generated from the liquid.
9. Describe how specific gravity and solubility data can affect means of firefighting for flammable liquids.
10. Discuss the effects of temperature and pressure on the evaporation rate of liquids.
11. Discuss the effect of pressure on the boiling point of a flammable liquid, and describe how atmospheric pressure changes can affect the safe handling and storage of flammable liquids.
12. Describe how viscosity can affect the generation of static charges in flammable and combustible liquids.
13. What are the basic safeguards for fire safety in dealing with flammable liquids?
14. What is the definition of a flammable gas?
15. What is the difference between a flammable gas and a flammable vapor? How do differences in the properties of flammable gases and vapors affect fire safety?
16. In a flammable gas storage area the valve on a cylinder of hydrogen that is under 2,500 psig pressure breaks, releasing hydrogen into the storage room. The dimensions of the room are 10 ft × 15 ft × 20 ft, and the hydrogen gas cylinder has a volume of 100 ft^3. Determine whether the contents of the room are flammable.
17. Twelve pounds of carbon is burned in a room that contains 3 lb moles of air. Assuming that 50% of the carbon forms carbon dioxide, and the other half forms carbon monoxide, determine whether the air in this room can support life. Note that the concentrations of both carbon monoxide and oxygen must be taken into account. Assume atmosphere with less than 19.5% oxygen cannot support life.
18. List at least three gases that can be produced during fires, and discuss their effects on safety.
19. Under what circumstances can water be used on a flammable liquid fire?
20. Describe how halogenated hydrocarbons extinguish a fire. What are the chemical structures of halon 1301 and halon 1211?
21. What factors must be considered before selection of a fire extinguisher?
22. What type of information must a visual inspection of fire extinguishers provide?

Chapter

6

Industrial Noise Control

Noise can cause hearing loss, create physical and psychological stress, and contribute to accidents by making it difficult to hear warning signals. An estimated 14 million U.S. workers are exposed to hazardous noise. Hazardous occupational noise exposures can be controlled, however, by engineering or administrative control measures such as reducing exposure time, use of quieter processes, equipment enclosure, and use of sound-absorbing materials.[1–4]

In order to understand and solve industrial noise problems, one must have a basic understanding of the theory and terminology of acoustics and noise control.

Sound is a wave phenomenon, produced as a result of small pressure variations in the atmosphere. These pressure variations may be caused by an object oscillating in a conducting medium such as air, with the conducting medium responsible for the transfer of vibrational energy. For example, air is capable of transferring the vibrational energy of sound at a speed of 1,128 ft/sec at 68°F. The propagation of sound waves is similar to transmission of water waves due to a disturbance created on the water surface.

Basic Terminology

An understanding of the following terms is essential to noise control.[1–3]

Amplitude. The amplitude refers to the maximum value of a sinusoidal wave, as shown in Figure 6.1. The amplitude of sound may be described in terms of either the quantity of sound produced at a given location away from the source or the overall ability of the source to emit sound. The amount of sound at a location away from the source generally is described by the sound pressure or sound intensity, whereas the ability of the source to emit sound is described by the sound cover of the source.

Frequency. The frequency of sound describes the rate at which complete cycles of high and low pressure regions are produced by the sound source. The unit of frequency is cycles per second (cps), also called hertz (Hz). The frequency range of the human ear is highly dependent upon the individual and the sound level. A normal-hearing young ear has a frequency range of approximately 20 to 20,000 cps at moderate sound levels. The frequency of a propagated sound wave heard by a listener will be the same as the frequency of the vibrating source if the distance between the source and the listener remains constant; however, the frequency detected by a listener will increase or decrease as the distance from the source is decreasing or increasing.

Speed of Sound. The speed of sound is defined as the distance through which a sound wave travels per second. The speed with which a sound wave can travel is a function of the type and physical conditions of the transferring medium. For example, the speed of sound in air, which is 1,128 feet per second at 68°F, increases with an increase in temperature. For gases other than

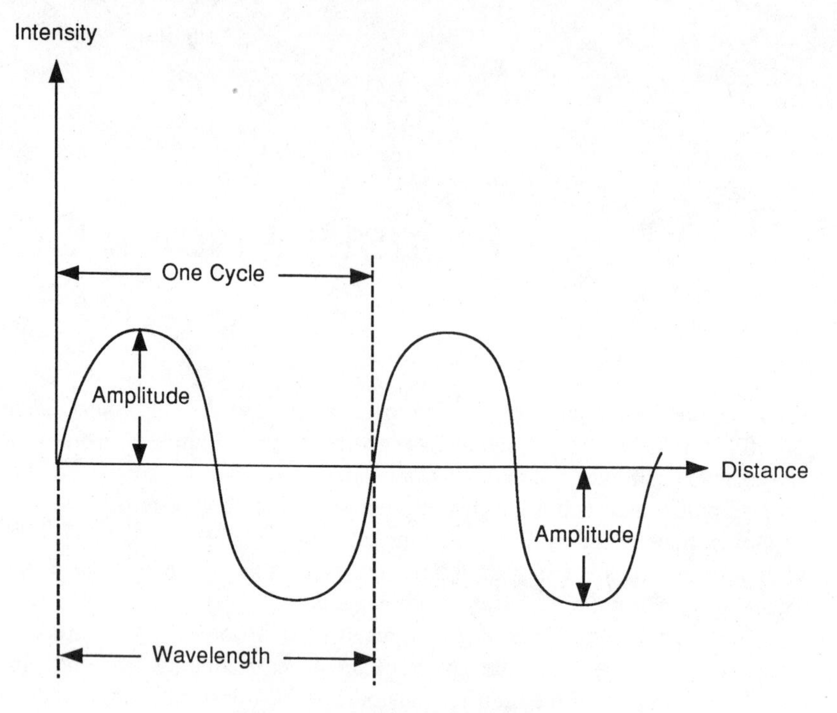

Figure 6.1. A sinusoidal wave.

air, the speed of sound can be calculated from the following relationship:

$$V = 233 \frac{C_p}{C_v} T/M$$

where:

V = Speed of sound, ft/sec
C_p/C_v = Ratio of heat capacity at constant pressure to heat capacity at constant volume, dimensionless
T = Absolute temperature, degrees Rankine (°F + 460)
M = Molecular weight of gas, Daltons

Wavelength. The sound wavelength is defined as the distance required to complete one wave cycle (see Figure 6.1). The relationship between the wavelength, frequency, and speed of sound can be described by the following expression:

$$F = \frac{V}{\lambda}$$

where F is the frequency, V is the speed, and λ is the wavelength. Figure 6.2 shows the relationship between frequency and wavelength for sound in air at normal temperature and pressure.[1]

Loudness. The impression of the sound amplitude to an observer is called the sound loudness, which is a function of the characteristics of the listener's ear.

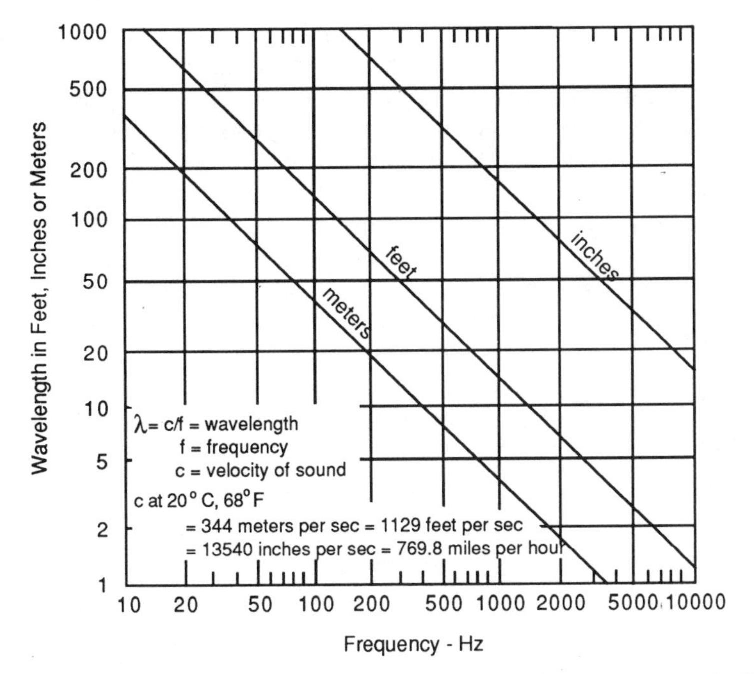

Figure 6.2. Relationship between frequency and wavelength for sound in air (Source: Reference 1)

Sound Intensity. The intensity of sound is defined as the amount of energy transmitted by the source per unit time through a unit area perpendicular to the direction of sound propagation. Intensity usually is expressed in units of joules per second per square meter (joules/sec/m^2).

Sound Power. The amount of total energy transmitted by a sound wave per unit time is called the sound power of the source. The sound power can be calculated from sound intensity using the following expression:

$$P = I \cdot A \cos \alpha$$

where:

P = Sound power, energy per time
I = Sound intensity, energy per time per area
A = Area
α = Angle of surface area with vertical axis

Sound Pressure. As air particles vibrate, momentary small fluctuations occur in the atmospheric pressure. These are the pressure changes that our ears detect as sound, or to which a microphone responds. The sound pressure changes are alternately positive and negative, with respect to atmospheric pressure, as the air is compressed and rarified.

In order to quantify sound behavior and effects, we must be able to apply numbers to the pressure changes. The best quantity to use is the average pressure. However, if we tried to average the sound pressure changes occurring at a particular point and over a particular time interval, we would find the average always equal to atmospheric pressure because all the positive pressure fluctuations are exactly counterbalanced by the negative ones. Thus, in place of a simple average, the instantaneous pressures first are squared, and then square roots are taken, before the average is determined. This procedure thus assigns a *positive*-valued quantity to a sound pressure, which is called the root-mean-square (rms) value of the sound pressure.

A very weak sound may have an rms sound pressure that is very small compared to atmospheric pressure; in fact, the rms sound pressure of a barely audible sound at 1,000 Hz (in the frequency region where humans hear best), in a very quiet environment, is about 0.0000000002 or 2×10^{-10} atmosphere, obviously a small pressure. A very loud sound could have an rms sound pressure of about 0.001 atmosphere.

Noise and Sound. The terms "noise" and "sound" are often used interchangeably, but generally sound is descriptive of useful communication or pleasant sounds such as music, whereas noise is used to describe discord or unwanted sound.

Period. The amount of time required for a complete wave cycle to take place is called the period. Frequency is the number of cycles per unit time, and period is the reciprocal of the frequency, normally measured in seconds.

$$T = \frac{1}{F}$$

where T is the period and F is the frequency.

Pure Tone. A pure tone refers to a sound wave with a single, simple sinusoidal change of level with time.

Random Noise. This type of noise is made up of components of many frequencies, whose instantaneous amplitudes occur randomly as a function of time.

Decibels. To simplify the sound pressure numbers while relating them to a meaningful scale, rms sound pressures are quoted in terms of decibels. (A meaningful scale is one that bears some relation to the apparent "loudness" of the noise.) Decibels, which are logarithmic values, are based on a reference starting point. The starting point, 0 decibel, is the rms sound pressure corresponding to the weakest audible sound (0.0000000002 atmosphere—the weakest sound that can be heard by a large proportion of people. All subsequent sound pressures are rms sound pressures and are referred to that standard reference pressure.

The decibel (abbreviated dB) is the unit for expressing the sound pressure level relative to 2×10^{-10} atmosphere. In the metric system, this reference pressure is 2×10^{-5} Newton/m^2. The unit termed the "pascal" is defined as 1 N/m^2; so the sound pressure level reference is expressed as 2×10^{-5} pascal or 20 micropascals. Thus, to be technically correct, one should say, "The sound pressure level is 75 decibels relative to 20 micropascals." As this is a universally recognized pressure base, it often is not stated, however, and one usually says, "The sound pressure level is 75 dB."

The word "level" is used to designate that the rms pressure is relative to the universal base sound pressure. The sound pressure level (SPL) for any measured sound is defined by:

$$\text{SPL (in decibels)} = 10 \log \frac{(\text{rms sound pressure measured})^2}{(20 \text{ micropascals})^2}$$

or:

$$= 20 \log \frac{(\text{rms sound pressure measured})}{(20 \text{ micropascals})}$$

In practice, a sound level meter is calibrated to read decibels relative to 20 micropascals; so a person is seldom aware of the rms pressure of the actual sound (that is, how many millionths of an atmosphere it is, or how many Newtons/m^2, or lb/in^2, or dynes/cm^2). Yet one is aware that very quiet sounds (a quiet whisper or the rustling of grass in a very slight breeze) may range from 10 to 20 dB, and very loud sounds (a nearby diesel truck, an overhead aircraft shortly after takeoff, or a loud clap of thunder) may range from 85 to over 130 dB. Instantaneous sound pressure levels of 160 dB can rupture the eardrum, and the risk of permanent hearing impairment increases as a function of sound levels above 80 dB.

Frequency Weighting. Anyone involved in noise control quickly learns a basic concept: the human response to sound is frequency-dependent. Humans hear best at frequencies around 500 to 5,000 Hz, for example, and, perhaps for this reason, they are most annoyed or disturbed by noise in that range. In addition, high sound levels and long exposure times to sounds in this same frequency range contribute to hearing loss. These considerations have ramifications on the effects of sound, so there is usually a need to know about the frequency distribution contained within a given sound being investigated, as well as a need to pay special attention to those frequencies having the greatest effects.

The typical sound level meter has three different frequency-weighting networks, identified as the A, B, and C scale networks. Their frequency responses are given in Figure 6.3. Extensive studies have shown that the high-frequency noise passed by the A-weighting network correlates well with human annoyance effects and hearing damage effects of the noise. Consequently, sound pressure levels, as measured with the A-scale filter, are used in various rating systems for judging the annoyance of noise and for evaluating the hearing damage potential of high sound levels and exposures. When the sound level meter is switched to the "A" position, the meter gives a single-number reading that adjusts the incoming noise at the microphone in accordance with this filter

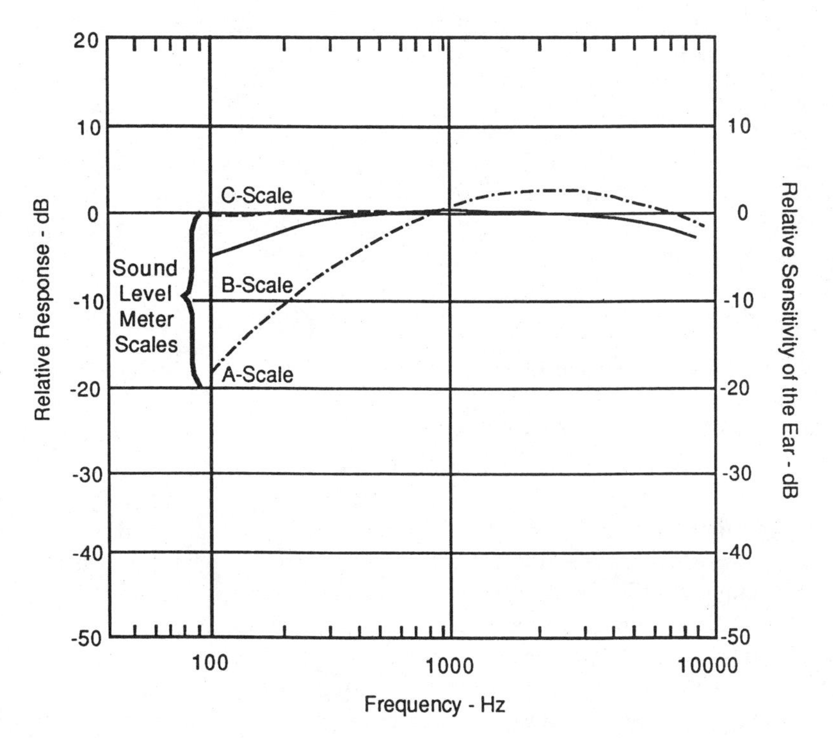

Figure 6.3. A, B, C scale sound frequency response. (Source: Reference 1)

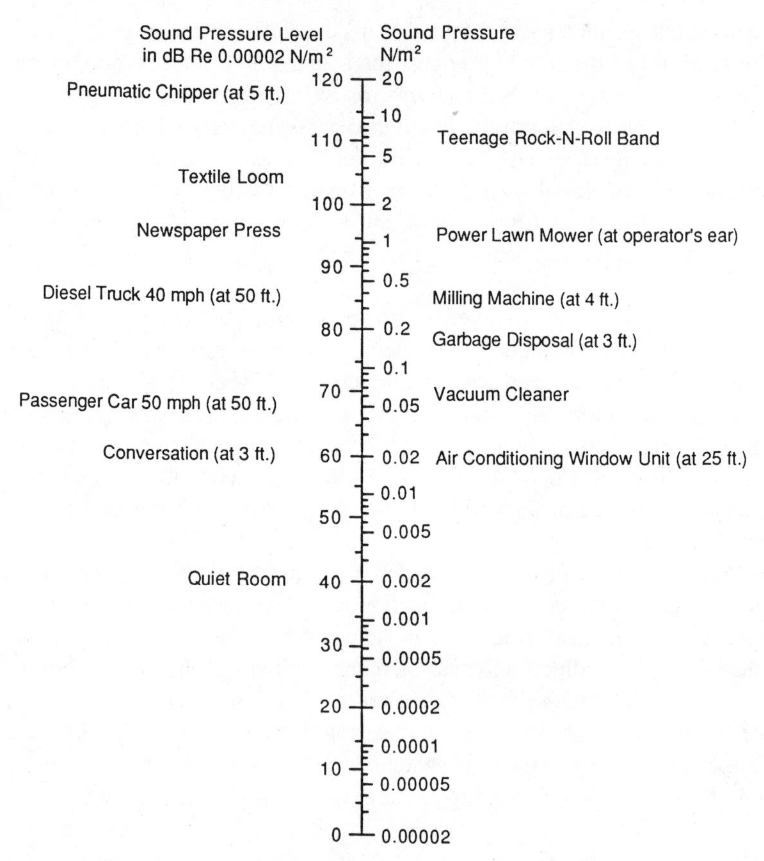

Figure 6.4. Relationship between A-weighted sound pressure levels in decibels and sound pressure in Newtons per square meter. (Source: Reference 3)

response and then indicates a numerical value of the total sound passed by this filter. The resulting value, called the A-weighted sound level, is expressed in units designated dBA.

Figure 6.4 shows the relationship between A-weighted sound pressure levels in decibels and sound pressure in Newtons per square meter (Newtons/m^2).

OSHA Regulations on Noise Control and Hearing Protection

Under the Occupational Safety and Health Act, every employer is legally responsible for providing a workplace free of excessive noise.

A final rule was released by the Department of Labor's Occupational Safety and Health Administration (OSHA) on March 8, 1983, entitled "Occupational Noise Exposure; Hearing Conservation Amendment." The amendment expands and supplements the requirements contained in 29 CFR 1910.95, the Occupational Noise Exposure Standard, by adding and revising paragraphs (c) through (p) and appendixes A through I. The amendment states that each employer shall administer a continuing, effective hearing conservation program whenever employee noise exposures equal or exceed an 8-hour time-weighted average (TWA) of 85 decibels on the A scale (slow response). These levels are referred to in the amendment as the action level.[5]

In many plant situations sound levels may vary during the day, so there is a need to account for time-varying noise in determining hazardous noise exposure. OSHA requires that exposure to time-varying noise be determined using the noise-dose concept. Each fractional noise dose can

Table 6.1. Exposure times allowed by OSHA for different sound levels.

Hours of Exposure	Sound Level dB(A)
8	90
6	92
4	95
3	97
2	100
$1\frac{1}{2}$	102
1	105
$\frac{1}{2}$	110
$\frac{1}{4}$ or less	115

be calculated by dividing the exposure time by the allowed time for a given sound level. The allowed times for different sound levels are summarized in Table 6.1.

OSHA standards establish limits on workplace noise exposure for given time periods. The action level for implementing a hearing conservation program is 85 dB(A) based on TWA, which requires a monitored audiometry program as a minimum. Increased exposures over time require more stringent requirements.

For sound levels not listed in the table, the following expression can be used to calculate the allowed time:

$$t = \frac{480}{2^{0.2(L_A - 90)}}$$

where t is the allowed time in hours, and L_A is the actual A-weighted sound level in dBA. Once the partial doses have been calculated, the total noise dose for the day can be calculated by summing all the partial doses:

$$D = \sum_{i=1}^{n} \frac{C_i}{t_i}$$

where D is the total noise dose, C is the exposure time, and t is the allowed time.

Table 6.2 demonstrates conversion from percent noise exposure or dose to the 8-hour time-weighted average sound level (TWA).

Monitoring

The monitoring section of the regulation states that whenever information is available that any employee's noise exposure may equal or exceed an 8-hour time-weighted average of 85 dB, then the employer shall develop and implement a noise monitoring program. The intent of the monitoring program is to identify employees for inclusion in the hearing conservation program and to enable the proper selection of hearing protectors. Either area sampling procedures or personal sampling procedures may be used to determine noise levels. Where circumstances such as high worker mobility, significant variations in sound level, or a significant component of impulse noise make area monitoring generally inappropriate, then the employer shall use representative personal sampling to comply with the monitoring requirements.

The regulation requires all noise from the range of 80 to 130 decibels to be measured, whether it is continuous, intermittent, or impulsive. All instruments used to measure employee noise exposure shall be calibrated to ensure measurement accuracy.

Table 6.2. Conversion from percent noise exposure to 8-hour time-weighted average sound level.

Dose or Percent Noise Exposure	TWA
10	73.4
15	76.3
20	78.4
25	80.0
30	81.3
35	82.4
40	83.4
45	84.2
50	85.0
55	85.7
60	86.3
65	86.9
70	87.4
75	87.9
80	88.4
85	88.8
90	89.2
95	89.6
100	90.0
150	92.9
200	95.0
300	97.9
400	100.0
500	101.6
600	102.9
700	104.0
800	105.0
900	105.8
999	106.6

Noise monitoring must be repeated whenever a change in production, process, equipment, or controls increases noise exposure to the extent that additional employees may be exposed at or above the action level.

Employers are required to notify each employee who is exposed at or above an 8-hour time-weighted average of 85 decibels of the results of the monitoring that has been accomplished, and employees have the right to observe the noise measurements thus obtained. Employees also have the right to observe noise measurements being conducted and must be notified of those areas exceeding the action level. Once a comprehensive noise measurement survey has been conducted, it need only be repeated when there has been a change in production, equipment, or controls to the extent that additional employees may be exposed at or above the action level, or the hearing protectors being used by employees are not adequate.

Audiometric Testing Program

Annual audiometric evaluations for situations that exceed the action level must be performed on all employees. The regulation indicates that the program shall be provided at no cost to employees and that specific criteria must be met in regard to who does the testing, the type of equipment, the calibration procedures, and responsibility for the overall program. New employees must be tested within six months of their first exposure at or above the action level.

Where industry utilizes mobile test van operations, the initial (baseline) audiogram can be obtained within one year of an employee's first exposure, provided that adequate hearing protection has been worn beyond the initial six months of the employee's first exposure. Prior to a baseline evaluation, employees must be free from exposure to workplace noise for at least 14 hours. In addition, employees should be told to avoid excessive noise exposures outside the workplace during the 14-hour quiet period.

Employees included in the audiometric monitoring must have their hearing checked annually, and it must be compared to the baseline evaluation. If employees show a standard thereshold shift as defined in the regulation, or a professional reviewer such as an audiologist or other specialist has determined a need for further evaluation, then the employer shall refer the employee for that evaluation as appropriate. Referral also can be made if the employer suspects a medical pathology of the ear is caused or aggravated by the wearing of hearing protectors. The regulation also indicates that the employee must be informed of test results and recommendations. The employer is encouraged to review the use of hearing protectors with employees, particularly if changes in hearing are determined to be a result of occupational noise exposures. Considerations and adjustments can also be made if an employee's hearing has improved.

Hearing Protectors

Employers must make hearing protectors available to all employees exposed to the action level of 85 dBA (TWA). Hearing protection is mandatory for any employee who is exposed to at least 90 dBA (TWA), or whose hearing has shifted, even if the employee is exposed to lesser levels of noise. It is the employer's responsibility to make sure that employees comply and to provide special training in the care and use of the protectors, as well as provide proper sizing and fitting procedures for their employees. The protectors must attenuate the employee's exposure to at least 90 dBA over an 8-hour time-weighted average. If employees still exhibit changes in hearing as a result of occupational noise, then the protectors should attenuate employee exposure to levels of 85 dBA or below.

Training Program

The employer must provide a training program for all employees exposed to the action level. Emphasis should be placed on the intent of the hearing conservation program as well as the effect of noise on hearing and how hearing loss can be prevented.

Recordkeeping

The employer is responsible for maintaining an accurate record of all employee exposures to hazardous noise as well as a record of all employee audiometric tests. The audiometric test records must, at a minimum, contain the following:

- Name and job classification of the employee
- The examiner's name
- Date of last calibration of the testing instrument
- Employee's most recent noise exposure assessment
- Record of background sound levels in testing rooms

Noise Measurements

Noise measurements normally are conducted for either compliance or diagnostic purposes. Compliance-type measurements must be conducted according to specific instructions set forth in laws and regulations. Diagnostic noise measurements, on the other hand, are taken to identify engineering or administrative control procedures to rectify hazardous noise problems.[1,2,6]

Sound Measuring Instruments

Noise measurement instruments are classified into sound level meters (SLM), integrating sound level meters (ISLM), and spectrum analyzers (SA).[6] The SLM, which is the most common instrument for the measurement of sound levels, can be Type 1 for precision measurements or Type 2 for general-purpose use. The ISLM instruments are designed for situations where there is a need to measure the average sound levels over a period of time. Spectrum analyzers identify the distribution of sound pressure levels in frequency.

Sound Level Meters (SLM)

The chief instrument for noise measurements is the sound level meter (SLM),[6] either Type 1 (precision) or Type 2 (general purpose), made in accordance with American National Standard S1.4 (1971), Specification for Sound Level Meters. The Type 2 instrument has broader tolerances on performance than the Type 1 instrument and is acceptable under the OSHA occupational noise exposure regulations. It usually is less bulky, lighter, and less expensive than the Type 1 SLM. A sound level meter typically consists of a microphone, a calibrated attenuator, a stabilized amplifier, an indicating meter, and the designated weighting networks. Figure 6.5 shows a typical sound level meter.

The microphone converts the sound energy into electrical signals, which are amplified and modified in such a way that they can be detected on an indicating meter. The SLM provides the capability of detecting a wide range of sound levels by use of a control device called an attenuator. This device can adjust the amplification of the signals received, usually in steps of 10 dB.

The amplifier in an SLM should have the capability of amplifying low-level sounds over a wide

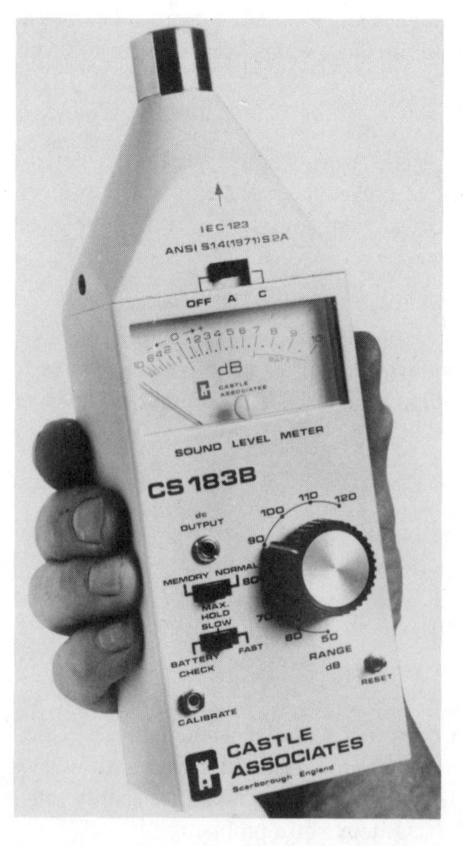

Figure 6.5. Sound level meter (SLM). (Source: Castle Associates)

frequency range. It also must minimize noise generation within the instrument to prevent interference with the sound level being measured.

The weighting network of an SLM can measure the response on either A-weighted, B-weighted, or C-weighted scales, depending on the sound frequency. The most commonly used of the three weighting scales is A-weighting. The A-weighted response is at a maximum at about 2,500 Hz, drops sharply below 1,000 Hz, and drops very rapidly at frequencies above 4,000 Hz. The B-weighted response is at a maximum at about 1,000 Hz and drops rapidly below 500 Hz. The B-weighted response is rarely used in sound level measurements. The C-weighted response is designed for frequencies in the range of 30 to 8,000 Hz.[6]

The indicating meter on an SLM, which is calibrated in decibels, receives an amplified and weighted signal from the weighting network and attenuators. The root-mean-square (rms) values for different components of the noise signal then are calculated. The logarithms of the average of the squares of sound pressures are taken for reporting sound levels in decibels (relative to 20 micropascals) (see above, section on "Basic Terminology").

All SLMs are sensitive to rough handling and should be treated with care. Microphones, especially, are subject to damage if mishandled. Instruction booklets provided with the units should be read carefully to determine how the instrument should be operated and under what conditions the readings are valid. The user should learn how to determine when battery power is too low, and how to ensure that the instrument is reading the sound environment and not internal electrical noise or an overloaded condition.

Some general advice about the use of sound level meters is summarized below:

- Wind or air currents can cause false readings. Use a wind screen with the microphone for any measurements when you can feel a wind or air current.
- Vibration of the meter can distort readings. Do not hold the meter directly against a vibrating machine.
- Magnetic distortion of the meter from adjacent power equipment can cause problems. Magnetic fields usually drop off quickly with distance from a motor or transformer. Move the SLM far enough away from the electric-magnetic equipment to be sure that the needle reading is attributable to the acoustic signal.
- Barriers or walls can obstruct sound and reduce sound levels or, by reflection, can increase sound levels. Avoid measurement positions where barriers or walls can alter the sound field, unless the position is clearly at the normal location of the operator.
- Avoid dropping the meter when it is hand-held; keep the safety cord wrapped around your wrist.

Integrating Meters

Two types of noise integrating meters are widely used: noise dosimeters and sound level monitors.

The noise dosimeter is capable of integrating sound levels and providing a measure of the fraction of the allowable daily dose. The integration of sound levels can be performed over a specified period of time. Because both sound level and exposure time are critical factors in occupational noise control, most dosimeters take both factors into account. For example, most dosimeters used in the United States provide for halving exposure time for every 5 dB(A) increase in sound level. This means that the dosimeter would show the same reading for a sound level of 80 dB(A) for 8 hours and a sound level of 85 dB(A) for 4 hours.

Sound level monitors are designed to provide the equivalent continuous sound level over a specified period of time. Some instruments can be programmed to take into account such factors as day–night average and community noise equivalent levels.

Spectrum Analyzers

An important factor in noise measurement is determination of noise frequency distribution, which is accomplished by the use of spectrum analyzers. When a sound signal is received by the instrument, it travels through different filtering media. Each filter medium is capable of separating a range of frequencies, called its band width. Although the ideal filter should have no response for frequencies outside its band width, actual filters can only approximate this ideal.[6]

Sound-Measuring Techniques

When the sound levels are known to change very little throughout the working day, a simple SLM reading suffices for characterizing the noise environment. However, the reading must be taken properly. The standard procedure is to locate the microphone at the ear position of concern, but with the worker at least 1 meter away. This is the free-field measurement that is preferred in American National Standard S1.13-1971, Methods for the Measurement of Sound Pressure Levels. For a general standing position, the preferred microphone height is 1.5 meters; for a seated worker, 1.1 meters.

When it is necessary to make sound measurements that will withstand scrutiny in the courts, several criteria are important:

- Measurements should be conducted by a qualified individual.
- The instruments and procedures must conform to ANSI standards.
- Instruments should be properly calibrated

Obtaining reliable data depends on periodic calibration of the instruments. The preferred calibrators deliver an acoustical signal of known frequency and sound pressure level. Some calibrators provide a variety of signals of different frequencies and levels. To ensure that the calibrators are correct, it is advisable to own two units, to make frequent intercomparisons of both units on the same sound level meter, and, annually, to have one of the calibrators recalibrated by the manufacturer or a reliable instrument laboratory, requiring that the calibration be traceable to the National Bureau of Standards.

To have minimum interference from the body of the observer, position the microphone at least 1 meter away from the observer, and position the observer to the side of the microphone (relative to the source of sound).

Generally, you first should explore the region of interest before obtaining the final sound level for compliance measurements. Directional effects sometimes can change the reading a few decibels in a short distance. One example is a noise source partially shielded by a machine structure, with the operator in and out of the acoustical shadow. Several readings may be needed to delineate the noise completely in the range of positions used by the worker in question.

It should be noted that certain limitations in noise measurement make assessment of the marginal situations difficult; these limitations include:

- Accuracy of instruments.
- Instrument performance differences: Two different instruments, both meeting laboratory standards for their response, may read field-encountered sounds differently.
- Representativeness of the exposure: Perhaps this is the most significant factor affecting variation in readings. Daily noise exposure patterns can vary significantly from day to day. This variation would be especially true in job-shop-type operations.

Decibel Addition

In many industrial situations employees may become exposed to noise from different sources. Under these circumstances the combined effect of noise from these sources must be taken into account for compliance and control measures. The following expression can be used to calculate the combined sound level from two random sources:

$$L_c = L_1 + 10 \log 10^{(L_2 - L_1)/10} + 1$$

where:

L_c = The combined sound level in decibels
L_1, L_2 = Individual sound levels in decibels

When calculations are performed according to the above equation, they normally agree to within 1 dB of the measured value on a sound meter. For example, consider two machines, each generating a noise level of 89 dBA. The combined noise level calculated from the above equation is 92 dBA, and we write:

$$89 \text{ dBA} + 89 \text{ dBA} = 92 \text{ dBA}$$

Table 6.3 can be used for decibel addition without the use of the above expression. An alternative method of decibel addition relies on a few simple rules:

1. When two decibel levels are equal or within 1 dB of each other, their sum is 3 dB higher than the higher individual level. For example, 80 dBA + 80 dBA = 83 dBA; 100 dB + 101 dB = 104 dB.
2. When two decibel levels are 2 or 3 dB apart, their sum is 2 dB higher than the higher individual level. For example, 87 dBA + 85 dBA = 89 dBA; 75 dBA + 77 dBA = 79 dBA.
3. When two decibel levels are 4 to 9 dB apart, their sum is 1 dB higher than the higher individual level. For example, 80 dBA + 86 dBA = 87 dBA; 32 dB + 36 dB = 37 dB.
4. When two decibel levels are 10 or more dB apart, their sum is the same as the higher individual level. For example, 80 dB + 95 dB = 95 dB.

Table 6.3. Obtaining decibel sum of two decibel levels.

Difference between Two Decibel Levels to be Added (dB)	Amount to be Added to Larger Level to Obtain Decibel Sum (dB)
0	3.0
1	2.6
2	2.1
3	1.8
4	1.4
5	1.2
6	1.0
7	0.8
8	0.6
9	0.5
10	0.4
11	0.3
12	0.2

Source: Reference 1.

In situations where several sound levels have to be added to obtain the combined effect, the following procedure can be used:

1. Add the two lowest decibel levels.
2. Add the result of step 1 to the next higher level.
3. Continue until all decibel levels have been added.

For example, assume that five machines are operating in a machine shop, and each is generating a sound level as summarized below:

Machine 1: 79 dBA
Machine 2: 85 dBA
Machine 3: 87 dBA
Machine 4: 95 dBA
Machine 5: 93 dBA

In order to calculate the combined sound level, we proceed as follows:

$$79 \text{ dBA} + 85 \text{ dBA} = 86 \text{ dBA}$$

$$86 \text{ dBA} + 87 \text{ dBA} = 90 \text{ dBA}$$

$$90 \text{ dBA} + 93 \text{ dBA} = 95 \text{ dBA}$$

$$95 \text{ dBA} + 95 \text{ dBA} = 95 \text{ dBA}$$

Therefore, the combined effect from all five machines operating simultaneously is equivalent to a sound level of 98 dBA.

Figure 6.6 depicts how several sound levels can be added to obtain the combined sound level.

Physiology of Hearing and Noise Effects in Humans

Before studying the various techniques used for occupational noise control, one must have a basic understanding of the physiology of hearing and the adverse effects of noise on hearing and health.[3,6,7]

The basic function of the human ear is to gather, conduct, and perceive sound waves from the

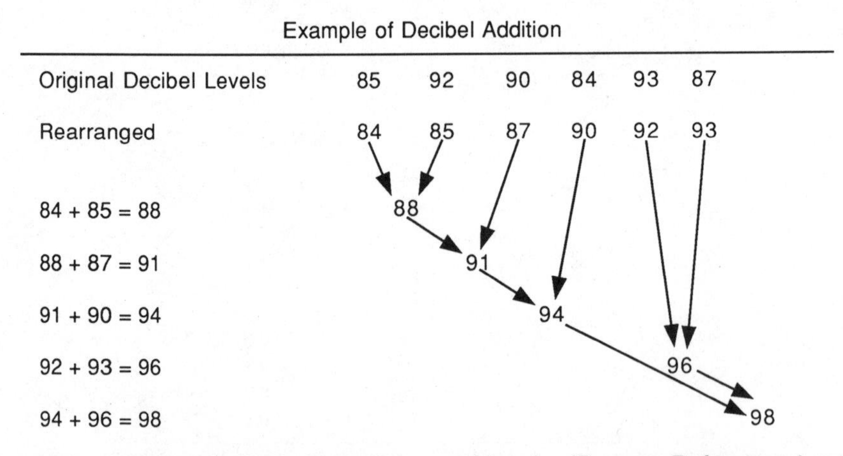

Figure 6.6. Combined effect of several sound levels. (Source: Reference 1 and 3)

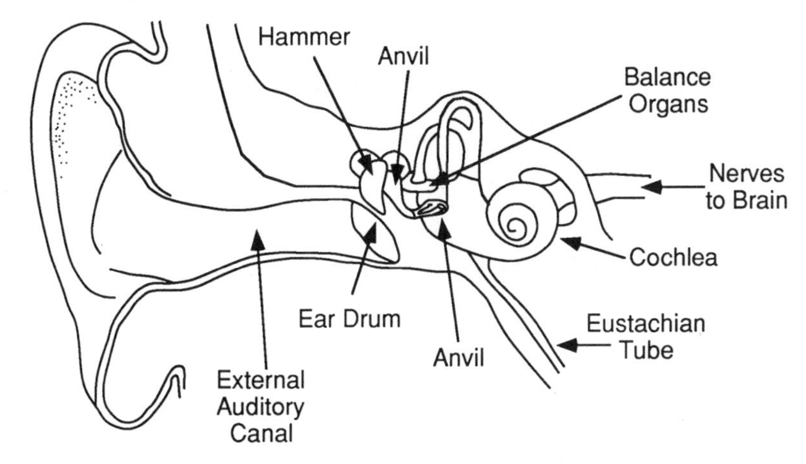

Figure 6.7. Schematic diagram of human ear.

environment.[3] Although the frequency range of the human ear can range from 16 to 30,000 Hz, few adults can sense sounds that exceed 11,000 Hz. Sound is a wave motion created as a result of pressure disturbances; the amplitude of the waves determines the intensity of the sound that reaches the ear. The sound energy is converted to electrochemical energy in the inner ear, and this energy is transmitted by the auditory nerves to the brain for interpretation.

The Human Ear

Anatomically, the human ear, diagrammed in Figure 6.7, can be divided into three major parts; the outer ear, an air-filled middle ear, and a fluid-filled inner ear.

Outer Ear

The visible portion of the ear, called the auricle, forms the entrance to the ear canal directing sound waves to the eardrum (also known as the tympanic membrane). The tympanic membrane, which separates the outer ear from the middle ear, is a cone-shaped membrane about a half-inch in diameter. The distance the eardrum moves in response to sound pressure waves is incredibly small, as little as one billionth of a centimeter.[1] Besides vibrating in response to sound waves, the eardrum protects the contents of the middle ear.[3]

Middle Ear

The middle ear is an air-filled cavity about 2 cm^3 in volume. Its major function is to transmit the oscillatory motion of the eardrum to the inner ear. The middle ear contains three bones, which are the smallest in the human body: the malleus (hammer), the incus (anvil), and the stapes (strirrup). The handle of the malleus attaches to the eardrum and articulates with the incus, which is connected to the stapes. The malleus and the incus vibrate as a unit in response to sound waves.[3] This assemblage of bones, known as the ossicles, transmits sound waves to the inner ear, and protects the inner ear by either amplifying or diminishing the sound waves.

In addition, there are two muscles attached to the stapes and the malleus, the tensor tympani and stapedius muscles. Their major function is to tighten up the eardrum and the motion of the ossicles, thereby reducing the efficiency of sound transmission. This phenomenon, known as the acoustic reflex, is carried out on command of the brain after a very loud sound reaches the eardrum.[7]

Inner Ear

The inner ear is a complex system of ducts and sacs that houses the end organs for hearing and balance. It consists of an outer bony portion and an inner membranous labyrinth. The center of the labyrinth connects the three semicircular canals and the cochlea. A watery fluid separates the bony area from the membranous labyrinth, and inside the membranous labyrinth are fluids called the endolymph and the cortilymph.[3]

The components of the inner ear have nothing to do with hearing but are responsible for certain senses. One component, the utricle, is responsible for the sense of acceleration and gravity. Another, an arrangement of three semicircular canals, provides humans with their sense of orientation in space and balance. The hearing part of the ear is called the cochlea. When sound waves travel through the external auditory canal, the foot of the stapes bone knocks against an oval window that is a wide opening in the cochlea, and sound is transmitted to the liquid inside. The round window lying just below the oval window is an elastic membrane, which is the final component that sound reaches in the human ear.[7]

Effects of Noise on Hearing and Health

Excessive noise can destroy hearing and has the potential of affecting other body organs including the heart. Although research on the effects of noise on human health is not complete, it is believed that exposure to excessive noise can cause high blood pressure and a quickened pulse rate. Noise also can cause an abnormal release of body hormones and the tensing of muscles.[3,4]

Hearing

The adverse effects of noise on hearing depend on both sound level and exposure time. Loss of hearing can be classified in the following categories:

1. Conductive impairment
2. Sensorineural impairment
3. Mixed impairment (both conductive and sensorineural)
4. Central impairment
5. Psychogenic impairment

A conductive hearing loss is produced by any condition that can interfere with the transmission of sound from the environment to the cochlea. A conductive loss can be due to the presence of wax in the auditory canal, a large perforation in the eardrum, or a buildup of fluid in the middle ear. Most conductive hearing losses can be treated by medical or surgical procedures.[3]

A sensorineural hearing loss is almost always irreversible. The sensory component of the loss involves the organ Corti, and the neural component implies degeneration of the neural elements of the auditory nerve. Exposure to excessive noise can cause sensorineural hearing loss.[3]

A mixed hearing loss occurs when there are components and characteristics of both conductive and sensorineural hearing loss in the same ear.

A central hearing loss implies difficulty in a person's ability to interpret what is heard. The abnormality is localized in the brain between the auditory nuclei and the cortex.

A psychogenic hearing loss indicates a nonorganic basis for an individual's threshold elevation. Two conditions in which such a loss may occur are malingering and hysteria.

The ear has no overload switch or circuit breaker, so it has no option but to receive all the sound that strikes the eardrum. In industry, excessive noise constitutes a major health hazard. Such exposure can cause both auditory and extra-auditory effects.

Auditory Effects

Noise induced hearing loss (NIHL) can occur unnoticed over a period of years. At first, excessive exposure to harmful noise causes auditory fatigue or a temporary threshold shift (TTS). This shift refers to the difference in one's hearing sensitivity measured before and after exposure to sound, and is called temporary because there is a return of the individual's pre-exposure hearing level after a period of hours away from the intense sound. However, repeated insults of excessive noise can transform a TTS into a permanent threshold shift (PTS). In fact, studies substantiate that the hearing sensitivity of factory workers in heavy industry is poorer than that of the general population.

Many factors influence the course of NIHL. The overall decibel level of the noise exposure obviously is important. If a noise exposure does not cause auditory fatigue, then such exposure is not considered harmful to one's hearing sensitivity.

Another consideration is the frequency spectrum of the noise. Noise exposure that has most of its sound energy in the high-frequency bands is more harmful to a worker's hearing sensitivity than low-frequency noises.

Another factor is the daily time distribution of the noise exposure. In general, noise that is intermittent in character is less harmful to the hearing than steady-state noise exposure. As the total work duration (years of employment) of worker exposure to hazardous noise increases, so too does the incidence and magnitude of the NIHL.

Finally, the susceptibility of the individual to hazardous noise must be considered because workers will not suffer identical hearing impairment if exposed to the same noise intensity over the same time period. Small percentages of workers will be highly susceptible or, on the other hand, refractory to the degrading effects of noise.

Extra-auditory Effects

The extra-auditory effects of noise result in physiologic changes other than hearing loss. We are familiar with the reflex-like startle response of an individual to a loud, unexpected sound. Less commonly noted are the cardiovascular, neurologic, endocrine, and biochemical changes secondary to intense noise exposure. Subjective complaints of nausea, malaise, and headache have been reported in workers exposed to ultrasonic noise levels. Vascoconstriction, hyperreflexia, fluctuations in hormonal secretions, and disturbances in equilibrium and visual function have been demonstrated in laboratory and field studies. These changes have been for the most part transient in character, and it remains to be clarified whether such noise exposure has long-lasting ill effects on the organism.

Noise Control

Once the sources of noise have been identified and the sound level has been measured, there might be a need for noise exposure reduction by either engineering or administrative controls.[1]

As noise is energy transmitted by wave motion, any engineering noise control considers either reduction of this energy at the source, diversion of the energy flow, or protecting the receivers from sound energy reaching them. Noise control measures can be divided into those requiring minimum or no equipment modifications and those necessitating major equipment or process modifications.

Techniques Involving Minimal Equipment Modification

The kinds of noise controls listed below can be effective in reducing noise exposures, but do not involve machine or process design changes. These alternatives are not necessarily simple or cheap, but one should consider them first, before exploring more complex solutions. The controls are:

Maintenance

Proper equipment maintenance can reduce the generation and release of unnecessary noise. For example, proper lubrication of parts reduces noise generation due to contact and friction.

Operating Procedures

In some processes the employee work exposure can be reduced by relocating the operator without interfering with the normal operation. For example, some operations can be performed efficiently from inside an operator booth.

Equipment Replacement

In some cases the solution to the noise problem might be to replace the noisy equipment with quieter equipment capable of performing the same tasks. For example, many manufacturers offer quiet electric motors or compressors.

Administrative Controls

Occasionally the use of administrative controls can reduce employee exposure to hazardous noise. For example, reducing the exposure time by rotating workers between high and low noise areas can reduce the daily noise dose.

Room Treatments

The presence of reflecting surfaces (walls, floors, ceilings, and equipment) in a workspace results in the buildup of sound levels in the reverberant field. By controlling the reflected sound (i.e., by preventing the reflections), reverberant field sound levels can be reduced by several decibels. Generally, reflections are prevented by the use of acoustically absorbent materials applied directly to wall or ceiling surfaces or suspended from the ceiling in the form of hanging baffles. The potential benefit of room treatment ranges anywhere from 0 dB (no benefit) to as much as 12 dB sound reduction.

Equipment Location

The sound level drops off as one moves away from a noise source. Outdoors, the sound level can be reduced by as much as 6 dB for every doubling of distance. Indoors, when workers are stationed close to noisy machines and where space permits, moving the noise sources (or the workers) may be beneficial. This situation often is encountered where operator-controlled production equipment is lined up in rows, and where operators may receive as much noise exposure from the machine behind then as from their own machine. If there is no room to spread out equipment, a likely alternative solution would be to shield the worker from the surrounding sounds.

Simple Machine Treatments

Vibration Isolation. Airborne noise can be produced by any solid vibrating member of a machine. The vibrating member alternately pushes and pulls against the air, creating small pressure changes that tend to radiate in all directions. The vibrating member may be driven into vibration by contact with a primary moving part or through some intermediate solid linkages in contact with the moving part. In such cases of forced vibration, techniques of vibration isolation may be applicable. In general, all vibration isolation techniques aim at disassociating the vibrating member from the force causing it to vibrate, generally by interposing a slightly compressed "springy" material between the forces and the member. An example would be supporting a panel on a machine by means of bolts that pass through neoprene grommets. Essentially, the panel is suspended from the machine by the neoprene.

Surface Damping. Frequently lightweight metal (or plastic) parts are set into bell-like vibration by multiple impacts (e.g., parts impacing on chutes) or by induced resonances caused by externally applied forces. The resulting free vibration can be effectively attenuated by application of externally applied damping materials. Damping treatments include application of specially treated aluminum tapes; application of troweled, painted, or sprayed-on materials; and application of constrained layer "sandwiches" of damping materials. In each case, the damping properties of these materials depend on the temperature, humidity, and chemical exposure.

Techniques Requiring Equipment to Be Added to Existing Machinery

Other forms of noise control may involve some kind of modification to the equipment, although some equipment changes that reduce noise exposure can be accomplished without redesigning the equipment. The modifications may change the machine noise emissions, may redirect the emissions, or may contain the emissions.[1,2]

Shields and Barriers

An acoustical shield is a solid piece of material placed between the worker and the noise source, often mounted on a machine. An acoustical barrier provides an obstacle to the travel of sound. Barriers can be quite useful, especially when they are used outdoors where they can reduce the sound level in a given direction by as much as 24 dB.

Both the barrier and the shield function by deflecting the flow of acoustical energy away from the worker. They are most effective when: (1) the worker is close to the noise source (positioned in the near field of the noise source), (2) the smaller dimension of the shield or barrier is at least three times the wavelength contributing most to the noise exposure received, and (3) the ceiling and other nearby reflecting surfaces are covered with sound-absorptive material.

Enclosures

Partial Enclosures. When a barrier is wrapped around a machine, with its top more or less open, it becomes a partial enclosure. Such an enclosure can be effective in reducing noise to workers nearby, but the noise escapes through the top and contributes to the reverberant sound in the workroom.

The spillover noise effects can be reduced by covering the inside of the enclosure with acoustically absorbent material. Also, suspended acoustically absorbent baffles may be placed over the openings to reduce the escaping noise.

Total Enclosure. If more than 12 to 15 dB of noise reduction is required, a total enclosure is needed so that noise is contained more fully. By virtue of their design, total enclosures can cause a heat buildup problem, which is handled by adding a ventilating blower, together with silencers for both supply and exhaust air.

As a general matter, enclosures should not touch any part of the machine and should be vibration-isolated from the floor. Nevertheless, the enclosure must be pierced for such services as electricity, air, steam, water, oil, or hydraulic power. These services can be regrouped, together with mechanical controls, to a convenient location and passed through a junction box that is later packed and sealed.

Silencers

Many types of noise control devices are termed "silencers." Duct silencers, for example, are cylindrical or rectangular structures fitted to the intake or discharge of air-moving equipment. These dissipative devices function by absorbing noise otherwise escaping from the intake or discharge. Duct silencers are lined internally with acoustical material.

The reactive muffler is another type of silencer used along piping or ductwork systems or at engine exhausts. These devices are designed to reflect pressure disturbances back toward the noise sources, thus functioning in a different fashion from dissipative silencers.

Techniques Requiring Equipment Redesign

Noise control at the source of the noise is highly desirable in many cases, especially when the need to retrofit or otherwise modify noise exposures is thereby eliminated. Usually, however, the expertise and resources necessary to redesign equipment on a large scale are beyond the means of the end user of a noisy product. Yet certain techniques may be useful to end users and may serve to eliminate the need for other forms of noise control. Treatments such as the following should be considered:

- Placing internal baffles in hoppers to encourage the product to slide, rather than fall, onto hopper surfaces.
- Changing chute slopes to encourage sliding rather than bouncing.
- Using soft material (e.g., neoprene) to reduce mechanical impacts.
- Replacing metal conveyors at transfer drops with canvas units, or reducing the height of the drops.
- Lining conveyor sides with plastic railing.
- Using timing mechanisms to space out conveyor line product flows, thereby preventing product impacts.
- Applying damping to the underside of conveyors, chutes, hoppers, and so on.

In other cases, it is possible to envision the use of alternative mechanisms to quiet noise emissions. Noisy hydraulic motors may be replaced with electric drives, and pneumatic parts ejectors may be replaced with mechanical mechanisms.

Noise Control Materials

Materials used in noise and vibration control fall into four general categories. Sound absorbers and sound isolators (discussed below) are used for control of airborne noise.[1,2,6] Vibration isolators and damping materials (discussed later, in the section on "Vibration") are used for control of hazardous vibration.

Sound Absorptive Materials

When a sound wave strikes a surface, part of the acoustical energy is absorbed by the surface, and part is reflected from the surface. Sound-absorptive materials have the capability of absorbing most of the sound energy that strikes them. Suitable materials are generally fibrous, lightweight, and porous. When a sound wave enters the pores of a porous material, its amplitude is successively reduced by the acoustical resistance to wave motion offered by the porous medium.

Sound-absorptive materials efficiency is affected by a number of factors. The amount of sound absorption at low frequencies is largely dependent on the depth of the air space, whereas in the frequency range of maximum absorption the resistance of the material to airflow is the critical factor. Examples of absorbent materials are acoustical ceiling tile, glass fiber, and foamed elastomers. These materials are employed in a wide range of applications such as: muffler linings; wall, ceiling and enclosure linings; wall fill; and absorbent baffle construction.[1]

Sound Absorption Coefficients. The sound absorption coefficient of a material is an indication of the amount of sound absorbed by the material at a given frequency. For example, a material with a coefficient of 0.75 absorbs 75% of the sound reaching it. It is common practice to report the absorption coefficients at six frequencies: 125, 250, 500, 1,000, 2,000, and 4,000 Hz. Table 6.4 summarizes the sound absorption coefficients of some common acoustic materials.[6]

The term "noise reduction coefficient" (NRC) refers to the arithmetic average of the absorption coefficient at frequencies 250, 500, 1,000 and 2,000 Hz. For example, a $\frac{1}{4}$-inch-thick polyurethane foam has an NRC of 0.20 (see Table 6.4).

Sound Transmission Loss Materials

The term "sound transmission loss" (STL) refers to the sound isolation properties of a material.[1] The STL is a measure of the amount of sound energy transmitted through a material relative to the amount flowing toward the material. Mathematically the STL is defined by the following expression:

$$\text{STL} = 10 \log \frac{E_i}{E_t}$$

where E_i is the incident energy and E_t is the transmitted energy. The STL increases with increasing frequency at a rate of about 5 dB for each doubling of frequency.

Material Selection

Both environmental and regulatory factors affect the selection of noise control materials. Environmental factors may include such phenomena as the presence of materials such as grease on machine parts and the temperature, whereas regulatory factors involve a variety of government-

Table 6.4. Sound Absorption Coefficients of Common Acoustic Materials.

	Frequency (Hz)					
Materials	125	250	500	1,000	2,000	4,000
Fibrous glass—hard backing (typically 4 lb/cu ft)						
1 inch thick	0.07	0.23	0.48	0.83	0.88	0.80
2 inches thick	0.20	0.55	0.89	0.97	0.83	0.79
4 inches thick	0.39	0.91	0.99	0.97	0.94	0.89
Polyurethane foam (open cell)						
$\frac{1}{4}$-inch thick	0.05	0.07	0.10	0.20	0.45	0.81
$\frac{1}{2}$-inch thick	0.05	0.12	0.25	0.57	0.89	0.98
1 inch thick	0.14	0.30	0.63	0.91	0.98	0.91
2 inches thick	0.35	0.51	0.82	0.98	0.97	0.95
Hairfelt						
$\frac{1}{2}$ inch thick	0.05	0.07	0.29	0.63	0.83	0.87
1 inch thick	0.06	0.31	0.80	0.88	0.87	0.87

Source: Reference 1.

regulated concerns, such as the use of lead-bearing materials near food processing areas. Both must be considered in materials selection.

Hearing Protection Equipment

Industrial environments that produce hazardous noise levels can be classified into moderate to high noise level areas and extremely high level noise areas.[7] When exposure to hazardous noise cannot be eliminated by engineering or administrative control measures, ear protection devices must be provided and used. When selected properly, these devices can lower the noise level in most manufacturing industries to within the legal specification limits. In addition to meeting the requirements for noise level reduction, ear protective devices must be comfortable and must be made of materials that do not irritate the skin and can be kept in sanitary condition over a long period of time.

Ear protective devices can be classified into two general categories:

- Ear plugs, or insert-type devices, which fit within the ear canal.
- Ear muffs, which are devices that fit around the ears and are supported either from a hard hat or from a head band that connects the individual muffs.

Insert Type

Ear plugs are designed to occlude the ear canal. They come in many varieties and are made of soft materials such as rubber, neoprene, wax, wool, and cotton. Some are disposable and discarded after use. Ear plugs that are reusable must be made of materials that can be easily disinfected. Some devices are made in several sizes to accommodate different-sized ear canals. Others come in one size that can be adapted by their natural expansion when inserted in the canal, or by removing one or more flanges on the unit. Some varieties are custom-molded, and they are supposed to provide the most comfortable and best fit.

In spite of their relative effectiveness in protecting the ears from hazardous noise, there are certain problems associated with ear plugs. In order to get a good seal, a certain amount of pressure must be applied, which can cause discomfort. Another problem associated with ear plugs is that they can easily absorb dirt or chemicals, which can cause infection of the ear canal. It is imperative to clean and disinfect reusable ear plugs before each use. Ear plugs also can lose the acoustic seal by simple jaw movements, which can change the shape of the ear canal.

Muff Type

Ear muffs come in a universal size and are available with foam- or liquid-filled cushions. Some devices fit only in one position (e.g., with the band over the head), whereas others are multipositioned and can be worn with the head band over the head, behind the head, or under the chin. Muffs cost more initially, but they are cleanable, and replacement parts are available.

For proper selection of ear muffs, the sound levels in the work area must be determined and the attenuation of the ear muff matched to these sound levels. Attenuation is a measure of the effectiveness of an ear protective device in eliminating hazardous high frequency noises. Before the selection of any hearing protective devices, the attenuation data provided by the manufacturer should be consulted to determine the most appropriate type. Figure 6.8 shows a typical ear muff protector. Ear muffs should be able to provide protection against high-frequency noise because most hearing damage occurs as a result of exposure to high-frequency noise.

Effective Use of Hearing Protective Devices

In industrial plants, efforts to encourage the use of protective equipment for employees and supervisors usually require an educational program on ear protection. There should be continual

Figure 6.8. Muff-type hearing protector. (Source: Mine Safety Appliances Company, Pittsburgh, PA.)

follow-up by supervisors to see that the program is accepted and that ear protection is worn when needed. For reminders, place signs in areas where protective equipment is mandatory. Supervisors should understand that if a plug or muff is uncomfortable, it may not be worn.

When they are used properly, hearing protectors can reduce potentially hazardous sound levels to nonharmful at-ear sound levels for most types of industrial noise environments. Laboratory measurements have shown that almost every hearing protector can provide 25 dB or more of attenuation. It should be recognized, however, that there may be significant differences between laboratory-measured performance and actual field performance.

Hearing protector performance is highly sensitive to the fit of the device being used. Any acoustical leakage around the device as a result of improper fit, seal breakage by eyeglass frames, or long hair, or loss of pressure on the cushions due to stretched supports, or improperly maintained cushions can so degrade hearing protector performance that only 10 dB or less of attenuation can be obtained. Unfortunately, workers tend to use hearing protectors improperly because looser-fitting devices are more comfortable than properly worn ones.

To ensure that hearing protectors serve as intended, they always should be provided as part of a more comprehensive hearing conservation program, including at a minimum, annual follow-up in the form of audiometric testing of individual hearing levels. Any successful hearing conservation program also should include education of the end users in the proper use of hearing protectors (as well as the potential hazards of improper use), and should provide (1) professional advice as to proper fit and (2) a wide variety of hearing protectors of all kinds (to account for individual preferences and differences in ear sizes). In addition, the program should be supported by management, to ensure company-wide cooperation. Finally, it is important to dispel some of the myths associated with the use of hearing protectors:

- Hearing protectors do not degrade a normal hearing person's ability to hear sounds or understand speech in high-noise environments. In fact, these devices can improve listening conditions. When hearing protectors are worn, all sounds are attenuated, and the signal-to-noise ratio remains the same at each frequency; the only difference is that the intensity of the sounds is reduced. However, because different frequencies are attenuated by different amounts, individual users will need to adjust to alterations in the sounds they hear.
- Hearing protectors do not appear to cause hygiene problems. Reusable devices can be cleaned and disposable devices replaced as required.

Certain problems associated with use of hearing protectors should be acknowledged:

- The devices may be uncomfortable, especially when first worn and especially in hot environments, where perspiration can cause ear muffs to slip or to irritate.
- The devices do make it more difficult to hear in low noise environments (i.e., under 80 dBA), and in intermittent noise environments workers naturally will want to remove the protectors during quiet periods.
- Workers with preexisting hearing impairment may lose some ability to hear certain sounds if the preexisting impairment complements the attenuation of the protector.
- Hearing protectors may make it difficult to localize a particular noise; that is, they can interfere with the ability to discriminate where a sound originates.

Vibration

Effects of Vibration

In industrial situations, exposure to vibration usually is accompanied by exposure to noise. Although the adverse effects of noise are substantially well researched and documented, the adverse effects of vibration on humans are the subject of many research programs. Even though this research is ongoing, studies have indicated that these effects can be rather serious. Humans exposed to excessive vibration can suffer from both physiological and psychological effects. Blurred vision, loss of motor control, and lack of ability to do one's job properly are some of the adverse effects of vibration.[3,7]

The most extensive investigations of industrial exposures of workers to vibration have been concerned with repeated exposure to low-frequency vibration transmitted through the upper extremities of the worker—that is, during the use of hand-held power tools that incorporate rapidly rotating or reciprocating parts. Unfortunately, most of these studies present only general descriptions of the clinical evidence of overexposure to vibration, and very few contain any controlled observations utilizing quantitative techniques; consequently there is large variability among the observations reported by different investigators. However, the main features of what can be called a vibration syndrome are evident.

The clinical evidence of overexposure to vibration during the use of hand tools can be conveniently grouped into the following types of disorders, in decreasing order of their appearance:[13]

- Raynaud's syndrome: Also known as ''dead fingers'' or ''white fingers,'' Raynaud's syndrome occurs manly in the fingers of the hand used to guide a vibrating tool. The circulation in the hand becomes impaired, and, when exposed to cold, the fingers become white and devoid of sensation, as though mildly frosted. The condition usually disappears when the fingers are warmed for some item, but a few cases have been sufficiently disabling that the affected individuals were forced to seek other types of work. In some instances, both hands are affected. Reynaud's syndrome has been observed in a number of occupations involving the use of fairly light vibrating tools; for example, in the metal trades, where air hammers are used for scarfing and chipping, and in stone-cutting, lumbering, and the cleaning departments of foundries, where there is a good deal of overtime work. Obviously, prevention of this condition is much more desirable than treatment. Preventive measures include directing the exhaust air from air-driven tools away from the hands so they will not become unduly chilled, using handles of a comfortable size for the fingers, and, in some instances, substituting mechanical cleaning methods for some of the hand methods that have produced many of the cases of ''white fingers.'' In many instances, simply preventing the fingers from becoming chilled while at work has been sufficient to eliminate the condition.

- Degenerative alteration of ulnar and axillar nerves, resulting in loss of the sense of touch, thermal sensations, and muscular weakness or even paralysis.
- Decalcification and fragmentation or deformation of bones.

Vibration also may cause equipment components such as nuts and bolts to break or crack as a result of fatigue, thereby producing a potentially hazardous situation.

Nature of Vibration

Vibration is defined as the oscillatory motion of a system around an equilibrium position. The system can be in a solid, liquid, or gaseous state, and the oscillation of the system can be periodic or random, steady-state or transient, continuous or intermittent.[3,7]

Periodic (Sinusoidal) Vibration

Vibration is considered periodic if the oscillating motion of a particle around a position of equilibrium repeats itself *exactly* after some period of time. The simplest form of periodic vibration is called pure harmonic motion, which can be represented by a sinusoidal curve as a function of time. Such a relationship is illustrated in Figure 6.9, where T is the period of vibration.

The motion of any particle can be characterized at any time by: (1) displacement from the equilibrium position; (2) velocity, or rate of change of displacement; and (3) acceleration, or rate of change of velocity. For pure harmonic motion, the three characteristics of motion are related mathematically.

Displacement. The displacement of a system from its equilibrium position under the influence of pure harmonic motion can be expressed mathematically by:

$$\chi = X \sin\left(2\pi \frac{t}{T}\right)$$

where:

χ = Distance from equilibrium position at any time
X = Maximum distance from equilibrium
t = Time
T = Period of vibration (see Figure 6.9)

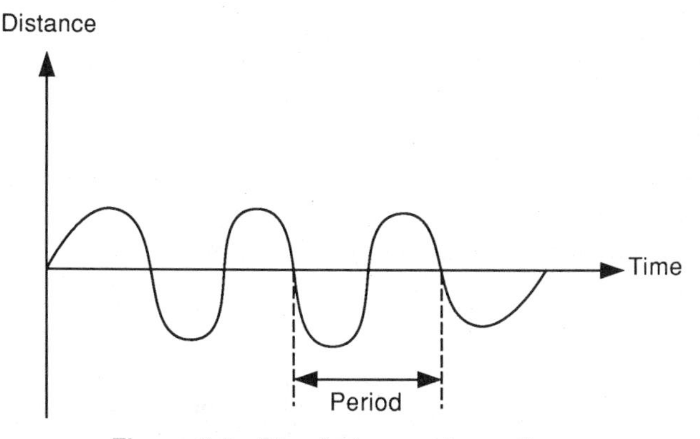

Figure 6.9. Simple harmonic motion.

Velocity. In many practical problems, displacement is not the most important property of the vibration. For example, experience has shown that the velocity of the vibrating part is the best single criterion for use in the preventive maintenance of rotating machinery.

Velocity is defined as the rate of change of displacement with time. The velocity of a vibrating system under pure harmonic motion can be expressed as follows:

$$v = \frac{d\chi}{dt} = \frac{d}{dt}\left(X \sin\left(2\pi \frac{t}{T}\right)\right)$$

$$v = V \sin\left(2\pi \frac{t}{T} + \frac{\pi}{2}\right)$$

where v is the instantaneous velocity, and V is the maximum velocity.

Acceleration. In many cases of vibration, especially where mechanical failure is a consideration, actual forces set up in the vibrating parts are critical factors. Because the acceleration of a particle is proportional to these applied forces, and because equal and opposite reactive forces result, particles in a vibrating structure exert forces on the total structure that are a function of the masses and accelerations of the vibrating parts. Thus, acceleration measurements are another means by which the motion of vibrating particles can be characterized.

Acceleration is defined as the rate of change of velocity with respect to time. Therefore, for a system under the influence of pure harmonic motion, one can write:

$$a = \frac{dv}{dt} = \frac{d^2\chi}{dt^2}$$

or:

$$= \frac{-4\pi^2}{T^2} X \sin\left(2\pi \frac{t}{T}\right)$$

or, equivalently:

$$a = A \sin\left(2\pi \frac{t}{T} + \pi\right)$$

where a is the instantaneous and A the maximum acceleration.

Random Vibrations

Random vibrations, the most commonly observed type of vibration, can be defined as nonperiodic motion that never exactly repeats itself. Because of its nonrepetitive nature, the theoretical treatment of this type of vibration is both time-consuming and difficult. Figure 6.10 shows an example of a random vibration.

Measurement of Vibration

Vibration measurement is important for several reasons. From a design standpoint, it is essential to measure the amplitude and frequency of a vibration to ensure that materials can withstand its stress–strain consequences. Another important factor is that if a machine is known to resonate at

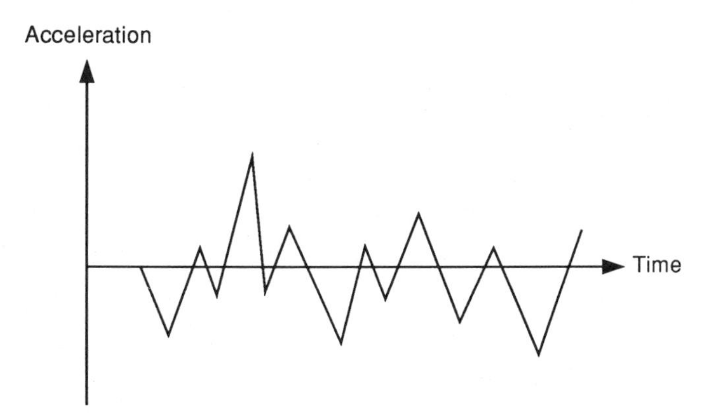

Figure 6.10. Typical random vibration.

a certain frequency, the machine operator can avoid operating the machine at a speed that causes resonance. Also, knowledge of the magnitude and frequency of the vibration is important in the selection and the installation of damping and isolating materials. Vibration monitoring also is useful in preventive maintenance operations; many times a change in vibration spectra is an indication of a fault developing within the system.[3,7]

The three major parameters that describe a vibration are displacement, velocity, and acceleration. Displacement measurements have been used on slow-moving machines because of the simplicity of measurements using optical devices and also because of relatively large displacement in these machines. In faster-moving machines, detection of displacement becomes more difficult, so the velocity of displacement is measured and used as the descriptive parameter. For high-frequency vibrations (above 1,000 Hz), it is necessary to measure the acceleration of the vibration because displacement and displacement velocity decrease very rapidly at high frequencies. Because high-frequency vibration has the potential of causing machine parts breakdown, its detection and measurement are critical in preventive maintenance.[7]

A wide variety of component systems, consisting of mechanical elements or a combination of mechanical, electrical, and optical elements, are available to measure vibration. The most common system uses a vibration pickup to transform the mechanical motion into an electrical signal, an amplifier to enlarge the signal, an analyzer to measure the vibration in specific frequency ranges, and a metering device calibrated in vibrational units.

Vibration Control

The two basic methods of vibration control employ damping materials and vibration isolators.

Damping Materials

Damping materials are used to reduce resonance effects in solids. Essentially, damping materials act as absorbents for solid-borne sound, converting vibrational energy into heat.

Vibration Isolators

Vibration isolators act on the same principle as isolators for airborne sound: introducing into the transmission path a material whose wave-transmitting properties are as different as possible from the medium carrying the wave. For vibration in solids, such materials are spring-like. Examples include resilient elastomer and metal springs, elastomer pads, and, in extreme cases, air springs. The weaker the spring, usually the greater the isolation.

Examples of Engineering Noise Control Principles

The following discussion is derived from reference 4:

1. *Changes in force, pressure, or speed will affect sound.* Sound is produced as a result of changes in these parameters, with great changes producing higher-level noises and small changes quieter ones. For example, a flat strip of metal can be bent noisily with a hammer or quietly with a pair of pliers. Another example is the cutting of cardboard by the fast action of a knife: the knife exerts a great force on the cardboard so that much noise is produced, but the cardboard can be cut much more quietly by running a blade gently across it.
2. *Vibration in solids and turbulence in fluids can produce sound.* The vibration of a violin box can create airborne sound; also the turbulent flow of fluids in piping systems can create sound, which can be transmitted to air. To control the sound generated by turbulent flow, the pipe can be covered with sound-absorbing material.
3. *Vibrations can produce sound even after traveling long distances.* Vibrations induced in solids and liquids can travel long distances and produce airborne sound. For example, vibrations from a train move along the rails and can be heard a considerable distance away. The best way to control such vibrations is to stop them as close to the source as possible, either by isolation or by using an appropriate damping material. For example, the vibrations from an elevator, which can be transmitted throughout a building, are controlled by isolating the elevator drive from the building structure.
4. *The time between changes affects sound frequency.* The frequency of a sound wave is determined primarily by the time lag between force or pressure changes that generate the sound. The longer the time between these changes, the lower the frequency of the sound. For example, consider two gears that have the same diameter but a different number of teeth. When these gears are rotated, the one with the lower number of teeth generates a lower-frequency sound.
5. *High-frequency sound is directional and can be reflected easily.* When high-frequency sound reaches a hard surface, it is easily reflected, much like light from a mirror, but high-frequency sound cannot travel through openings or around corners easily. This principle can be used to control high-frequency noise by placing a hard surface in its path.
6. *Low-frequency noise can easily travel around objects and through openings.* Low-frequency noise can radiate at the same level in all directions; it can travel around corners and through openings or holes and then continue to radiate in all directions. This principle indicates that the noise from equipment such as compressors that generate low-frequency noise can be controlled only by a complete enclosure of damped material lined with a suitable absorbent.
7. *High-frequency noise is greatly reduced by distance.* High-frequency sound is reduced much more effectively than low-frequency sound by passing through air. This principle can be demonstrated by considering a ship siren. Although the sound of the siren may seem extremely loud to those on board ship, it is much less loud at a distance. This principle is used in noise control in many industrial situations. For example, consider a roof fan that generates low-frequency noise and disturbs the residents of houses outside plant. This fan can be replaced by another one of similar capacity but with a larger number of fan blades. The new fan will generate higher-frequency noise than the old one and may not cause disturbances.
8. *Low-frequency noise is less disturbing than high-frequency noise.* The human ear is less sensitive to low-frequency noise than it is to high-frequency noise. For example, consider two passing trains that make equal amounts of noise in terms of decibels. The train that moves faster is more disturbing to the human ear because it generates a higher-frequency

noise than the slower one. In industrial applications, it might be possible to change an item of equipment that generates high-frequency noise for one that generates a lower-frequency noise.

9. *Small vibrating surfaces generate less noise than larger surfaces.* It is possible for an object with a small surface area to vibrate intensely without creating a disturbing noise. For example, an electric shaver lying on a bathroom counter will generate a loud noise compared to one carried by its cord. This principle can be used in noise control measures for several work-related situations. For example, consider the control panel of a hydraulic system that generates too much noise; the panel can be detached from the system and its vibrating surface reduced, so that it becomes quieter.
10. *Perforated surfaces generate less noise than solid surfaces.* In many situations large vibrating surfaces cannot be avoided. The use of a densely perforated surface can greatly reduce the sound level. For example, consider a flywheel that has a solid cover over its belt and drive. Sound radiation from the flywheel can be greatly reduced by installation of a perforated cover.
11. *Irregular objects create less wind tone than regular ones.* When air passes an object at a certain speed, a disturbing tone may be produced. This phenomenon can be prevented by making the object irregular. For example, at certain air speeds loud noises can be generated around smokestacks. This can be prevented by mounting a spiral strip of sheet metal on the smokestack. Regardless of the wind's direction, it will encounter an irregular object.
12. *Piping systems with less change in flow direction produce less noise from turbulence than more convoluted systems produce.* This principle can be used in the design of piping systems to minimize the amount of noise by reducing the number of unnecessary bends in lines.
13. *Low-velocity mixing produces the least amount of exit noise.* When a flowing gas mixes with a nonmoving gas, pressure differences are created that may result in noise production. A lower outflow speed will produce a lower sound level. For gas speeds below 325 ft/sec, reducing the speed by half results in lowering of the sound level by 25 decibels. For example, consider the exhaust air from an air-driven machine that, because of its turbulence and high speed, generates a loud noise. The exhaust can be replaced by another one that has a larger cross-sectional area, causing lower velocities and less turbulence.
14. *Sound sources should not be placed near corners.* The closer a sound source is placed to radiating surfaces, the more noise is radiated to a given distance. The worst place would be near corners, where the sound source is close to three radiating surfaces. The best place for sound sources would be away from the walls. For example, the amount of noise in a machine shop can be reduced considerably by placing the machines as far away from the walls as possible.

References

1. NIOSH, *Industrial Noise Control Manual*, U.S. Department of Health, Education, and Welfare, NIOSH, Cincinnati, Ohio (1978).
2. Fader, B., *Industrial Noise Control*, 1st Ed., John Wiley & Sons, New York (1981).
3. NIOSH, *The Industrial Environment, Its Evaluation and Control*, NIOSH, Cincinnati, Ohio (1973).
4. Industrial Accident Prevention Association (IAPA), *Noise Control, A Guide for Employers and Employees*, IAPA, Toronto, Ontario.
5. The Office of the Federal Register, *Code of Federal Regulations*, Office of Federal Register, Washington, D.C. (1985).
6. Harris, C. M., *Handbook of Noise Control*, 1st Ed., McGraw-Hill Book Co., New York (1979).
7. Cheremisinoff, P. N., and P. P. Cheremisinoff, *Industrial Noise Control Handbook*, 1st Ed., Ann Arbor Science Publishers, Ann Arbor, Mich. (1977).

8. American Society of Safety Engineers (ASSE), *Readings in Noise Control and Hearing Conservation*, ASSE, DesPlaines, Ill. (1985).
9. Bragdon, C. R., *Noise Pollution, a Guide to Information Sources*, 1st Ed., Gale Research Company, Detroit, Mich. (1979).
10. Gasaway, D. C., *Hearing Conservation: A Practical Manual and Guide*, 1st Ed., Prentice-Hall, Englewood Cliffs, N.J. (1985).
11. Archibald, C. J., *Noise Control Directory*, 1st Ed., Fairmont Press, (1979).
12. Acoustical Society of America (ASA), *Test-Site Measurement of Maximum Noise Emitted by Engine Powered Equipment*, ASA, (1975).
13. Halliday, D., and R. Resnick, *Physics*, 1st Ed., John Wiley & Sons, New York (1967).
14. Beranek, L. L., *Acoustics*, 1st Ed., McGraw-Hill Book Co., New York, (1954).
15. National Safety Council (NSC), Administration and Human Relations Aspects, Data Sheet, NSC, Chicago.
16. Anticaglia, J., and A. Cohen, Extra-auditory effects of noise as a health hazard, *Am. Ind. Hyg. Assoc. J.*, Vol. 31, p. 277, Westmont, N.J. (1970).
17. Kryter, K., *The Effects of Noise on Man*, Academic Press, New York (1970).
18. Chamber of Commerce of the United States, *Analysis of Worker's Compensation Laws*, Chamber of Commerce, Washington, D.C. (1978).
19. National Safety Council (NSC), An Industrial Hearing Conservation Program, Data Sheet, NSC, Chicago.
20. National Safety Council (NSC), Procedures of a Sound Survey, Data Sheet, NSC, Chicago.
21. U.S. Department of Commerce, *Handbook of Noise Rating*, U.S. Department of Commerce, Washington, D.C. (1974).

Exercises

1. What is the amplitude of a sound wave? What are the two different ways of expressing a sound wave amplitude?

2. What is the relationship between the frequency of a sound wave as detected by a listener and the distance between the listener and the sound source?

3. What is the ratio of the speed of sound in air to the speed of sound in hydrogen gas at 68°F? Calculate the wavelength of a sound wave traveling through hydrogen at 68°F with a frequency of 1,000 Hz.

4. What is the rms value of a sound pressure, and how can it be calculated from individual sound pressure data?

5. Calculate the sound pressure level, in decibels, for a source that generates a sound pressure of 0.0005 atmosphere.

6. What is the A-weighted sound level, and what is its importance in occupational noise control?

7. A worker in a machine shop is exposed to noise according to the following table:

Hours of exposure	*Sound level, dBA*
1	83
2	80
1	83
4	75

Determine whether this worker is exposed to hazardous noise according to OSHA regulations.

8. What are the minimum OSHA requirements for an audiometric testing program?

9. What are the employer's responsibilities for recordkeeping in conjunction with OSHA's occupational noise control regulations?

10. What are the major noise measurement instruments, and what type of noise measurements can be conducted using each type of instrument?

11. What are the two types of noise integrating meters, and what type of noise measurement data can each provide?

12. What noise measuring instrument is suitable for determination of the noise frequency distribution? Briefly describe the operation of this instrument.
13. What criteria are used when a noise measurement needs to withstand scrutiny in a court of law?
14. In a warehouse, three forklifts and a grinding machine operate, and each produces a sound level according to the following table:

Forklift 1	80 dB
Forklift 2	79 dB
Forklift 3	81 dB
Grinding machine	80 dB

    Determine whether this warehouse is in compliance with OSHA regulations, and if not, recommend procedures to bring it into compliance with noise regulations.
15. What are the major functions of the eardrum?
16. What are the ossicles in the middle ear, and what is their major function?
17. What is acoustic reflex, and what muscles in the middle ear are responsible for it?
18. What are the different categories of hearing loss?
19. What is a sensorineural hearing loss?
20. Describe noise induced hearing loss (NIHL) and occupational noise exposure factors that affect NIHL.
21. What are the extra-auditory effects of occupational hazardous noise exposure?
22. Mention four noise reduction techniques that do not require machine or process design change.
23. What is an acoustical shield? And how can it be used to reduce occupational noise exposure?
24. What are the four types of materials used in noise control?
25. What are the two classes of ear protective devices?
26. What is the proper procedure for the selection of ear muffs?
27. What is Raynaud's syndrome, and how can it be prevented?
28. Why is it important to measure vibration in industrial equipment?
29. What are the three major characteristics that describe a vibration?
30. What are the two basic methods of vibration control?

Chapter

7

Electrical Safety

More than a thousand Americans are killed each year by electrical shock, and thousands more are burned or maimed. A survey taken in the state of California on electrical work injuries showed that more than 90% of the fatalities occurred when a person who was grounded made contact with a "hot" wire or an energized equipment housing. Line-to-line contact accounted for fewer than 10% of the deaths. Electrical safety requires an understanding of what electricity is, how electrical energy is transferred, and how the path through which electrical current travels can be controlled.[1]

Electricity can be defined as the flow of electrons along a conductor. Electrons are negatively charged particles distributed in orbits around the nuclei of atoms, which are the smallest units of an element that can exist either alone or in combination. In an atom, the negative charge of the electrons is neutralized by the positive charge of particles called protons, so that the atom is electrically neutral. For example, an atom of hydrogen has one electron and one proton, as shown in Figure 7.1. The electrons rotate around the nucleus and are held in orbit by the attractive forces it exerts on them, but they can be freed and removed from their orbits if enough energy is supplied to overcome the attractive forces of the nucleus. This energy can take the form of heat, pressure, or magnetic forces. Once electrons are removed from their orbits, they can drift randomly in no particular direction; these electrons are called free electrons. If an external force is applied so that their movement is controlled and directed in a given direction, they are responsible for producing an electrical current in that direction, as shown in Figure 7.2.

Some materials, such as metals, have loosely bonded electrons, and the amount of thermal energy available at room temperature is sufficient to generate free electrons. Materials that have a relatively large number of free electrons at room temperature, which are called conductors, are capable of conducting electricity (the movement of electrons in a given direction). On the other hand, materials that do not have a large number of free electrons at room temperature (such as plastics), which are called insulators, are incapable of conducting electricity. Materials that fall in between the two extremes are termed semiconductors.

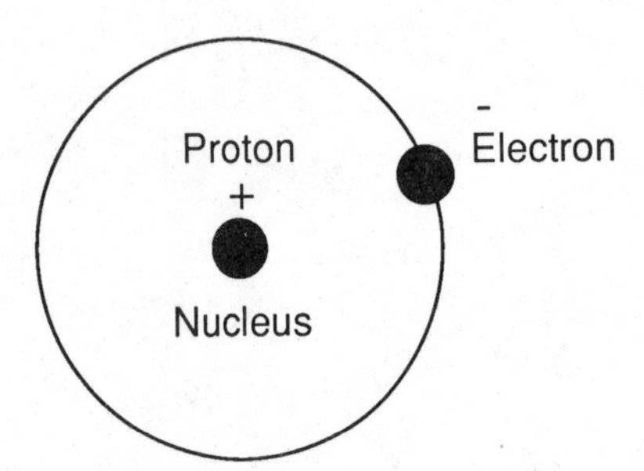

Figure 7.1. Hydrogen atom.

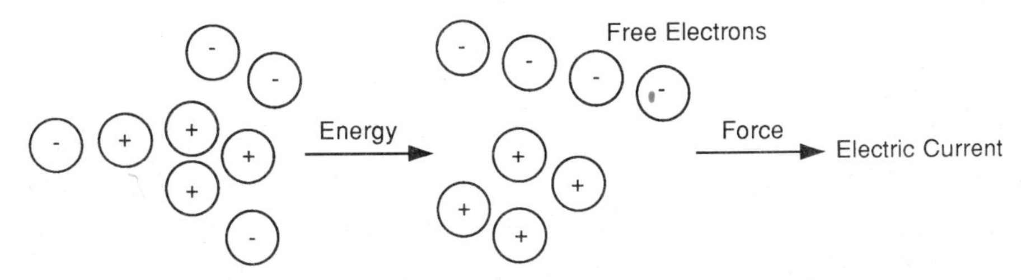

Figure 7.2. Movement of electrons by application of an external force produces an electric current.

Voltage, Electric Current, and Ohm's Law

The flow of electrons takes place from the negatively charged end of a circuit to its positively charged end. The factor of most concern in an electrical shock is the amount and path of such current flow through the human body.

In order for electrons to move between two points, a potential difference must exist. The potential difference between two points in a circuit is measured in terms of volts. The higher the potential difference, the easier it is for the electrons to move from one point to another, that is, the higher the electrical current.[1,2]

The flow of electrons also is governed by the resistance offered by the conducting material. The higher this resistance, the lower the current.

The current flow in a circuit is measured in terms of amperes. One ampere, by definition, is the flow of 6.28×10^{18} electrons per second past a given point in a circuit. Sometimes it is necessary to use smaller units of measurement for the current flow, the most commonly used units being the milliampere (0.001 ampere) and the microampere (one millionth of an ampere).[1,2]

The relationship among potential difference, electrical current, and resistance can best be described by Ohm's law:

$$V = RI$$

where V is the potential difference in volts, I is the current flow in amperes, and R is the conductor resistance in ohms. By definition, one ohm is the resistance of a conductor that maintains a current flow of one ampere under a potential difference of one volt.

Electrical Shock

Electrical shock occurs when the human body becomes part of an electrical circuit that has a sufficient potential difference to overcome the body's resistance. Electrical shock is a common hazard encountered by people involved in the installation, maintenance, and operation of electronic equipment. Although potential difference determines whether the body's resistances will be overcome, the damaging factor in electrical shock is the current flow.[1]

The electrical current that flows through the body from a "hot" wire or a defective piece of equipment to ground is known as a ground fault current, and is different from the short circuit that results when two or more circuit conductors are crossed. Most of the resulting deaths and injuries might be prevented by the use of a ground fault circuit interrupter (GFCI), also known simply as a ground fault interrupter (GRI).

Electrical Resistance of the Human Body

As moisture enhances electrical conductivity, the electrical resistance of the human body varies with the amount of moisture on the skin. Studies have revealed that hand-to-hand resistance can vary between 100,000 ohms and 600,000 ohms, depending upon skin moisture. Also a relatively high voltage can break down the outer layers of the skin, thereby reducing the resistance. Once the skin resistance is broken, current can flow through the blood and other organs, which have a substantially lower resistance than that of skin.

Table 7.1 summarizes the human body's resistance to electrical current flow.

Resulting Effects of Current on the Human Body

Using Ohm's law and assuming a hand-to-hand resistance of 2,400 ohm's, a current of 50 mA (0.050 ampere) would flow between the hands if one hand were in contact with a 120-volt source and the other hand were grounded.

Based on the research of Charles F. Dalziel, professor emeritus at the University of California, Berkeley, the effects of 60 Hz (hertz, cycles/sec) alternating current on the human body are generally accepted to be as follows:[1]

- 1 milliamp or less: No sensation, not felt.
- More than 3 mA: Painful shock.
- More than 10 mA: Local muscle contractions, sufficient to cause ''freezing'' to the circuit for 2.5% of the population.
- More than 15 mA: Local muscle contractions, sufficient to cause ''freezing'' to the circuit for 50% of the population.
- More than 30 mA: Breathing difficulty; can cause unconsciousness.
- 50 to 100 mA: Possible ventricular fibrillation of the heart.
- 100 to 200 mA: Certain ventricular fibrillation of the heart.
- Over 200 mA: Severe burns and muscular contractions; heart more apt to stop than fibrillate.
- Over a few amperes: Irreparable damage to body tissues.

The path the current takes through the body affects the degree of injury. A small current that passes from one hand to the other hand through the heart is capable of causing severe injury or death. However, there have been cases where larger currents caused an arm or leg to burn off without going through the vital organs of the body. In many such cases the person was not killed; had the same current passed through the vital organs of the body, the person easily could have been electrocuted. Another factor affecting the outcome of an electrical current passing through the body is the duration of the current flow: the longer this duration, the more devastating the

Table 7.1. Human Resistance to Electrical Current.

Body Area	Resistance (ohms)
Dry skin	100,000–600,000
Wet skin	500–1,000
Internal body (hand to foot)	400–600
Ear to ear	about 100

Source: Reference 1.

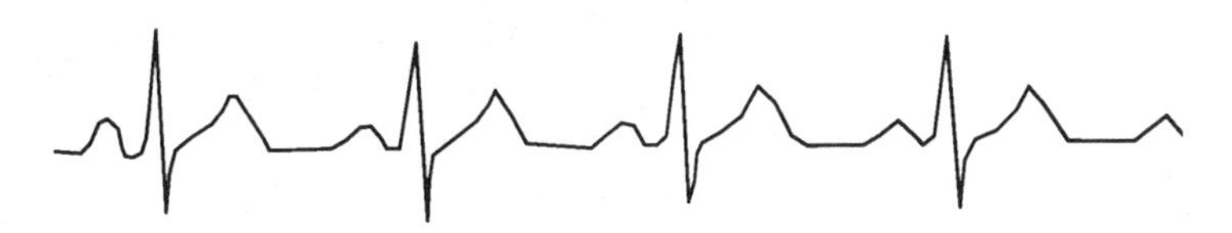

Figure 7.3. Electrocardiograms of the healthy rhythm of the heart and of ventricular fibrillation. (Source: Reference 1)

results can be. An electrical current passing through the body can cause severe injury or death by:

1. Contracting the chest muscles, resulting in breathing difficulty and death due to asphyxiation.
2. Affecting the central nervous system, resulting in malfunction of vital body functions such as respiration.
3. Interference with the normal rhythm of the heart beat, resulting in ventricular fibrillation, which is defined as ''very rapid uncoordinated contractions of the ventricles of the heart resulting in loss of synchronization between heartbeat and pulse beat.'' Once ventricular fibrillation occurs, it will continue and death will ensue within a few minutes. Resuscitation techniques, if applied immediately, may save the victim. (Figure 7.3 shows electrocardiograms illustrating the healthy rhythm of the heart and ventricular fibrillation.)
4. Affecting the heart muscle, resulting in severe heart muscle contraction and cessation of heart action.
5. Destruction of body tissues by heat generated when the current overcomes tissue resistance.

Recognition and Prevention of Shock Hazards

The damaging factor in electric shock is current flow (see preceding section), so every attempt should be made to minimize the flow in case of accidental contact with an energized conductor.[2,3] Referring to Ohm's law, $V = RI$, we see that to minimize the current at a given voltage, the resistance to flow, R, must be maximized. For example, the resistance of a wet finger is about 500 ohms, and that of a dry one is 100,000 ohms; thus, if a voltage of 120 volts is maintained, the wet finger can draw a current of 240 mA, whereas the dry finger can draw only 1.2 mA. Therefore, electricity and water must be regarded as incompatible. No one should attempt to install, operate, or repair electrical equipment without protective equipment when the body's resistance has been reduced even by perspiration!

To work with electricity safely, workers must recognize shock hazards and protect themselves against them. Any condition that can generate leakage of current from an electrical apparatus or conductor should be considered a potential shock hazard. Some of the more obvious shock hazard conditions are:

- Working with electrical equipment that lack the Underwriters Laboratories (UL) label.
- Using electrical equipment with broken or cracked insulation.
- Working with electrical equipment that is not adequately grounded.
- Working with electrical equipment on damp floors, or in high-humidity areas.
- Using metallic ladders to work on electrical equipment.
- Working on any electrical circuit without ensuring that the power has been shut off.

An excellent way to prevent shock hazards is to organize a preventive inspection and maintenance program for electrical equipment and appliances in the workplace. All electrical equipment should be inspected and tested on a regular basis to detect any unsafe condition, such as broken insulation, a potential shock hazard. The preventive maintenance program should rectify all unsafe conditions detected by the inspection program.

Several methods are available for shock prevention, including insulation, isolation, grounding, use of ground fault interrupters, and lockouts and enclosures. The following paragraphs summarize and briefly discuss these methods.

Insulation

Use of insulation is one of the best techniques available for preventing electrical shock.[2–6] The primary function of the insulating material is to prevent leakage of electrical current from the conductor. The second function of insulation, which is mechanical in nature, is to physically separate the conductor from the worker. Heat is generated as a result of electrical current within the conductor; so the third function of the insulation is to dissipate this heat and prevent overheating. The insulating material also should have a relatively long service life, which means good thermal, physical, and chemical stability.

Electrical insulating materials can be divided into organic and inorganic materials, and most of the organic insulating materials are combustible. Plastic materials such as polyvinyl chloride and polyethylene are the most widely used organic materials for electrical insulation.

Although the main function of electrical insulation is to prevent the leakage of current from the conductor, insulation can fail for a number of reasons. The environment can adversely affect the chemical structure of plastic insulating compounds, and in many cases, degrade the plastic compound into smaller molecules, with the loss of insulating, mechanical, and thermal properties. For example, the oxidation of polymeric compounds by oxygen or ozone can result in the degradation of the polymer chain and a substantial reduction in the average molecular weight of the polymer. As most of the electrical, chemical, mechanical, and thermal properties of polymers are governed by molecular weight and molecular weight distribution, oxygen or ozone can cause the insulating material to fail. Another environmental factor that can adversely affect the performance of insulation is ultraviolet radiation. When absorbed by some polymeric materials such as polyvinylchloride (PVC) or polyethylene, ultraviolet rays can supply sufficient energy to break the relatively weak covalent bonds within the polymer molecule and cause substantial reduction in the average molecular weight of the polymer. In the case of PVC, the chemical reactions also produce hydrogen chloride, which can further degrade the polymer. For example, these environmental factors cause the formation of conjugated double bonds in PVC, which is responsible for the discoloration of materials made from that polymer. The problem of ultraviolet absorption for many plastics and rubber compounds can be eliminated by the addition of an inhibiting compound such as carbon black, which provides ultraviolet light protection.

Another factor that can adversely affect the properties of electrical insulation is heat aging. Because heat is always produced as a result of electrical current flow within conductors, it is important to understand the effects of long-term heat exposure on the properties of insulation and to know the maximum exposure temperature for the material. The Underwriters Laboratories (UL) has developed a temperature index from which the upper use temperature of a polymer can be estimated. This index is obtained by exposing various samples of polymeric material to elevated temperatures for various time durations and by observing the adverse effect of temperature on the mechanical, thermal, and electrical properties of the polymer. For example, the upper use temperature for Nylon 6 is 65°C using the UL index.

Humidity can adversely affect the performance of electrical insulation. Water vapor can condense and penetrate the insulation, causing the electrical resistance of the insulation to be reduced substantially.

Insulation that survives environmental damage may fail from abuse. For example, cords may be subject to abrasion or impact damage, or may even be chewed by rodents.[2,3]

Isolation of Personnel

Every attempt should be made to isolate personnel who work on energized electrical equipment.[2-5] When ladders are used for work on electrical wiring, it is imperative that they be made of nonconducting materials such as wood or fiberglass. Plastic gloves designed for electrical work should be worn where there is a potential for contact with a live conductor. These gloves should be inspected periodically to ensure that they are free of holes and tears, which would adversely affect their conductive properties. Gloves used for electrical work should be stored in a cool, dark, and dry area to prevent deterioration of the polymeric material and loss of the gloves electrical insulating properties.

Personnel working around energized equipment can also be insulated by floor mats designed for electrical work. These mats should be kept dry, free of holes, and protected from conducting materials such as metal chips, nails, screws, and so on.

Grounding and Use of Ground Fault Interrupters

Grounding of electrical circuits and electrical equipment is done to protect personnel against shock, to safeguard against fire, and to protect against damage to electrical equipment. Grounding involves maintaining a path to the earth for the flow of excess electrons that have been generated through a fault in the system or the equipment. There are two kinds of grounding, electrical circuit or system grounding and electrical equipment grounding. Electrical grounds can be designed and installed, or they can be unintentional. In a designed ground the path for the flow of electrons to the earth usually is a large-diameter copper wire that may be attached to cold water underground piping or another grounded conductor. An unintentional ground may be formed through wet floors, furniture, and so on. Although unintentional grounds have a relatively large resistance compared to intentional grounds, they can produce severe shock in a person who touches a ''hot'' conductor while touching an unintentional ground.[1,3]

System Grounding

Electrical system grounding is accomplished when one conductor of the circuit is intentionally connected to the earth.[3] This is done to protect the circuit should a lightning strike or other high voltage occur. Grounding a system also stabilizes the voltage in the system so that expected voltage levels are not exceeded under normal conditions.

All alternating current circuits operating at 50 volts or more must be grounded to protect against power surges. For example, in most places the local utility company will ground the neutral wire at the transformer. This system ground will protect the wiring and appliances in a house or building from any accidental power surges due to lightning. The grounded conductor must once again be grounded when it is connected to any electrical appliance within the structure. This grounding usually is accomplished by connecting the conductor wire to an underground piping system or any other type of conductor buried in the ground.

It should be kept in mind that not all systems require grounding. In some instances, grounding a system might make it inherently unsafe. For example, in many manufacturing processes and plants, the operation of strategic equipment might be unsafely interrupted by the presence of a system ground. The equipment in manufacturing processes might operate under pressure and temperature processing conditions posing potential physical and/or health hazards, and the interruption of such equipment by a first fault to the ground could create such hazards as explosion, fire, or release of toxic gases to the atmosphere. However, in such cases it is imperative that if a system ground is not used, trained electrical technicians using fault current indicators be present to quickly rectify any unsafe condition.

OSHA regulations mandate system grounding for all 3-wire direct current (d.c.) systems usually used in heavy industrial applications such as large mills and extrusion machines. Also 2-wire d.c. systems that operate between 50 and 300 volts must be grounded. Common 2-wire d.c. systems include battery charging machines, electroplating, and some crane operations. Grounding is accomplished by making an electrical connection to ground at the source.

Equipment Grounding

Equipment grounding is accomplished when all metal frames of equipment and enclosures containing electrical equipment or conductors are grounded by means of a permanent and continuous connection or bond.[1-3] The equipment grounding conductor provides a path for a dangerous fault current to return to the system ground at the supply source of the circuit, should an insulation failure take place. If installed properly, the equipment grounding conductor is the current path that enables protective devices such as circuit breakers and fuses to operate when a fault occurs.

When insulation on the conductor is in good working condition, there is no need for equipment grounding. However, as was noted before, insulation can fail for a variety of reasons. When the insulation on a live conductor fails, a potential difference equal to the line voltage can develop across the metallic part of the equipment enclosure. This would pose a serious shock hazard to a grounded person who might touch the enclosure.

Electrical equipment with open heating coils such as toasters or electrical heaters should not be grounded. Grounding such equipment would increase the shock hazard. The exposed metallic part of the following equipment must be grounded unless specifically labeled ''double insulation'':

- Portable electric tools such as drills or saws.
- Communication receivers and transmitters.
- Electrical equipment located in damp locations such as garages or basements.
- TV antenna towers.
- Electrical equipment in hazardous locations such as a flammable liquid storage area.
- Electrical equipment operated in excess of 150 volts to ground.

One must exercise special care when working with equipment such as capacitors and transformers. Due to the stored charge there is a potential shock hazard associated with this type of equipment even when the equipment is turned off and unplugged. It is advisable to discharge the item by means of a shorting stick until repairs are complete.

The size of the equipment grounding conductor can play an important role in the safe handling of electrical appliances, as it is this conductor that must carry the fault current. Unless sufficient current can reach the circuit protective device, the circuit will not be protected. The wrong size grounding conductor also can pose the danger of overheating or the development of hot spots, which can lead to fires and/or equipment damage. The National Electric Code (NEC) should be consulted for proper size selection for the grounding conductor. Electrical equipment should be inspected on a regular basis to ensure that an electrically continuous equipment grounding path is maintained from the metal enclosure.

Ground Fault Interrupter (GFI)

Fatal shock can occur when the human body becomes part of the electrical current path between an ungrounded conductor and the ground. A ground fault interrupter (GFI), also known as a ground fault circuit interrupter (GFCI), is a fast-acting device that can detect the flow of current to the ground and act in milliseconds by opening the circuit. Although the person who becomes part of a ground fault circuit may feel an uncomfortable shock, it is unlikely that the shock will become fatal, simply because the person is not in the circuit for a sufficient time.

It is important to realize that GFIs only can provide protection against line-to-ground contact and provide no protection against line-to-line contact. For example, if someone touches both the ''hot'' wire and the neutral wire in a circuit, the GFI will do nothing, and the person could receive a fatal shock. It also is imperative to understand that in spite of all the success that GFIs have had in protecting against line-to-ground contact, they are no substitute for proper grounding and general safety precautions.

Lockouts and Enclosures

Before making any attempt to do maintenance work on electrical equipment, one must ensure that the power to the equipment has been shut off.[2,4,5] For smaller electrical appliances, the worker can simply unplug the device. However, larger electrical equipment used in manufacturing facilities normally is connected permanently to the power source, and the power to such equipment can be disconnected only through switches or circuit breakers. Many fatal shocks to electrical workers have occurred when maintenance work was being performed and a coworker turned on the power. Therefore, it is extremely important to lock the circuit breakers with one's own lock and to test the equipment to ensure that the right circuit breaker has been shut off. Some circuit breakers cannot be locked, and during maintenance a sign should be installed on such breakers indicating that work is being performed on the equipment. Figure 7.4 shows a typical warning sign.

Electrical enclosures provide an excellent way to protect personnel against accidental contact with live uninsulated conductors and should always be considered. However, workers must take extreme care when using metallic enclosures to ensure that they are adequately grounded. Temporary restraints such as ropes also help in preventing unauthorized personnel from entry into an area where there might be a potential shock hazard.

Electrical Equipment for Hazardous Locations

Hazardous locations may contain flammable liquids, gases, or vapors, or combustible dusts in sufficient quantities to produce an explosion or fire hazard. In hazardous locations, specially designed equipment and special installation techniques must be used to protect against the explosive and flammable potential of these substances.[1,3,5]

Hazardous locations are classified as Class I, Class II, or Class III, depending on the type of hazardous substance that may be present. In general, Class I locations are those in which flammable vapors and gases may be present; Class II locations are those in which combustible dusts may be found; Class III locations are those in which there are ignitable fibers.

Each of these classes is divided into two hazard categories, Division 1 and Division 2, depending on the likelihood of the presence of a flammable or ignitable concentration of a substance. In general, the electrical installation requirements for Division 1 locations are more stringent than for Division 2 locations. Division 1 locations are designated as such because a flammable

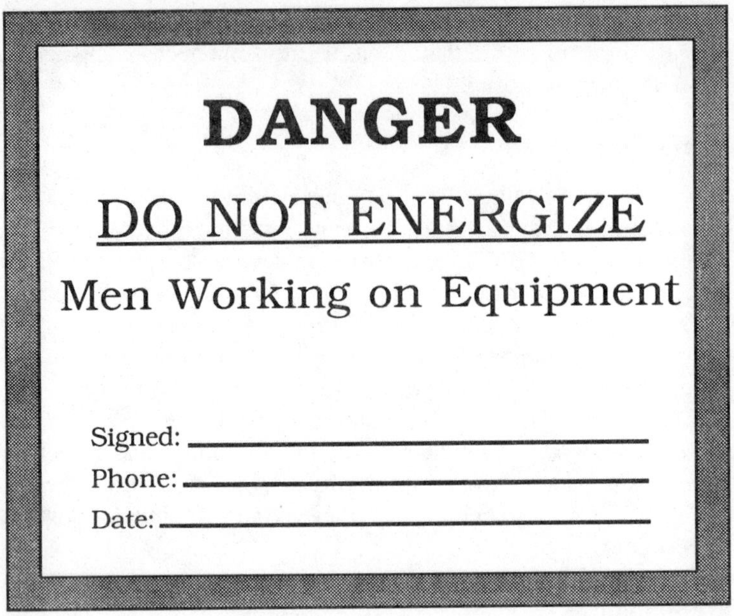

DANGER

DO NOT ENERGIZE

Men Working on Equipment

Signed: ______

Phone: ______

Date: ______

Figure 7.4. Typical warning sign for electrical equipment lockout.

gas, vapor, or dust is normally present in hazardous concentrations. In Division 2 locations, hazardous quantities of these materials are not usually present, but may occasionally exist either accidentally or when material in storage is handled.

Additionally, Class I and Class II locations are subdivided into groups of gases, vapors, and dusts having similar properties.

Table 7.2 summarizes the classification of different hazardous locations.

Equipment Design

General-purpose electrical equipment can cause explosions and fires in areas where there are flammable vapors, liquids, and gases and combustible dusts or fibers, which require special electrical equipment designed for the specific hazard involved.[4,5] Explosion-proof equipment should be designed for flammable vapor, liquid, and gas hazards and dust-ignition-proof equipment for combustible dust. Nonsparking equipment, intrinsically safe equipment, and purged and pressurized equipment also may be needed. In some cases, general-purpose or dust-tight equipment is permitted in Division 2 areas.

Electrical equipment may arc, spark, or produce heat under normal operating conditions. For example, circuit controls, switches, and contacts may arc or spark. Motors and lighting fixtures may heat up, producing temperatures high enough to cause ignition. Electrical equipment installed in these areas should control the sparking, arcing, and heating.

Installations in hazardous locations must be intrinsically safe and be designed to provide protection from hazards arising from the combustibility or flammability of vapors, liquids, gases, dusts, or fibers that might be present.

Equipment and wiring approved as intrinsically safe are acceptable in any hazardous (classified) location for which they are designed. Intrinsically safe equipment is not capable of releasing sufficient electrical or thermal energy to cause ignition of a specific flammable or combustible atmospheric mixture in its most easily ignitable concentration.

To avoid contaminating nonhazardous locations, the transport of flammable gases and vapors

Table 7.2. Summary of Class I, II, III Hazardous Locations.

Classes	Groups	Divisions 1	Divisions 2
I—Gases, vapors, and liquids (Art. 501 NEC)	A: Acetylene B: Hydrogen, etc. C: Ether, etc. D: Hydrocarbons, fuels, solvents, etc.	Normally explosive and hazardous	Not normally present in an explosive concentration (but may accidentally exist)
II—Dusts (Art. 502 NEC)	E: Metal dusts (conductive* and explosive) F: Carbon dusts (some conductive* and all explosive) G: Flour, starch, grain, combustible plastic or chemical dust (explosive)	Ignitable quantities of dust normally are, or may be, in suspension; or conductive dust may be present	Not normally suspended in an ignitable concentration (but may accidentally exist). Dust layers are present
III—Fibers and flyings (Art. 503 NEC)	Textiles, woodworking, etc. (easily ignitable, but not likely to be explosive)	Handled or used in manufacturing	Stored or handled in storage (exclusive of manufacturing)

*Note: Electrically conductive dusts are dusts with a resistivity less than 10^5 ohm-centimeter.
NEC: National Electric Code.
Source: References 4 and 5.

by equipment must be prevented. Interconnections between circuits must not produce an unexpected source of ignition to non-intrinsically safe equipment. Separation of intrinsically safe and non-intrinsically safe wiring may be necessary to ensure that the circuits in hazardous or classified locations remain safe.

Equipment approval must be based on the class, division, and group of the hazardous or classified location. There are two types of equipment specifically designed for hazardous or classified locations: explosion-proof and dust-ignition-proof. Explosion-proof apparatus is intended for Class I locations, and dust-ignition-proof equipment is primarily intended for Class II and Class III locations. Equipment approved specifically for hazardous locations carries an Underwriters Laboratories (UL) label, which indicates the class, division, and group for the location where it may be installed (see Figure 7.5). Equipment approved for use in a Division 1 location may be installed in a Division 2 location for the same class and group.

Generally, equipment installed in Class I locations must be approved as explosion-proof. It is impractical to keep flammable gases outside of enclosures; so arcing equipment must be placed in enclosures that will withstand an explosion. This minimizes the risk of an external explosion when a flammable gas enters the enclosure and is ignited by the arcs. Not only must the equipment be strong enough to withstand an internal explosion, but the enclosures must be designed to vent the resulting explosive gases. This venting must ensure that gases are cooled to a temperature below the ignition temperature of the hazardous substance involved before being released into the hazardous atmosphere.

When an internal explosion occurs, it tends to distort the shape of the enclosure from rectangular to elliptical, as exaggerated in Figure 7.6. Adequate strength is a major requirement for the design of an explosion-proof enclosure. To prevent failure of the enclosure, openings should be designed to relieve the pressure. All joints and flanges must be held to narrow tolerances.

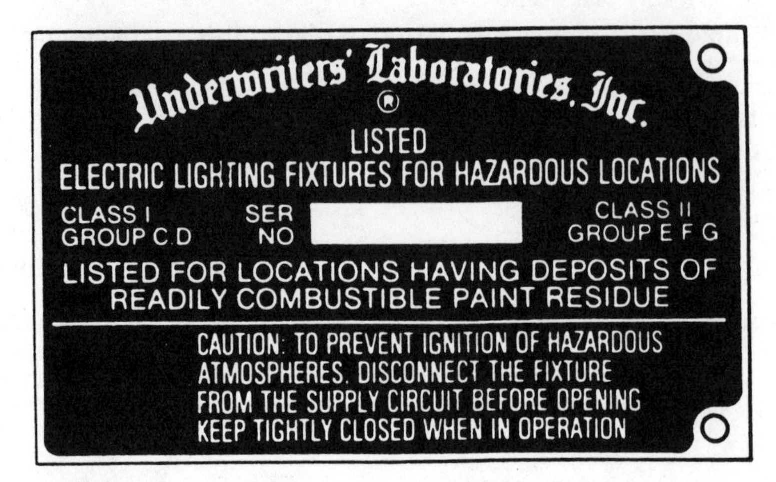

Figure 7.5. Label showing approval for use in hazardous (classified) locations. (Source: References 4 and 5)

Because the explosion characteristics of hazardous substances vary with the specific material involved, each group requires special design considerations. For Class I hazardous locations, there are four chemical atmosphere groups: A, B, C, and D (see Table 7.3). Design characteristics for these four groups require containment of maximum explosion pressure, maximum safe clearance between parts of enclosures, and operation at a temperature below the ignition temperature of the atmospheric mixture involved.

Dust-Ignition-Proof Equipment

In Class II, Division 1 locations, equipment generally must be dust-ignition-proof. Section 502-1 of the NEC defines dust-ignition-proof as equipment ''enclosed in a manner that will exclude ignitable amounts of dust or amounts that might affect performance or rating and that, where

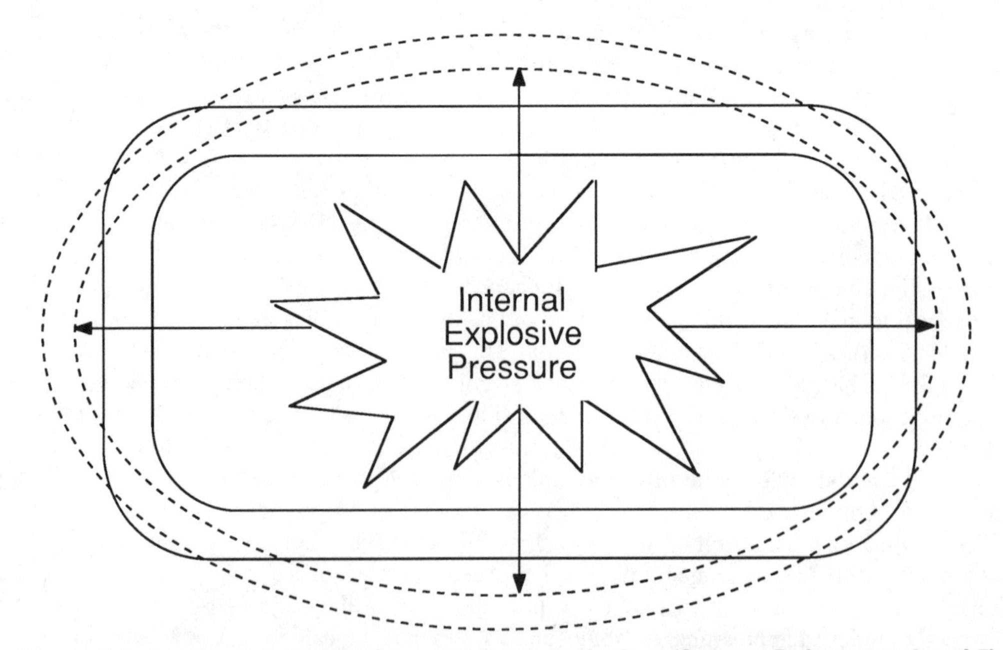

Figure 7.6. Equipment designed to contain an explosion. (Source: References 4 and 5)

Table 7.3. Chemical Atmospheres by Group—Class I.

Group A:	Acetylene
Group B:	Acrolein (inhibited), arsine, butadiene, ethylene oxide, hydrogen, manufactured gases containing more than 30% hydrogen (by volume), propylene oxide, propylnitrate
Group C:	Acetaldehyde, allyl alcohol, N-butyraldehyde, carbon monoxide, crotonaldehyde, cyclopropane, diethyl ether, diethylamine, epichlorohydrin, ethylene, ethylenimine, ethyl mercaptan, ethyl sulfide, hydrogen cyanide, hydrogen sulfide, morpholine, 2-nitropropane, tetrahydrofuran, unsymmetrical dimethyl hydrazine (UDMH 1, 1-dimethyl hydrazine)
Group D:	Acetic acid (glacial), acetone, acrylonitrile, ammonia, benzene, butane, 1-butanol (butyl alcohol), 2-butanol (secondary butyl alcohol), N-butyl acetate, isobutyl acetate, diisobutylene, ethane, ethanol (ethyl alcohol), ethyl acetate, ethyl acrylate (inhibited), ethylene diamine (anhydrous), ethylene dichloride, ethylene glycol monomethyl ether, gasoline, heptanes, hexanes, isoprene, isopropyl ether, mesityl oxide, methane (natural gas), methanol (methyl alcohol), 3-methyl-1-butanol (isoamyl alcohol), methyl ethyl ketone, methyl isobutyl ketone, 2-methyl-1-propanol (isobutyl alcohol), 2-methyl-2-propanol (tertiary butyl alcohol), petroleum naphtha, pyridine, octanes, pentanes, 1-pentanol (amyl alcohol), propane, 1-propanol (propyl alcohol), 2-propanol (isopropyl alcohol), propylene, stryene, toluene, vinyl acetate, vinyl chloride, xylenes

Source: Table 500-2, Article 500—Hazardous (classified) locations, 1981 National Electrical Code, National Fire Protection Association, Boston, Mass. Also, References 4 and 5.

installed and protected in accordance with this code, will not permit arcs, sparks, or heat otherwise generated or liberated inside the enclosure to cause ignition of exterior accumulations or atmospheric suspensions of a specified dust on or in the vicinity of the enclosure.''

Dust-ignition-proof equipment is designed to keep ignitable amounts of dust from entering the enclosure. In addition, dust may accumulate on electrical equipment, causing overheating of the equipment, as well as the dehydration or gradual carbonization of organic dust deposits. Overheated equipment may malfunction and cause a fire. Dust that has carbonized is susceptible to spontaneous ignition or smoldering. Therefore, equipment must be designed to operate below the ignition temperature of the specific dust involved.

In Class II hazardous locations, there are three groups: E, F, and G (see Table 7.4). Special designs are required to prevent dust from entering the electrical equipment enclosure. Assembly joints and motor shaft openings must be tight enough to prevent dust from entering the enclosure. In addition, the design must take into account the insulating effects of dust layers on equipment and must ensure that the equipment will operate below the ignition temperature of the dust involved. If conductive combustible dusts are present, the design of equipment must take the special nature of these dusts into account.

In general, equipment that is approved as explosion-proof may not be acceptable for use in Class II locations. For example, because grain dust has a lower ignition temperature than that of many flammable vapors, equipment approved for Class I locations may operate at a temperature that is too high for Class II locations. On the other hand, equipment that is dust-ignition-proof for Class II generally is acceptable for use in Class III locations because the same design considerations are involved for both.

Table 7.4. Chemical Atmospheres by Group—Class II.

Group E:	Metal dust, including aluminum, magnesium, and their commercial alloys, and other metals of similarly hazardous characteristics having resistivity of 10^2 ohm-centimeter or less
Group F:	Carbon black, charcoal, coal, or coke dusts
Group G:	Flour, starch, grain dust, or combustible plastic or chemical dusts having resistivity greater than 10^8 ohm-centimeter

Source: References 4 and 5.

Marking

Approved equipment must be marked to indicate the class, group, and operating temperature range. Furthermore, the temperature marked on the equipment must not be greater than the ignition temperature of the specific gases or vapors in the area.[2,4,5]

There are, however, four exceptions to this marking requirement. First, equipment that does not produce heat (for example, junction boxes or conduits) and equipment that does produce heat but has a maximum surface temperature of less than 100°C (212°F) is not required to be marked with the operating temperature range. The heat normally released from this equipment cannot ignite gases, liquids, vapors, or dusts.

Second, any permanent lighting fixtures that are approved and marked for use in Class I, Division 2 locations do not need to be marked to show a specific group, as these fixtures are acceptable for use with all of the chemical groups for Class I (Groups A, B, C, and D).

Third, fixed general-purpose equipment in Class I locations, other than lighting fixtures, that is acceptable for use in Division 2 locations does not have to be labeled according to class, group, division, or operating temperature. This type of equipment does not contain any devices that might produce arcs or sparks and thus is not a potential ignition source.

Fourth, for Class II, Division 2 and Class III locations, fixed dust-tight equipment, other than lighting fixtures, is not required to be marked. In these locations, dust-tight equipment does not present a hazard so it need not be identified.

Equipment installed in hazardous or classified locations must be of a type and design that will provide protection from hazards arising from the combustibility and flammability of vapors, liquids, gases, dusts, or fibers. The employer is responsible for demonstrating that the installation meets this requirement. Guidelines for installing equipment under this option are contained in the NEC in effect at the time of installation of that equipment. These guidelines must be met, and the employer must demonstrate that the installation is safe for the hazardous or classified location.

Class I Hazardous Locations

Article 501 of the NEC contains installation requirements for electrical wiring and equipment used in Class I hazardous areas. The requirements as they pertain to Class I, Division 1 hazardous locations are summarized in Table 7.5. The requirements for Class I, Division 2 locations are summarized in Table 7.6.

Class II Hazardous Locations

Article 502 of the NEC is concerned with the installation requirements for electrical wiring and equipment used in Class II hazardous areas. The requirements as they pertain to Class II, Division 1 and Division 2 locations are summarized in Table 7.7.

Class II locations are hazardous because of the presence of combustible dust. As discussed previously, these dusts are broken down into three groups: E, F, and G. The dusts are also divided into two categories: conductive (having resistivity less than 10^5 ohm-centimeter) and nonconductive. Where conductive dusts are present, there are only Class II, Division 1 locations. Group E dusts are conductive, some group F dusts are conductive, and Group G dusts are nonconductive.

In general, equipment in Class II, Division 1 locations should be dust-ignition-proof, whereas equipment in Division 2 locations need only be dust-tight. Additionally, equipment should be able to function at full rating without causing excessive dehydration or carbonization of organic dust deposits. Maximum operating surface temperatures are given in Table 7.8. Because some Group G chemical and plastic dusts have ignition temperatures approaching or below those given

Table 7.5. Summary of Equipment Requirements for Class I, Division 1 Hazardous Locations.

A. Meters, relays, and instruments, such as voltage or current meters and pressure or temperature sensors, must be in enclosures approved for Class I, Division 1 locations. Such enclosures include explosion-proof and purged and pressurized enclosures. See NEC Section 501-3(a).

B. Wiring methods acceptable for use in Class I, Division 1 locations include: threaded rigid metal or steel intermediate metal conduit and Type MI cable. Flexible fittings, such as motor terminations, must be approved for Class I locations. All boxes and enclosures must be explosion-proof and threaded for conduit or cable terminations. All joints must be wrench-tight with a minimum of five threads engaged. See NEC 501-4(a).

C. Sealing is required for conduit and cable systems to prevent the passage of gases, vapors, and flame from one part of the electrical installation to another through the conduit. Type MI cable inherently prevents this from happening by its construction; however, it must be sealed to keep moisture and other fluids from entering the cable at terminations. See NEC Section 501-5.
   (1) Seals are required where conduit passes from Division 1 to Division 2 or nonhazardous locations.
   (2) Seals are required within 18 inches of enclosures containing arcing devices.
   (3) Seals are required if conduit is 2 inches in diameter or larger entering an enclosure containing terminations, splices, or taps. See Figure 85 for a description of seals.

D. Drainage is required where liquid or condensed vapor may be trapped within an enclosure or raceway. An approved system of preventing accumulations and a system to permit periodic drainage are two methods used to control condensation of vapors and liquid accumulation. See NEC Section 501-5(f).

E. Arcing devices, such as switches, circuit breakers, motor controllers, and fuses, must be approved for Class I locations. See NEC Section 501-6(A).

F. Motors shall be: (1) approved for use in Class I, Division 1 locations; (2) totally enclosed with positive pressure ventilation; (3) totally enclosed inert-gas-filled with a positive pressure within the enclosure; or (4) submerged in a flammable liquid or gas. The last kind of installation is permissible, however, only when there is pressure on the enclosure that is greater than atmospheric pressure, and the liquid or gas is only flammable in air. This type of motor is not permitted to be energized until it has been purged of all air. The latter three types of motors must be arranged to be de-energized should the pressure fail or the supply of liquid or gas fail—as with the submerged type. Types (2) and (3) may not operate at a surface temperature above 80% of the ignition temperature of the gas or vapor involved. See NEC Section 501-8(a).

G. Lighting fixtures, both fixed and portable, must be explosion-proof and guarded against physical damage. See NEC Section 501-9(a).

H. Flexible cords must be designed for extrahard usage, contain an equipment grounding conductor, be supported so that there will be no tension on the terminal connections, and be provided with seals where they enter explosion-proof enclosures. See NEC Section 501-11.

I. Receptacles and attachment plugs for use with portable equipment must be approved explosion-proof devices and provided with an equipment grounding connection. See NEC Section 501-12.

J. Signaling, alarm, remote control, and communications systems are required to be approved for Class I, Division 1 locations regardless of voltage. See NEC Section 501-14(A).

K. Equipment grounding is required of all non-current-carrying metal parts of the electrical system. In addition, locknuts and bushings must not be relied upon for electrical connection between raceways and equipment. If locknuts and bushings are used, bonding jumpers are required. See NEC Section 501-16.

NEC: National Electrical Code, NFPA 70.
Source: References 4 and 5.

in the table, equipment used with such dusts should have even lower operating surface temperatures.

Class III Hazardous Locations

Class III hazardous locations are areas where ignitable fibers and flyings are present. In general, equipment acceptable for use in Class II, Division 2 locations also is acceptable for installation in Class III locations. Equipment in Class III locations should be able to operate at full rating

Table 7.6. Summary of Class I, Division 2 Hazardous Locations.

A. Meters, instruments, and relays in Class I, Division 2 locations must be in approved explosion-proof enclosures. However, general-purpose equipment may be used if circuit-interrupting contacts are immersed in oil or enclosed in a hermetically sealed chamber or in circuits that do not release enough energy to ignite the hazardous atmosphere. See NEC Section 501-3(b).

B. Wiring methods: Generally, threaded rigid or intermediate conduit or types PLTC, MI, MC, MV, TC, or SNM cable systems must be used. Boxes and fittings are not required to be explosion-proof unless they enclose arcing or sparking devices. See NEC Section 501-4(b).

C. Seals are required for all conduit systems connected to explosion-proof enclosures. Seals also are required where conduit passes from hazardous to nonhazardous areas or from Division 1 to Division 2 areas. See NEC Section 501-5(b).

D. Drainage is required where liquid or condensed vapor may be trapped within an enclosure or along a raceway. See NEC Section 501-5(f).

E. Most arcing devices are required to be in explosion-proof enclosures. These include items such as switches, circuit breakers, motor controllers, and fuses. However, general-purpose enclosures may be used for Class I, Division 2 locations if the arcing and sparking parts are contained in a hermetically sealed chamber or are oil-immersed. See NEC Section 501-6(b).

F. Motors, generators, and other rotating electrical machinery suitable for use in Class I, Division 1 locations also are acceptable in Class I, Division 2 locations. Other motors must have their contacts, switching devices, and resistance devices in enclosures suitable for Class I, Division 2 locations (see note E, above). Motors without brushes, switching mechanisms, of similar arc-producing devices also are acceptable. See NEC Section 501-8(b).

G. Lighting fixtures in Class I, Division 2 locations must be totally enclosed and protected from physical damage. If normal operating surface temperatures exceed 80% of the ignition temperature of the gas, liquid, or vapor involved, then explosion-proof fixtures must be installed. See NEC Section 501-9(b).

H. Flexible cords in Divisions 1 and 2 are required to: (1) be suitable for extrahard usage, (2) contain an equipment grounding conductor, and (3) be connected to terminals in an approved manner, (4) be properly supported, and (5) be provided with suitable seals where necessary. See NEC Section 501-11.

I. In general, receptacles and attachment plugs must be approved for Class I locations. See NEC Section 501-12.

J. For signaling systems and other similar systems, see NEC Section 501-14.

K. Equipment grounding is required of all non-current-carrying metal parts of the electrical system. In addition, locknuts and bushings must be relied upon for electrical connection between raceways and equipment. If locknuts and bushings are used, bonding jumpers are required. See NEC Section 501-16.

NEC: National Electrical Code, NFPA 70.
Source: References 4 and 5.

without causing excessive dehydration or carbonization of accumulated fibers or flyings. The maximum operating surface temperature is 165°C (329°F) for equipment that is not subject to overloading, and 120°C (248°F) for equipment that may be overloaded.

Table 7.9 summarizes some of the requirements for installations in Class III locations.

Static Electricity

Static electricity is a surface phenomenon due to electron exchange between two contacting surfaces (see Figure 7.7). When the two surfaces are in close proximity to each other, the negative charge of excess electrons on one surface is neutralized by the positive charge of the electron-deficient surface. When the two surfaces are separated, the accumulated charge has no place to go, and is called static electricity. Insulating materials such as plastics hinder the movement of electrons; so the more insulating a material is, the more difficult it is to establish neutrality on separation of the two contacting surfaces, and the greater the degree of charge accumulation. Although friction is not a requirement for the generation of static electricity, it enhances the electron transfer process and increases the magnitude of charge accumulation.

The accumulated charge on a given surface, because of its high potential, can flow to a nearby

Table 7.7. Summary of Class II Hazardous Locations.

A. Wiring methods for Class II, Division 1 locations: Boxes and fittings containing arcing and sparking parts are required to be in dust-ignition-proof enclosures. For other than flexible connections, threaded metal conduit or Type MI cable with approved terminations is required for Class II, Division 1 locations. See NEC Section 501-4(a). In Class II, Division 2 locations, boxes and fittings are not required to be dust-ignition-proof but must be designed to minimize the entrance of dust and prevent the escape of sparks or burning material. In addition to the wiring systems suitable for Division 1 locations, the following systems are suitable for Division 2 locations: electrical metallic tubings, dust-tight wireways, and types MC and SNM cables. See NEC Section 502-4(b).

B. Suitable means of preventing the entrance of dust into a dust-ignition-proof enclosure must be provided where a raceway provides a path to the dust-ignition-proof enclosure from another enclosure that could allow the entrance of dust. See NEC Section 502-5.

C. Switches, circuit breakers, motor controllers, and fuses installed in Class II, Division 1 locations must be dust-ignition-proof. In Class II, Division 2 areas, enclosures for fuses, switches, circuit breakers, and motor controllers must be dust-tight. See NEC Section 502-6.

D. In Class II, Division 1 locations, motors generators and other rotating electrical machinery must be dust-ignition-proof or totally enclosed pipe-ventilated. In Class II, Division 2 areas, rotating equipment must be one of the following types: (1) dust-ignition-proof, (2) totally enclosed pipe-ventilated, (3) totally enclosed nonventilated, or (4) totally enclosed fan-cooled. Under certain conditions, standard open-type machines and self-cleaning squirrel-cage motors may be used. See NEC Section 502-8.

E. In Class II, Division 1 locations, lighting fixtures must be dust-ignition-proof. Lighting fixtures in Class II, Division 2 locations must be designed to minimize the accumulation of dust and must be enclosed to prevent the release of sparks or hot metal. In both divisions, each fixture must be clearly marked for the maximum wattage of the lamp, so that the maximum permissible surface temperature for the fixture is not exceeded. Additionally, fixtures must be protected from damage. See NEC Section 502-11.

F. Flexible cords in Divisions 1 and 2 are required to: (1) be suitable for extrahard usage, (2) contain an equipment grounding conductor, (3) be connected to terminals in an approved manner, (4) be properly supported, and (5) be provided with suitable seals where necessary. See NEC Section 501-12.

G. Receptacles and attachment plugs used in Class II, Division 1 areas are required to be approved for Class II locations and provided with a connection for an equipment grounding conductor. In Division 2 areas, the receptacle must be designed so the connection to the supply circuit cannot be made or broken while the parts are exposed. This is commonly done with an interlocking arrangement between a circuit breaker and the receptacle. The plug cannot be removed until the circuit breaker is in the off position, and the breaker cannot be switched to the on position unless the plug is inserted in the receptacle. See NEC Section 502-13.

H. For signaling systems and other similar systems, see NEC Section 501-14.

I. Equipment grounding is required of all non-current-carrying metal parts of the electrical system. Locknuts and bushings must not be relied upon for electrical connection between raceways and equipment enclosures. If locknuts or bushings are used, bonding jumpers are required. See NEC Section 502-16.

NEC: National Electrical Code, NFPA 70.
Source: References 4 and 5.

Table 7.8. Maximum Safe Surface Temperatures for Equipment in Class II Hazardous Locations.

Class II Group	Equipment that is not Subject to Overloading		Equipment (such as motors or power transformers) that May be Overloaded			
			Normal Operation		Abnormal Operation	
	°C	°F	°C	°F	°C	°F
E	200	392	200	392	200	392
F	200	392	150	302	200	392
G	165	329	120	248	165	329

Source: References 4 and 5.

Table 7.9. Summary of Class III Hazardous Locations.

A. In Class III hazardous locations, wiring must be within a threaded metal conduit or be of type MI or MC cable unless flexibility is required. Fittings and boxes are required to provide an enclosure that will prevent the escape of sparks or burning material. See NEC Section 503-3.

B. Switches, circuit breakers, motor controllers, and similar devices used in Class III hazardous locations must be within tight metal enclosures that are designed to minimize the entry of fibers and flyings and must not have any openings through which sparks or burning materials might escape. See NEC Section 503-4.

C. Motors, generators, and other rotating electric machinery must be totally enclosed nonventilated, totally enclosed pipe-ventilated, or totally enclosed fan-cooled. The windings of totally enclosed nonventilated motors are completely enclosed in a tight casing and are cooled by radiation and conduction through the frame. Enclosed pipe-ventilated motors have openings for a ventilating pipe, which conveys air to the motor and then discharges the air to a safe area. In totally enclosed fan-cooled motors, the windings are cooled by an internal fan that circulates air inside the enclosure. Under certain conditions, self-cleaning textile motors and standard open-type machines may be used. See NEC Section 503-6.

D. Lighting fixtures must have enclosures designed to minimize the entry of fibers, to prevent the escape of sparks or hot metal, and to have a maximum exposed surface temperature or less than 165°C. See NEC Section 503-9.

NEC: National Electrical Code, NFPA 70.
Source: References 4 and 5.

conducting object, and, if the magnitude of the charge is large enough, can generate a spark. Static electricity sparks have been responsible for many fires in flammable liquids.

The hazards of static electricity can range from a minor shock on cold, dry days to devastating fires and explosions. The minor shock a person receives from a static electricity discharge is never fatal, but the natural reaction to this shock can lead to a serious accident if the person is working on a ladder or near moving machinery.

Some flammable liquids such as jet fuel and fuel oil can generate and retain large amounts of electrostatic charges as a result of contact between fluid layers when they are pumped through pipelines. When these flammable charged liquids are stored in tanks or transported, a spark can develop in the vapor space above the liquid surface (see Figure 7.8). The spark may supply enough energy to raise the vapor temperature to ignition, and if the vapor–oxygen mixture is within the explosive range, a devastating explosion and fire can result.[2,3]

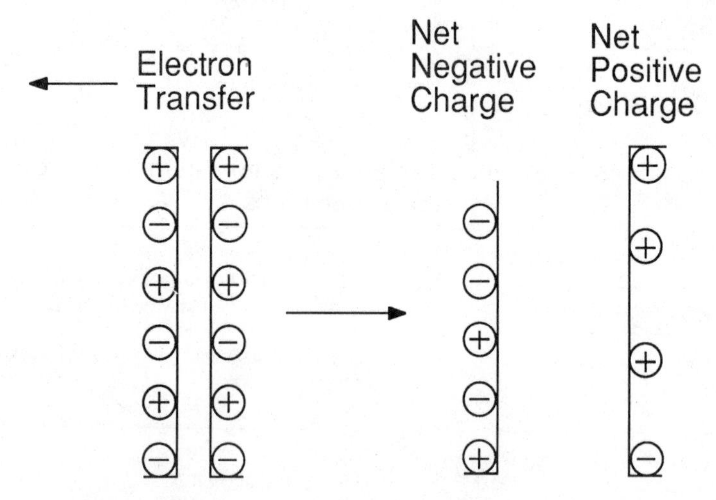

Figure 7.7. Static electricity is a surface phenomenon.

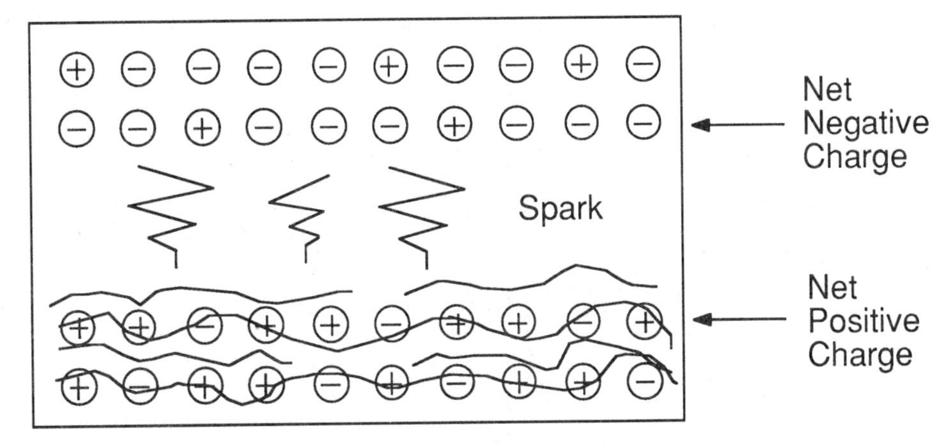

Figure 7.8. Static electricity generation in flammable liquids.

Electrostatic Hazard Control

The principal hazard in dealing with static electricity is the ignition of flammable liquids, vapors, and dusts. Although it is preferable to eliminate such atmospheres, a flammable atmosphere may exist in some industrial situations. Examples include "empty" drums or tanks of flammable liquids, or the generation of an explosive atmosphere in a flammable liquid storage area as a result of leaks or spills. Although it is preferrable to avoid the generation of static electricity altogether, in many industrial operations it is necessary to transfer and transport flammable liquids that can accumulate large amount of electrostatic charges. As noted above, the principal hazard associated with static electricity is not the accumulation of charges, but the transfer of these charges to an object with lower potential. There are several ways to control and eliminate the hazards associated with static electricity by providing a conducting path for the accumulated charges to discharge safely. These methods include bonding and grounding, use of antistatic and conductive materials, humidification, and ionization of the surrounding atmosphere.[1-3]

Bonding and Grounding

When two conducting bodies are connected by a conductor, charges can flow freely between the two bodies and eliminate potential differences and the possibility of sparking. The mechanism of connecting two conducting bodies by means of a conductor is known as bonding. Grounding, on the other hand, provides a conducting path between the charged object and the earth (see Figure 7.9). The earth can act as an infinite source to which electrons can flow or be removed; so any excess charges can flow to the earth safely when the charged object becomes grounded.

It should be noted that although bonding eliminates the possibility of sparking between bonded objects, it does not eliminate the possibility of sparking between the charged object and other objects or the ground.

Many flammable liquids can build up electrostatic charges when agitated or during transfer; so it is imperative to ensure that proper bonding and grounding procedures have been followed before any attempt is made to transfer a flammable liquid. When flammable liquids are transferred from a storage container to another container, a piece of noninsulated wiring must be used for bonding the two containers. The use of insulated wiring is not recommended, as it cannot be inspected for loss of electrical continuity as a result of corrosion or breakdown. If grounding is used, the grounding connections must be inspected to ensure electrical continuity and provision

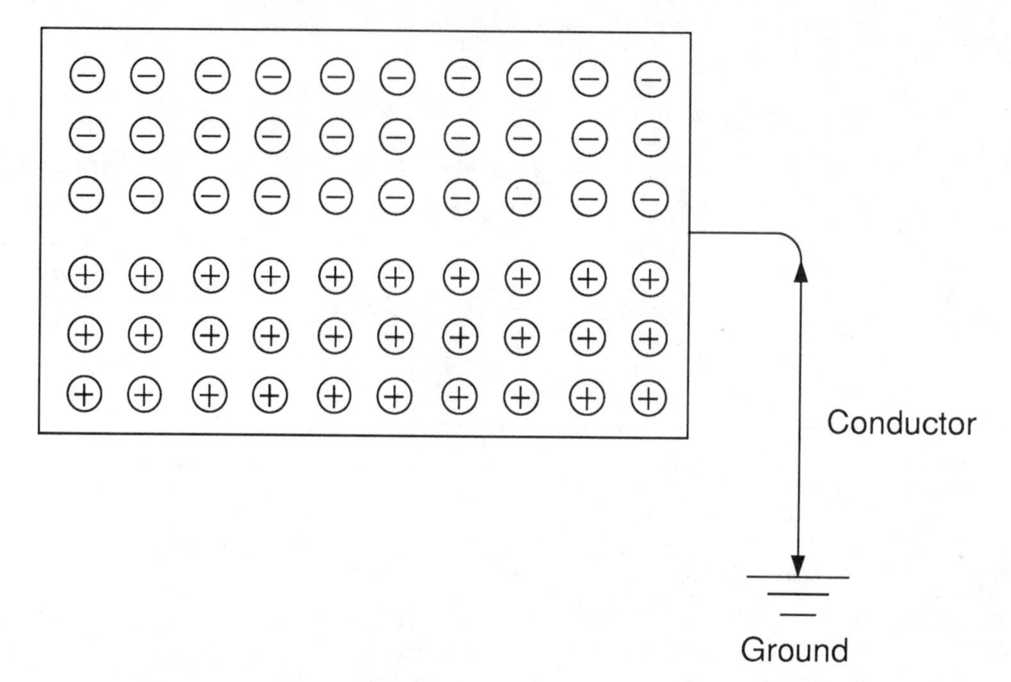

Figure 7.9. Grounding eliminates the buildup of static charges.

of a low-resistance path to the ground. Aboveground storage tanks containing flammable liquids need to be grounded only if they are on concrete or some other nonconducting support.

In extremely hazardous locations where there is danger of fire or explosion due to static charges, personnel may need to be grounded as well as equipment. The grounding of personnel can be accomplished by a number of methods, including the use of conductive floors, conductive clothing such as shoes, or clothing containing metallic fiber (see Figure 7.10).

Bonding and grounding also should be used during the transfer of flammable liquids to tank cars and in moving belts made of nonconductive materials.

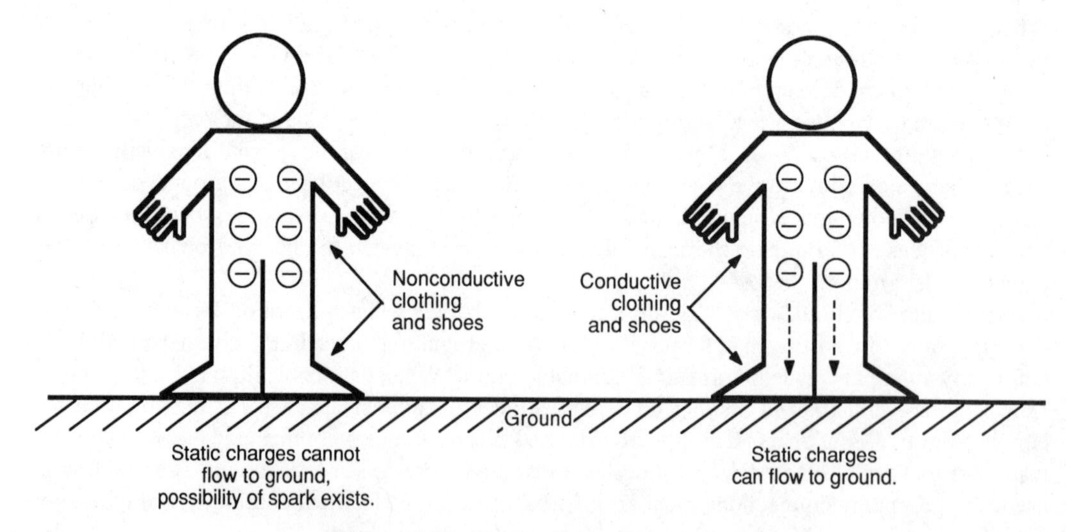

Figure 7.10. Grounding of personnel.

Antistatic Conducting Materials

The primary objective in electrostatic hazard control is to provide a safe conducting path for the flow of accumulated charges. Natural fibers such as cotton or wool have a tendency to absorb moisture, which would drastically reduce the resistance to flow of electrons and reduce the degree of charge accumulation.

Antistatic compounds also can increase the surface conductivity of the charged material. These are mostly chemical surfactants with high molecular polarity, which can substantially increase the surface conductivity.[2]

Humidification

Relative humidity is defined as the ratio of the partial pressure of water vapor in air to the equilibrium vapor pressure of water at the surrounding temperature. It should be noted that if the same amount of water in air is present at two different temperatures, the relative humidities are different because the equilibrium vapor pressure of water is a strong function of temperature. This is an important factor because it dictates the rates of condensation and vaporization on the surface of a charged nonconducting material, thus greatly affecting any increase in conductivity of the charged surface due to condensation. When the humidity is high, enough water molecules can condense on the surface of a nonconducting material to make it conductive, thus providing a path for charged particles to flow to the earth and reducing the possibility of sparking. Although the rates of accumulation of charges between the two contacting surfaces are the same in the summer and winter months, the electrostatic effects are much more pronounced on cold, dry winter days because low temperature and low humidity make the charged surface extremely nonconductive. On the other hand, static electricity normally is not a problem when the relative humidity is 70% or more. Humidification is not a widely accepted method for the control of electrostatic hazards as it can generate an uncomfortable working environment and has the potential of adversely affecting equipment.

Ionization

In certain situations when grounding or bonding is not practical, the air surrounding a charged object can be ionized to provide a conducting path for the flow of charges. Typical ionizing devices include: electronic static bars, radioactive static eliminators, and static combs.

Lightning

The convective updrafts and downdrafts acting on cloud particles with different charge polarities can result in charge separation and distribution within the cloud. Positive charges accumulate in the upper portion of the cloud, and negative charges accumulate in the lower portion. This charge separation can create potential differences of up to 100 million volts with current flows of up to 300,000 amperes. The electric field created as a result of this charge separation includes the earth and can cause charge separation on the surface of the earth by induction. The neutralization of charges can take place in two different ways: the negative charges can flow from the lower part of the cloud to the electron-deficient upper portion, causing an ''intracloud flash''; or neutralization can take place by flow of electrons and negative ions from the lower portion of the cloud to the induced charges on the surface of the earth—a phenomenon called a ''cloud to earth flash.''[2]

Lightning Hazards

The flow of static charges from a cloud follows the path of least resistance to the earth. This path can be a natural one, such as tall trees or mountaintops, or it can be man-made, such as tall buildings, TV and radio antennas, and so on; and sometimes it can include a human being. When the human body provides the path for the flow of lightning charges to the earth, electrocution—responsible for 150 fatalities annually in the United States—is almost inevitable. Lightning hazards can be summarized as follows:[2]

- Electrocution of unprotected persons.
- Hazards to airplanes as a result of either intracloud flash or cloud to earth flash.
- Damage to structures.
- Damage to electrical equipment.

Lightning Hazard Control Measures

The underlying concept in the control of lightning hazards is to intercept the flash at a safe point and provide a path of safe and low resistance from that point to the earth. For example, on structures of all kinds metallic rods are installed on the high points of the structure, with a low-resistance wire connecting the rod to the ground. It is important to bond all metallic structures and pipes to this system to prevent development of a potential difference. The following safety measures must be followed during lightning storms:[2,3]

- Avoid standing in high points. Remember that the flow of lightning current takes the path of least resistance to the earth.
- If you are in an automobile, stay inside. The car provides one of the best protections against lightning.
- Stay indoors; but if you have to be outside, wear rubber clothing.
- Avoid staying underneath tall trees. If you are in a forest, stay under a short tree away from the tallest trees, preferably in low-lying areas such as ditches, canyons, and so on.
- Do not stand near open windows and doors because the resistance to current flow has been reduced there, and the lightning current can flow to the ground through your body.
- During a storm do not use plug-in equipment such as hair dryers.
- Avoid using the telephone, as lightning might hit the telephone lines outside your house or building.
- Avoid working on or near conducting surfaces such as fences.
- If you are in a small boat, lie down. Remember that your body could be the least resistant path for electric current in an open area.
- Do not handle flammable liquids and gases during a lightning storm.

Safety Precautions for Electrical Hazards

The use of proper insulation, grounding and bonding techniques, lockouts, isolation, and the right kind of ground fault interrupters can greatly contribute to eliminating the hazards associated with electricity. However, as in any other operation, safe work practices and procedures are a critical element in working safely with electrical equipment. The following can help to provide a safe working environment in your operation.[2-5]

- Before performing any maintenance work on any electrical equipment, be sure that the power has been shut off and proper lockout tagout procedures have been followed.
- Do not work on any electrical equipment if you are not qualified to do so, or if you are not sure what you are doing.
- Use proper protective equipment when working with electricity. A hard hat made of conductive material, a metallic ladder, or a rubber glove with holes in it would not give you much protection against shock in case of contact with a bare conductor.

- The fact that the power to a system has been shut off does not necessarily mean that the system is de-energized. Many capacitors can store electric charges for a long period of time. Always test the system to be sure that it is de-energized.
- In working with electricity and electrical equipment, always remember Murphy's law that "anything that can go wrong, will go wrong."
- Do not wear conducting materials such as rings, watches, bracelets, and so on, when working with electricity.
- When working with electrical equipment, be sure that proper grounding and bonding procedures have been followed.
- Inspect insulation on wiring to ensure that it is in good working order. Insulation can fail, leaving no protection against fatal shock.
- When working with electrical equipment, never stand on wet or damp floors. This would provide a low-resistance path to the ground in case of accidental contact. Use rubber mats that are free of holes and conducting materials such as nails and screws.
- When replacing fuses, be sure that the right-capacity fuse is being used. Do not use fuses of a higher load rating as a replacement.
- In case of electrical fire, use the right kind of fire extinguishing agent made of nonconductive materials. You could get a fatal shock by using water.
- Use the proper enclosure for bare conductors.
- In hazardous locations such as a flammable liquid storage area, use only explosion-proof devices and nonsparking switches and tools.
- Provide lightning protection on all structures.
- Train some employees in basic first aid and CPR for emergencies.
- Never work on energized electrical equipment without protective equipment, especially when your fingers are wet or your body resistance has been reduced by perspiration.

References

1. OSHA, Instruction CPL2.35 CH2, Office of Compliance Programming, Washington, D.C. (1982).
2. Lacy, E. A., *Handbook of Electronic Safety Procedures*, 1st Ed., Prentice-Hall, Englewood Cliffs, N.J. (1977).
3. National Safety Council, *Accident Prevention Manual for Industrial Operations*, 9th Ed., National Safety Council, Chicago (1986).
4. OSHA, *An Illustrated Guide to Electrical Safety*, U.S. Department of Labor, Washington, D.C. (1983).
5. American Society of Safety Engineers (ASSE), *An Illustrated Guide to Electrical Safety*, ASSE, Des Plaines, Ill. (1983).
6. Hilado, C. J., *Flammability Handbook for Electrical Insulation*, 1st Ed., Technomic Publishing Co., Westport, Conn. (1982).

Exercises

1. Describe how an electric current can be produced.
2. Determine the amount of current, in milliamperes, flowing through the body of a person who is grounded and in contact with a copper conductor in a 110-volt line. The resistance of copper wiring in 1,100 ohms.
3. Discuss the factors affecting the outcome of an electric current passing through the body.
4. Discuss the different ways that an electric current can cause severe injury or death in humans.
5. Describe at least three methods for shock prevention.
6. Describe how environmental factors can affect the properties of electrical insulation.
7. What is the main objective of system grounding? Why, in some instances, does system grounding create an unsafe condition?
8. Briefly describe equipment grounding, and state the need for this type of grounding.

9. What is a ground fault interrupter (GFI), and how can it provide protection against fatal shocks?
10. What is an intrinsically safe electrical device?
11. What are the two types of electrical equipment specifically designed for classified locations?
12. What is the major safety requirement for electrical equipment that is to be installed in a Class I location?
13. Define a dust-ignition-proof electrical device. Which class and division of hazardous locations require the use of this type of electrical equipment?
14. Describe how electrostatic charges can develop between two contacting surfaces.
15. Describe how static electricity can cause a fire in flammable liquids.
16. What are the four principal methods for control of electrostatic hazards?
17. Explain why bonding does not eliminate the possibility of a spark between charged objects and the ground or another objects.
18. Under what conditions must aboveground flammable storage tanks be grounded?
19. Describe the mechanism through which an intracloud flash and a cloud to earth flash take place.

Chapter

8

Determining the Cause of Accidents and Conducting Effective Safety Audits

Thousands of accidents occur throughout the United States every day, most due to failure or improper interaction among equipment, people, supplies, and the environment.

An accident can be defined as any unplanned event that can cause loss of human life, injury, and/or property damage. Accidents usually are complex occurrences, composed of many individual events. Generally, most accidents are the result of an unsafe condition or an unsafe act. According to statistics published by the National Safety Council, 98% of industrial accidents are caused by unsafe conditions and unsafe acts.[1] Natural disasters such as earthquakes have contributed to only 2% of industrial accidents. In this regard, the systematic elimination of unsafe conditions and unsafe acts in the workplace can reduce the probability of an accident by 98%. Because accidents do not just happen but are caused, it is important for management to devote as much concern to conditions before an accident as after it; it is more critical to realize what can cause an accident than what has caused it.

Basic, Indirect, and Direct Causes

Accident causes can be divided into basic causes, indirect causes, and direct causes. Although unsafe conditions and unsafe acts can trigger an accident, they are only symptoms of failure; the basic causes of accidents usually can be traced to poor management policies and decisions. Unsafe acts and conditions due to poor management can be categorized as indirect causes of accidents, along with personal factors such as lack of skill, poor vision, or use of drugs or alcohol. Unplanned releases of energy and/or hazardous materials are direct causes of accidents resulting from unsafe conditions and unsafe acts (see Figure 8.1).[2]

In order to prevent the occurrence or to minimize the impact of an accident, each of the three categories of accident causes must be analyzed and properly managed.

Management Safety Program

Most accident prevention programs and activities involve methods and procedures for the elimination of unsafe conditions and acts. Although this should be an important part of any accident prevention program, it is equally important to understand that unsafe acts and conditions are the symptoms of poor management.

The management in any organization must have a safety policy clearly describing management's intent for safety in areas such as the following:[3]

- *Production relative to safety:* This is an important area for management to address. For example, an effective management program must value safety as highly as production, quality, and employee morale; otherwise safety rules can easily be ignored for the sake of production or product quality.

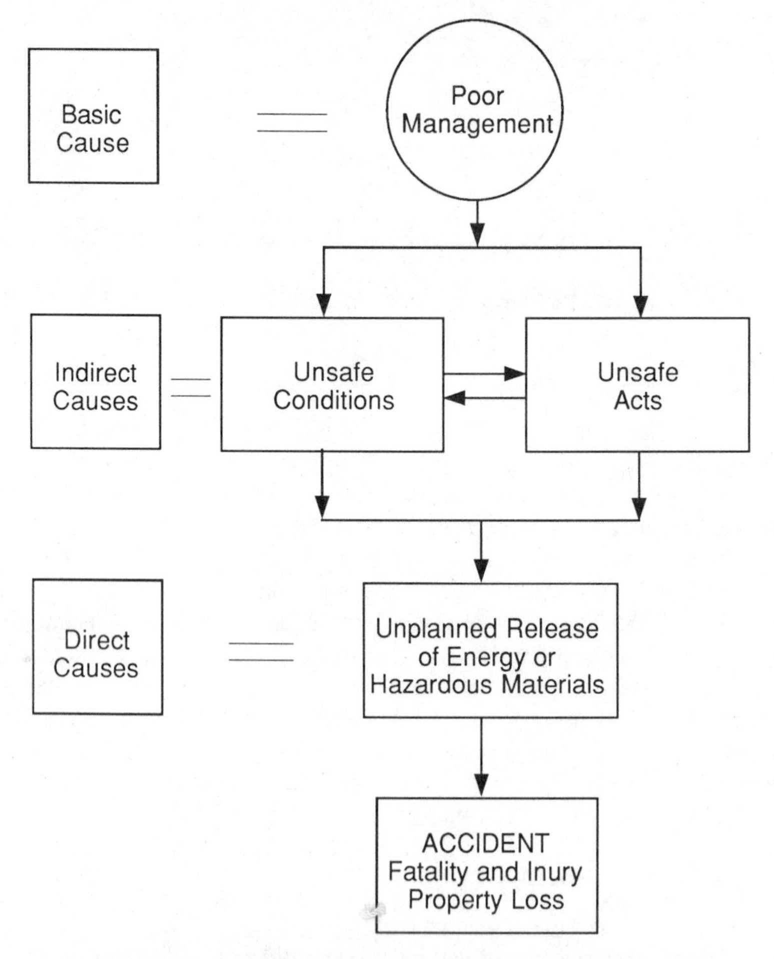

Figure 8.1. Causes of accidents. (Source: Reference 2)

- *Staffing procedures:* Management must ensure that an adequate staff is in place to carry out safety policies and rules and to provide a means for the smooth upward and downward flow of safety information.
- *Assignment of responsibility, authority, and accountability:* It is extremely crucial for management to clearly define the responsibility, authority, and accountability of each member of the organization in regard to safety. Those individuals who are responsible for the creation or the enforcement of safety rules (such as safety officers or members of an inspection team) must have the authority to carry out their duties and have the clear support of management.
- *Employee selection, training, and supervision:* Management should clearly define its policies in regard to the selection, training, and supervision of all personnel in general, and of personnel who are responsible for safety in particular. The latter should receive enough training to acquire the knowledge and expertise needed to carry out their duties.
- *Communication procedures:* It is management's responsibility to set forth policies for the flow of safety information within the organization. Often a safety program may sound effective but fail for lack of proper communication among management, safety personnel, and other employees.
- *Inspection procedures:* Management should clearly define how and by whom safety inspections will be conducted, emphasizing how the recommendations that may result from a safety

inspection will be implemented within the organization. For example, if a number of unsafe conditions have been identified in a safety inspection, management should clearly establish procedures and a time frame for the rectification of those conditions.

- *Equipment, supplies, and facilities design:* An effective management safety program should set forth policies and procedures for the safe design of new equipment and any design changes needed in existing equipment, supplies, or facilities. Many accidents have occurred because of an unsafe design or a change in the design of otherwise safe equipment. For example, management might require that any design changes in equipment be approved by the organizations' safety committee or safety office.
- *Standard and emergency procedures:* An effective safety program requires a system that clearly defines the organization's standard operating and emergency procedures. The standard operating procedures must contain guidelines for the safe use of equipment, safe work practices, and personal protective equipment. The emergency procedures must clearly identify the steps to be followed during an emergency.

Symptoms of Human Failure

As noted earlier, most industrial accidents are caused by unsafe acts and unsafe conditions. It should be understood, however, that unsafe acts and conditions do not cause accidents by themselves. They are often the result of poor management, inadequate controls, lack of knowledge, improper assessment of hazards, or other personal factors. A sound accident prevention program must set forth a system for the identification and the rectification of unsafe practices and conditions.[1,3–5]

Unsafe Acts

An effective method for the identification of unsafe acts is to actually observe a worker performing a task. The job can be broken down into several steps, and each step can be analyzed for maximum worker safety. Unsafe acts fall into the category of human failure, and they usually result in an accident or the creation of an unsafe condition. For example, a worker who places an open container of gasoline (flash point = 40°F) in a warehouse is committing an unsafe act; and as a result of that unsafe practice, an unsafe condition (potential for a fire) is created.

The most effective way to deal with an unsafe act is to *stop* the act immediately and *instruct* the employee on how the job can be performed in a safe manner.[5] It is equally important to have a *training* program in place to ensure that each employee is properly trained in how to perform his or her tasks safely. In any organization a few employees may enjoy committing unsafe practices; so a system must exist to *discipline* such employees for repeated and willful commitment of unsafe acts.

Because unsafe acts have been a major contributing factor to many industrial accidents, many organizations have a system in place for recording such acts. Figure 8.2 shows information that might be included on a typical unsafe act form.

It is imperative for supervisors and managers to conduct regular inspections of their facilities and focus their attention on the identification of unsafe practices and on the reasons for their occurrences. The following examples of unsafe acts highlight different categories of human failure:[4,5]

- *Operating without qualification or authorization:* There are few instances when an employee might operate an apparatus or machine for which he or she is not properly trained. It is important that all equipment be operated by authorized personnel who are familiar with its dangers and operating procedures. In the operation of complex equipment or in situations where an instrument is used by several individuals (such as instruments in college labora-

CALIFORNIA STATE UNIVERSITY, LONG BEACH
SCHOOL OF ENGINEERING

UNSAFE ACT REPORT

Date: ______________________

Location where unsafe act took place: Bldg.__________ Room__________

Department: ______________

Name of person committing unsafe act: ______________________

Name of supervisor notified: ______________________

Unsafe act identified by: ______________________

Description of unsafe act:

Potential hazard of the unsafe act:

cc: Department Chair
Faculty Member
Dean
Safety Director
File

Figure 8.2. Unsafe act form.

tories), it is necessary to prepare a written operating procedure and list any personal protective equipment that might be needed.

- *Lack or improper use of PPE:* There may be occasions when a worker does not have or is not using the personal protective devices required for the safe performance of a given task.
- *Failure to secure equipment:* Many items of equipment are subject to automatic startup or unexpected movement. Examples are some air compressors and ladders. Failure to secure such equipment prior to its use constitutes an unsafe act.

- *Operating equipment at unsafe speed:* Often workers will try to finish jobs too quickly, perhaps operating a vehicle or a machine at unsafe speeds. Workers also may take shortcuts that can contribute to accidents. Management must ensure that such activities are not encouraged.
- *Failure to warn:* If equipment has the potential of automatic startup or shutdown, or if equipment is about to be moved, adequate warning signals must be given. Also floors or other walking surfaces that might pose a danger should have adequate warning signs.
- *Bypass or removal of safety devices:* Most equipment comes with safety devices such as guards, fuses, and so on. Some workers have a tendency to remove or bypass such devices for their own comfort.
- *Using defective equipment:* Equipment often will become defective as a result of wear and tear or improper use. For example, a piece of cable with defective insulation can generate a fatal electric shock.
- *Use of tools for other than their intended purpose:* One of many examples is the use of chiesels to open drums.
- *Working in hazardous locations without adequate protection or warning:* Examples include such things as cleaning a hazardous spill without appropriate PPE or working in a traffic lane without lights or a warning.
- *Improper repair of equipment:* Among the many examples are welding done on a container of flammable liquid, working on live electrical equipment, and working on a machine that can accidentally be turned on without proper lockout and tagout.
- *Horseplay:* This type of activity can result in a serious accident or injury and should not be allowed on company property, whether the worker is on or off the job.
- *Wearing unsafe clothing:* Examples include the use of gloves around moving machine parts or wearing watches, rings, or other metallic parts while working with electricity.
- *Taking an unsafe position:* Among the examples are improper lifting, hanging from a beam while performing a job, and reaching a height that requires overextension of the body.

Unsafe Conditions

Unsafe conditions usually arise because of poor planning or an unsafe act.[1,4–6] It is imperative that every organization devise a system to identify and rectify such conditions. When an unsafe condition is identified, every effort must be made to *remove* the condition. If a hazard cannot be eliminated immediately, *guards* such as screen and enclosures must be installed to protect personnel from the hazard.[5] During the time that the unsafe condition exists, appropriate *warning* signs must be installed to alert personnel to the dangers. It is equally important that when an unsafe condition is identified, a *recommendation* in regard to its rectification be provided. Finally, a *follow-up* inspection should be carried out to ensure that the condition has been corrected. Figure 8.3 shows a typical unsafe condition report form. The following are examples of unsafe conditions:[1]

- Inadequate supports or guards
- Defective tools, equipment, or supplies
- Congestion of the workplace
- Inadequate warning systems
- Fire and explosion hazards
- Poor housekeeping
- Hazardous atmospheric condition (gases, dust, fumes, vapors)
- Excessive noise
- Poor illumination
- Poor ventilation
- Radiation exposure

CALIFORNIA STATE UNIVERSITY, LONG BEACH
SCHOOL OF ENGINEERING

No.__________

UNSAFE CONDITION REPORT

Date of Notification ____________________ Department ____________________

Location Where Unsafe Condition Exists ____________________

Unsafe Condition Identified by ____________________

Description of Unsafe Condition ____________________

Suggestion(s) ____________________

Corrective Action (Response required within 30 days of notification.)
LACK OF COMPLIANCE MAY RESULT IN CLOSURE OF THIS LABORATORY.

Date of Corrective Actions ____________________

Corrective Action Taken by ____________________

Signature ____________________

PLEASE RETURN TO THE DEAN'S OFFICE UPON COMPLETION OF CORRECTIVE ACTION

DISTRIBUTION

1- Copy to be returned upon completion (white)
2- Department (green)
3- Dean (canary)
4- Safety Director (pink)

Figure 8.3. Unsafe condition report form.

Accident Prevention

Accidents can be prevented when all employees work together to prevent them.[1,7,8] This has been proved in organizations of all sizes. Some of the more obvious preventive measures involve the identification and elimination of unsafe acts and conditions, but they are just part of a series of steps needed to establish a meaningful accident prevention program. Most organizations approach safety by establishing a safety policy; carefully selecting and training workers, supervisors, and managers; periodically reviewing all procedures; instituting inspection procedures; and correcting deficiencies. However, accident prevention actually should be addressed on three levels.

At the highest level, an attempt must be made to determine the basic causes of each accident, by conducting special surveys of (a) the hazards that exist (hazard analysis), (b) the procedures used to conduct each job (job safety analysis, or JSA), and (c) incidents and accidents that occur (incident/accident investigation). These surveys are used to establish a meaningful safety policy, create safety awareness, and determine the personal and environmental factors that require attention.

At the second level, that of indirect causes, an attempt must be made to eliminate unsafe acts and conditions. This can best be done by first keeping accurate records of all incidents and accidents and then periodically evaluating those records to determine trends and conditions that must be corrected. Again, this must be a joint effort of all who are involved: workers, supervisors, managers, and top executives. In particular, safe procedures must be developed; suitable education and training programs must be instituted (both formal and informal, including periodic safety meetings, on and off the job); the work environment and procedures must be improved so that all employees are motivated to work safely; personnel must be assigned carefully; equipment and facilities must be designed properly; meaningful inspections must be conducted periodically; equipment and facilities must be maintained diligently.

At the third level, that of direct causes, special attention must be given to the protection of people and property, should an unplanned release of energy or hazardous material occur. Where possible, quantities of available energy or hazardous material must be reduced; when all else fails, each worker must be protected with suitable equipment and structures (personal protective equipment, cabs, canopies, barricades). In addition, arrangements must be made to furnish first aid, medical attention, and transportation to a medical facility.

Figure 8.4 depicts the elements of a successful accident prevention program.

Effective Safety Audits

A safety audit is a management tool that can be used to measure the effectiveness of an organization's safety and health program in meeting its goals and objectives.[9,10] A properly conducted safety and health audit should accomplish the following:[10]

- Determine if the organization's safety and health program is meeting its objectives and goals.
- Establish a basis for facility organization, employee participation, and personnel accountability in safety matters.
- Evaluate the effectiveness of the organization's safety and health program regardless of the strengths and weaknesses of other areas within the organization.
- Detect and correct any operations, procedures, and/or equipment that is in violation of federal, state, or local laws, regulations, and standards.
- Identify the strengths and weaknesses of the current safety and health program.
- Facilitate the formulation of an improvement plan that can easily be communicated to all levels of management within the organization.

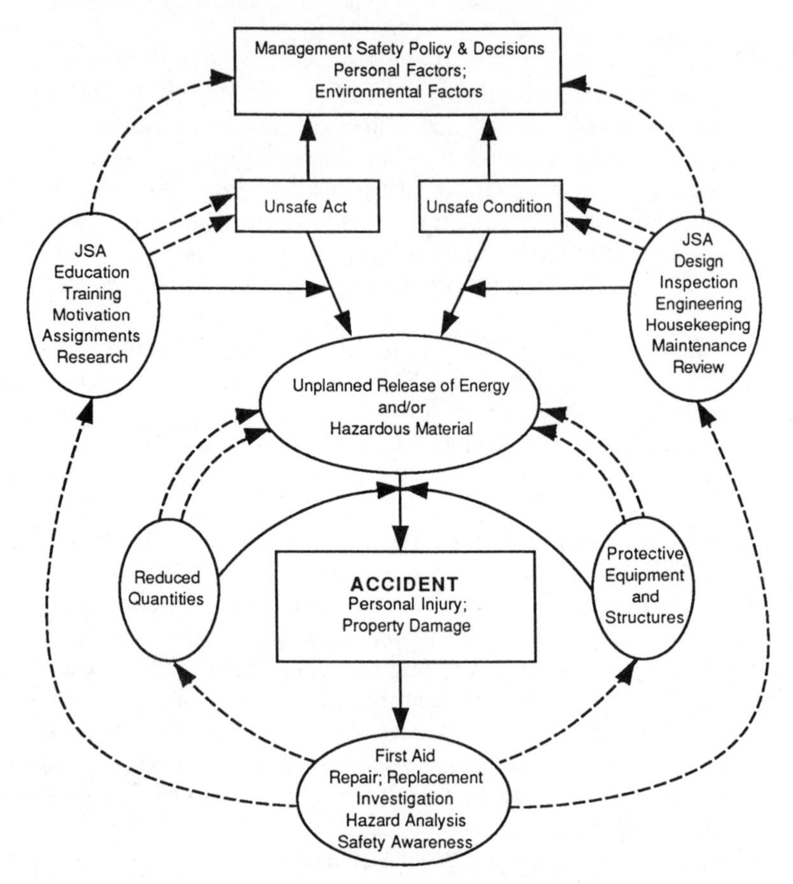

Figure 8.4. Elements of an accident prevention program. (Source: Reference 1)

Types of Audits

Safety and health audits can be classified as: comprehensive, limited, formal, and informal.

A comprehensive audit is conducted for the purpose of reviewing all aspects of an organization's safety and health program. This rather time-consuming process, when conducted properly, can identify all problem areas that might adversely affect the success of the safety and health program.

A limited audit focuses attention on a particular area or program. For example, an audit that concentrates on management safety awareness or fire protection and prevention falls into this category.

A formal inspection normally follows a detailed and structured evaluation system. This type of audit usually results in a well-documented detailed written report addressing specific safety problems within the organization.

An informal audit is conducted for the purpose of identifying specific safety problems in a given area and usually results in an interim status safety report.

A safety audit can be either internal (self-review) or external (independent review).

An internal review can be conducted either by an inspection team composed of different personnel within the organization or by supervisors and managers. The advantages of an internal audit include the auditors' knowledge of management styles within the organization and the low cost associated with this type of audit. An internal audit could have the disadvantage of being biased because it is a self-evaluation.

An external audit normally is conducted by outside consultants hired by the organization. The major advantage of an external audit is its objectivity; that is, it provides an opportunity for independent evaluation. This type of audit also provides better expertise in terms of identification and rectification of safety problems. The disadvantages of an external audit include the auditors' lack of familiarity with facility and personnel, the audit's relatively high cost, and the possibility of no or minimum follow-up.

Qualifications of an Auditor

In order to conduct an effective safety and health audit, the auditor must have the following qualifications:

- Be properly educated and trained in the area of audit (e.g., electrical, chemical, etc.).
- Have previous work experience in the area.
- Be technically competent.
- Have adequate investigative and analytical skills.
- Be objective and independent.
- Have adequate communication skills.

The Audit Process

A successful audit process starts with a written program that identifies its objectives, scope, auditors, program components, evaluation criteria, and reporting and follow-up provisions.

The next step is to organize a meeting between facility management and auditors. This meeting should address such issues as audit timetables and expected results.

The third step in an audit process involves preparation of auditors to review program components, evaluation criteria, scheduling requirements, and historical and potential loss performance.

The next step includes organization and implementation of data collection and analysis techniques, followed by the preparation of a written report addressing both the strengths and the shortcomings of the organization's safety and health program. The report must also include, where appropriate, any suggested corrective actions.

The final step in the audit process is the formulation and implementation of an improvement plan, followed by periodic status evaluations and follow-up reports.

Standards for Evaluation

The following questions can be used as guideline in an effective safety audit program.

Historical Performance

A. What has been the accident trend in the past?
B. How does the facility compare with similar facilities or with the industry average?
C. Which shift, area, or department has the majority of incidents or accidents?
D. What are the high-cost injuries (e.g., worker's compensation data)?
E. What are the job-related injuries or illnesses?

High-Risk Exposures

A. What are the areas that pose a high risk due to either the frequency or the severity of accident occurrence?
B. How long have they existed?
C. Is reducing the risk practical or economical?

Laws, Codes, Standards, and Regulations

A. Is the facility in compliance with federal, state, or local regulations such as OSHA, RCRA, SARA, NFPA, and so on?

Management

A. Does management exhibit an active interest in the conduct of an effective safety and health program?
B. Does management have a basic understanding of environmental and safety legislation and its impacts on the operation of the facility?
C. Does adequate funding exist for short- and long-term safety programs?
D. Does the organization hold any regular safety meetings?
E. Does management communicate properly with safety and health personnel?
F. Does management receive the correct safety and health information, such as reports of accidents, unsafe conditions, and unsafe acts?
G. Does management have proper plans for the safety training of personnel?
H. Is management involvement in safety issues visible to all company personnel?
I. Does management have any procedures in place to measure the company's safety performance?
J. Are there any plans for provision of PPE and its use by employees?
K. Has management established any methods or procedures to systematically identify and rectify unsafe conditions and acts?

Safety Program Components

1. Policy

A. Does the safety and health policy of the organization communicate management's commitment and philosophy to all employees?
B. Have safety responsibilities for each level within the organization been identified?
C. Are safety goals and objectives identified in the organization's policy?

2. Responsibility

A. Have specific safety and health responsibilities been identified for all levels within the organization?
B. Has the accountability of different levels in regard to safety been identified? What measures are used to determine the safety program's effectiveness (e.g., safety performance evaluations)?

3. Hazard Evaluation and Control

A. Are safety inspections successful in identifying unsafe acts and conditions?
B. Are employees encouraged to report unsafe conditions and acts to management?
C. Are the unsafe conditions and acts corrected in a timely manner?
D. Are safety-related work orders monitored to assure completion?
E. Is there a procedure for dealing with hazards that are not corrected in a timely manner?
F. Does the organization have a preventive maintenance program?
G. Is there a procedure for the selection, provision, and use of personal protective equipment?

4. Accident Reporting and Investigation

A. Are employees reporting work-related injuries or illnesses?
B. Are there procedures for investigating incidents and accidents no matter how minor they are?
C. Are both direct and indirect causes of accidents reported?

5. Training

A. Are new employees trained in the organization's safety policy and rules?
B. Are personnel adequately trained in how to identify hazards and how to protect against them?
C. Is there a procedure for conducting a job safety analysis for high-risk jobs?
D. Does management have proper safety training?
E. Does the organization have adequate training programs to comply with training requirements of the regulatory agencies (e.g., a hazard communication standard)?

6. Safety Meetings

A. Does the organization hold regularly scheduled safety meetings?
B. Are the topics discussed suitable for the organization's safety goals and objectives?
C. Is there any procedure to measure the effectiveness of the safety meetings?
D. Do the meetings generate effective communication between management and personnel?

7. Safety Committee(s)

A. Have the goals and duties of each safety committee within the organization been specified?
B. Do the safety committees have the clear support of management?
C. Have all departments within the organization been represented on the safety committee(s)?

8. Work Practices and Procedures

A. Is the level of housekeeping adequate?
B. Is there a program in place to control high-risk areas?
C. Are personnel aware of the organization's safety rules and policies?
D. Is there a program in place to enforce the use and maintenance of personal protective equipment?
E. Is there a procedure for monitoring employees and areas to identify unsafe acts and unsafe conditions?
F. Have special work practices been established for handling hazardous materials and wastes?
G. Is working safely a job requirement?

9. Industrial Hygiene

A. Does the organization have a program to identify potentially hazardous exposures such as excessive noise, radiation, chemicals, and so forth?
B. Has a chemical inventory been completed?
C. Are copies of material safety data sheets (MSDSs) available for all hazardous chemicals used within the organization?

D. Does the organization have a program in place to measure the concentration of hazardous materials, to ensure that employees are not exposed to levels above the permissible exposure limit (PEL)?
E. Are the results of hazardous material exposure monitoring effectively communicated to employees, medical personnel, and management?
F. Are existing controls for hazardous material exposure effective (e.g., an engineering control such as ventilation to reduce exposure to a corrosive vapor to below PEL levels)?
G. Have high-risk exposure areas been evaluated for physical and health hazards?

10. Medical

A. Does the organization have a medical program to monitor the effect of employee exposure on health?
B. Have specific facility exposures been communicated to medical personnel?
C. Does the organization's medical facility have the capability for treating injuries and/or illnesses that are expected in the area?
D. Is the organization located within a reasonable distance of an outside medical facility?
E. Is there a need to train few employees on first aid procedures?
F. Are medical records kept in an orderly manner, and are they available to employees?

11. Fire Prevention and Control

A. Have potential fire hazards within the facility been identified?
B. Have employees been trained on potential fire hazards (e.g., flammables, combustibles, oxidizers, etc.)?
C. Is there a system to monitor the readiness of the firefighting equipment (e.g., fire extinguishers, sprinkler systems, etc.)?
D. Do potential fire hazard areas (such as a flammable liquid or gas storage) have approved electrical equipment and proper warning signs?
E. Is there a program to train employees in the use of different types of firefighting equipment such as portable fire extinguishers?
F. Are the fire brigade members properly outfitted and trained?

12. Communications

A. Are the records of occupational incidents and accidents available to affected employees?
B. Does the organization comply with OSHA, RCRA, and SARA provisions for record keeping and posting?
C. Are employees aware of the provisions set forth by regulatory agencies for record keeping and posting?

13. Employee Relations

A. Is the employees' union supportive of the company's safety and health program?
B. Is there a system to reward employees for safe work practices?
C. Is there a system to handle safety rule infractions properly?
D. Are employee safety complaints handled in a proper manner? Are they filed with OSHA?

14. Regulatory Requirements

A. Has the company received any citations from any regulatory agency (OSHA, EPA)?
B. Is there a system to identify and correct areas that are not in compliance?
C. Is management aware of the regulatory requirement for compliance?

15. Contractor Safety

A. Does the organization require contractors to comply with safety rules and policies? Is this stated in the contract?
B. Is there a system to make contractors aware of the company's safety rules and policies?
C. Are contractors familiar with the company's specific hazards and emergency response procedures?
D. Is there a procedure to monitor and review the contractors' safety performance?

16. Regulatory Recordkeeping

A. Is documentation available to support the facility's activities (e.g., permits, training records, etc.)?
B. Are work-related injuries and illnesses properly reported?
C. Is EPA, DOT, OSHA documentation available (e.g., manifests, inventory, etc.)?

17. Emergency Preparedness

A. Does the facility have a written contingency plan?
B. Have different emergency scenarios been identified?
C. Has a leader who takes total control during an emergency been identified? Is he or she backed up by one or more alternates?
D. Has the facility's emergency response team been adequately trained?
E. Are there procedures to monitor the adequacy of the emergency equipment?
F. Are there emergency drills to ensure that employees are familiar with emergency procedures?

18. Employee Motivation and Involvement

A. Does the facility have an effective promotional safety program?
B. Do employees have a voice in the structure of the promotional safety program?
C. Are the rewards/prizes appropriate and effective?

19. Purchasing

A. Are there procedures established to ensure that only approved safety equipment and devices are purchased?
B. Before ordering any safety equipment, are the safety personnel within the organization consulted to ensure that the right type of equipment is purchased?
C. Are different manufacturers of safety devices consulted?

20. Employee Placement

A. Is there a program to examine employees' compatibility with the assigned tasks? (For example, employees who cannot wear a respirator because of a physical condition such as asthma should not be assigned to tasks that require respirator use.)
B. Are employees trained in the physical and health hazards of their job and means of protecting against them?

21. Off-the-Job Safety

A. Is there any interest on the part of management in off-the-job safety? (Some companies use this performance to prove that their employees are safer on the job than they are at home!)
B. Is there any training program for off-the-job safety (e.g., safe driving, safe use of household tools, etc.)?
C. Does the off-the-job safety program involve members of the employee's family?
D. Are there any incentives such as prizes for off-the-job safety performance?

22. Safety and Health Manual

A. Does the facility have a written safety manual that describes the company's safety policy, the individual's responsibilities, and safety procedures and rules?
B. Is the manual content communicated properly to all employees?
C. Is the manual used as a working resource?

Safety and Health Department

A. Does the safety and health department have adequate staffing to meet the safety goals and objectives set forth by the facility management?
B. Does the safety and health manager have the clear support of the organization's top management?
C. Does the safety and health manager have the managerial, technical, and communication skills required to conduct an effective safety and health program?
D. Are the safety and health department staff properly trained in safety issues and safety challenges confronting the organization? Is funding available for professional development such as safety seminars or courses?
E. Is there a program in place to monitor the effectiveness of the safety and health department?

Safety Audit Evaluation Methods

The main objective of a safety and health audit is to identify deficiencies within an organization's safety program. Upon completion of a safety audit, a report of findings must be prepared and distributed among the appropriate members of the organization. There are basically two formats used for a safety evaluation report: the narrative format and the performance grading format.

The Narrative Format

The narrative evaluation format addresses questions raised during the audit about different program components, in terms of effectiveness, strengths, weaknesses, and recommendations. For example, a sample safety evaluation report for Company X using the narrative format might contain the following information:

Safety and Health
Evaluation Report for
Company X

Re: Safety and Health Policy

The safety and health policy of Company X has not been revised since 1980. Although the policy indicates management's desire for a safe and healthy work environment, it does not specify management's goals and responsibilities. The policy is unsigned, and most employees are unaware of its existence. Recommendations for improvement include:

1. State management's commitment to safety.
2. State management's goals and objectives in regard to safety.
3. Identify the responsibilities of management, supervisors, and employees.
4. Have the policy signed by top management.
5. Communicate the policy to all employees.
6. Review and revise it periodically.

Performance Grading Format

This method is based on specific scoring requirements for questions examined during the audit. Typical scoring methods include but are not limited to:

- A, B, C, D, F
- Poor, fair, good, excellent
- Yes, no
- Percentage (90%, 70%, 50%, etc.)

Tables 8.1 through 8.3 summarize different ranking techniques for the evaluation of a company's safety policy.

Formulating the Improvement Plan

A successful safety audit program should provide for recommending an improvement plan. This plan should present recommendations for resolving problems identified during the safety audit. The improvement plan should at least include the following:

- The safety program component.

Table 8.1. Evaluation of the Safety Policy of Company X Using the Percentage Method.

Area	Ranking
Is there a written safety policy?	(10%) yes
Has the policy been revised in the past three years?	(10%) yes
Is the policy signed by top management?	(10%) yes
Are basic responsibilities included in the policy?	(10%) yes
Are safety goals and objectives identified?	(30%) yes
Is management's commitment to safety clearly indicated?	(20%) yes
Are personnel familiar with the policy?	(10%) yes
Total points out of possible 700 points = 100 (seven questions)	

Table 8.2. Evaluation of the Safety Policy of Company X.

Poor	Fair	Good	Excellent
• No written policy	Policy has not been reviewed or revised	Management indicates desire for a safe and healthy work environment	
• Employees unaware of existence of safety policy • Safety policy not signed by top management • Responsibilities not identified in the policy			

Table 8.3. Evaluation of the Safety Policy of Company X.

Area	Grading
Is there a written policy?	A, B, C, D, F
Does the safety policy identify responsibilities?	A, B, C, D, F
Does the safety policy indicate goals and objectives?	A, B, C, D, F
Does the safety policy indicate management's commitment to safety?	A, B, C, D, F
Is the safety policy communicated properly?	A, B, C, D, F
Has the safety policy been signed by top management?	A, B, C, D, F

- A brief description of the problem(s) identified.
- Actions required to solve the problem.
- The estimated time required for the corrective action to be completed.
- Individual(s) who are responsible and will be held accountable for implementing the corrective action(s).

Table 8.4 shows a sample improvement plan formulated for a company's safety and health policy.

Table 8.4. Sample Improvement Plan for Safety and Health Policy of Company X.

Program Component	Problem	Solution	Timetable	Responsibility
Safety policy	Last review 8 years ago	Revise policy	3rd quarter 1989	Safety manager
Safety policy	Responsibilities not defined	Revise policy	4th quarter 1989	Safety manager
Safety policy	Management commitment vague	Revise policy	4th quarter 1989	Safety manager
Fire protection	Inadequate ventilation in inflammable storage area	Install new ventilation	3rd quarter 1989	Safety committee

References

1. U.S. Department of Labor, *Accident Investigation*, U.S. Department of Labor, Mine Safety and Health Administration, National Mine Health and Safety Academy (1987).
2. ANSI, *American National Standard Method of Recording Basic Facts Relating to the Nature and Occurrence of Work Injuries*, ANSIZ16.2-1962, American National Standard Institute Inc., New York (1969).
3. Bird, F. E., Jr., *Management Guide to Loss Control*, Institute Press, Atlanta Ga. (1974).
4. Boley, J. W., *A Guide to Effective Industrial Safety*, 1st Ed., Gulf Publishing Co., Houston, Tex. (1977).
5. Revelle, J. B., *Safety Training Methods*, 1st Ed., John Wiley & Sons, New York (1980).
6. Henrich, H. W., *Industrial Accident Prevention*, 1st Ed., McGraw-Hill Book Co., New York (1959).
7. National Safety Council, *Supervisor Safety Manual*, National Safety Council, Chicago (1973).
8. National Safety Council, *Accident Facts*, National Safety Council, Chicago (1974).
9. King, R. W., and J. Magid, *Industrial Hazard and Safety Handbook*, 1st Ed., Newnes-Butterworths, Boston, Mass. (1979).
10. Weigand, M. A., *How to Conduct Effective Safety Audits*, Continuing Safety Education Seminars, American Society of Safety Engineers, Des Plaines, Ill. (1988).

Exercises

1. Define ''accident,'' and outline different categories of accident causes.
2. A flammable liquid storage warehouse went into flames and the fire destroyed three neighboring buildings. The damage is estimated at 5 million dollars. List the possible basic, indirect, and direct causes of this accident.
3. You have recently been appointed as the manager of the process development department in an oil company's research center. The department has a poor safety record that includes several injuries and many citations issued by OSHA. Summarize the steps that you would immediately undertake to eliminate unsafe acts and conditions within your department.
4. Perform an informal inspection of your work area, and identify at least three unsafe conditions.
5. What steps should be undertaken when an unsafe act is discovered?
6. What steps should be undertaken when an unsafe condition is discovered?
7. List three major goals of an effective safety and health audit.
8. What are the different types of safety and health audits? Briefly describe the major advantages and disadvantages of each.
9. List major qualifications of a safety and health auditor.
10. What are the major components of an effective safety and health improvement plan?

Chapter

9

System Safety Analysis in Process Design

Introduction to System Safety

The system safety approach is a thorough and systematic means of addressing workplace hazards. The formal hazard evaluation techniques employed are the result of increasing demands for more precise hazard assessment, as the complexity of production, construction, and processes makes it impossible to informally assess risk.[4,10]

The procedures used in system safety engineering are among the most effective and advanced methods of preventing system failures that result in accidents. The system safety approach identifies the hazards associated with a system that are due to process, equipment, or human interactions; and it must be applied from the earliest stages of process development to the shipment of products and the disposal of wastes.

System safety procedures begin by defining a system and focusing on how accidents can occur within that system as a result of equipment failure, human error, environmental conditions, or a combination of these factors. The preventive measures used include the design of equipment and/or the development of procedural safeguards to mitigate the hazards, as early identification, analysis, control, or elimination of the hazards in a process can eliminate the need for major design changes later. Of course, costs of the early detection of hazards in a process and their elimination or control must be included in determinations of the overall economic feasibility of a process, especially in the commercialization of new and extremely hazardous technology. The cost of equipment and/or design changes to control or eliminate hazards could change the overall economic picture of a project.[28,29]

The system safety approach to the safe design and operation of potentially hazardous processes can be illustrated by a system that allows for equipment, human, and environment interactions (Figure 9.1). Investigators try to identify hazards in that system that could result from equipment failure, human error, environmental conditions, or a combination of these effects. In this context, the hazards generally fall into one of four categories: physical, chemical, biological, and ergonomic (Figure 9.1). After the hazards are identified within the system, the next step is to determine whether the risks associated with the hazards are acceptable, and, if not (as with the loss of human life), at what cost the risks can be eliminated or reduced. In determining the risks associated with the hazards, the severity of a hazard and its probability of occurrence should be taken into account. For example, the release of a highly toxic chemical from a storage facility has such an adverse impact on the surrounding community that it would generate an unacceptable risk even if the probability of such a release were very low.[28]

Risk Evaluation

Once the system components and their failure modes have been identified, the acceptability of risks taken as a result of such failures must be determined. The risk assessment process will yield

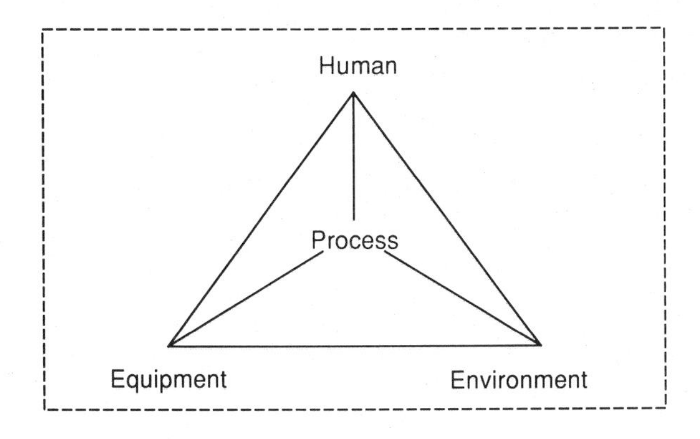

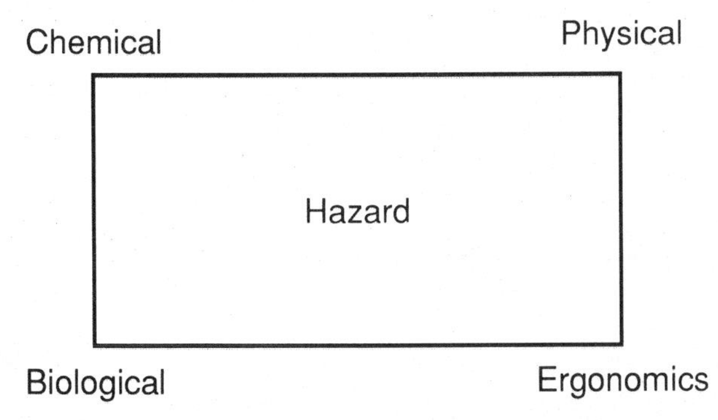

Figure 9.1. System interactions for use in hazard evaluation procedures.

more comprehensive and better results when reliable statistical and probability data are available. In the absence of such data, the results are a strong function of the engineering judgment of the design team. The important issue is that both the severity and probability (frequency) of the accident must be taken into account.

Table 9.1 summarizes one method of probability and severity assessment that can be applied to a system component failure. Both probability and severity have been ranked on a scale of 0 to

Table 9.1. Risk data summary.

Probability	Severity									
	0.1	0.2	0.3	0.4	0.5	0.6	0.7	0.8	0.9	1.0
0	0.1	0.2	0.3	0.4	0.5	0.6	0.7	0.8	0.9	1.0
0.1	0.2	0.3	0.4	0.5	0.6	0.7	0.8	0.9	1.0	1.1
0.2	0.3	0.4	0.5	0.6	0.7	0.8	0.8	1.0	1.1	1.2
0.3	0.4	0.5	0.6	0.7	0.8	0.9	1.0	1.1	1.2	1.3
0.4	0.5	0.6	0.7	0.8	0.9	1.0	1.1	1.2	1.3	1.4
0.5	0.6	0.7	0.8	0.9	1.0	1.1	1.2	1.3	1.4	1.5
0.6	0.7	0.8	0.9	1.0	1.1	1.2	1.3	1.4	1.5	1.6
0.7	0.8	0.9	1.0	1.1	1.2	1.3	1.4	1.5	1.6	1.7
0.8	0.9	1.0	1.1	1.2	1.3	1.4	1.5	1.6	1.7	1.8
0.9	1.0	1.1	1.2	1.3	1.4	1.5	1.6	1.7	1.8	1.9
1.0	1.1	1.2	1.3	1.4	1.5	1.6	1.7	1.8	1.9	2.0

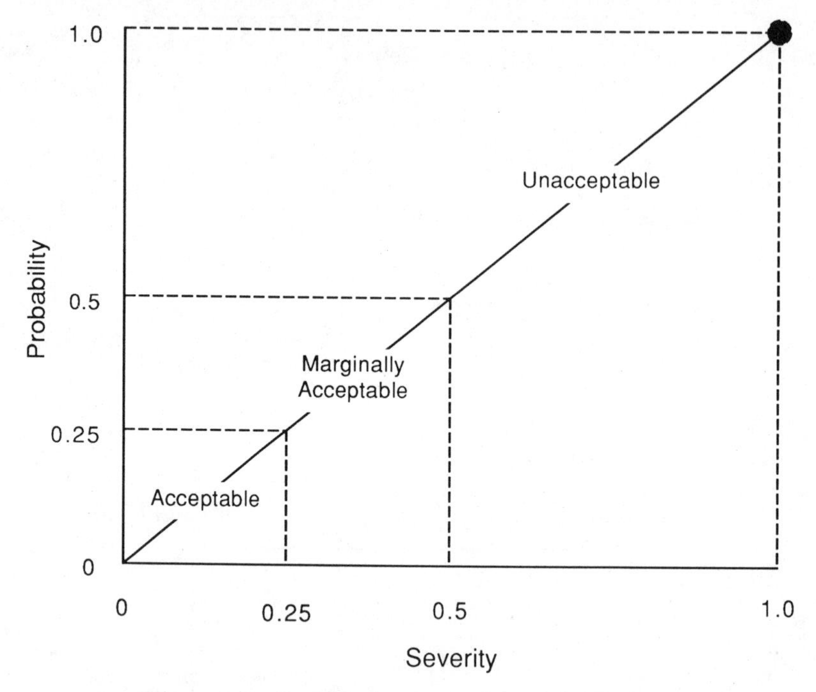

Figure 9.2. Graphic representation of risk data.

1, with table entries being the sum of probability and severity. Although the acceptability of risk is a management decision, one can divide the situations presented by Table 9.1 into unacceptable, marginally acceptable, and acceptable regions. Figure 9.2 is an example of the graphic representation of risk data.[28,30]

Need for Incorporation of System Safety into the Design Process

Members of the engineering profession need to be aware of both their legal liability and their professional responsibility for the public welfare. An examination of public toleration of engineering failures and the public's understanding of the dialectic science–engineering connection will show the truth of this statement. When a man was put on the moon, that was considered to be a scientific achievement; but the Challenger disaster was seen as an engineering/management failure.[31]

Although the exact definition of the engineer's moral and legal responsibilities vis-à-vis the public is debatable, there will be increasing public pressure on industry in future years to enhance the quality of life by advancing technology safely and without interference with valuable natural resources. In recent years this pressure has been transmitted to industry through strict occupational safety and environmental health laws and regulations. Also, court attitudes toward work- and environmental-related accidents and injuries have changed dramatically, as evidenced by a court ruling on June 13, 1985, when three company executives were found guilty of murder and were sentenced to 25 years in prison and payment of $10,000 fine over the cyanide poisoning death of an employee.[32]

In this new era of hazardous technological development, new dimensions must be added to the role of the engineer to ensure that his or her technical, legal, and moral responsibilities to the public are fulfilled. Today's engineer must be equipped with appropriate tools to apply the scientific laws of nature to the design and operation of hazardous technologies safely and with minimum interference with the environment. Many technological organizations also have come to the hard realization that safe work practices and policies will dramatically translate themselves into profit by eliminating or reducing the number of accidents in the workplace.

Even though the public has put considerable pressure on industry, this pressure has not been transmitted proportionately to the nation's engineering schools, which are responsible for the training and the education of its engineers. If the public's expectations of engineers are to be fulfilled, safety must be regarded as a science, and safety topics must be incorporated into the regular curriculum of the engineering schools. Also, the philosophy of the system safety approach must be incorporated into all phases of development of hazardous technologies, from concept to design, to pilot, to plant, to semiworks, and to commercial operation.[33,34]

System Safety Techniques

Several hazard evaluation procedures have been developed that, when applied properly to a given system, can identify hidden system failure modes and recommend procedures for their rectification. Many occupational and environmental safety problems are realized because of an emergency. Often, however, once the emergency is over, the problem is considered resolved. Although solving safety problems that have occurred is a responsibility of the design engineer, the true role of the design team must be to prevent accidents from happening in the first place. When integrated with engineering design, these hazard evaluation procedures provide the design engineer with the tools needed to identify and modify those components of the system that could cause an accident.[1,23,28]

In order to properly apply system safety techniques to the design and operation of potentially hazardous technologies, the design engineer must have a clear understanding of the system and be able to prepare a written response to questions such as the following:

1. What is the intended function of the system?
2. What are the raw materials, intermediates, and final products and by-products?
3. What steps are taken to convert the raw materials to final products (e.g., chemical reactions, physical operations, etc.)?
4. How does the system interact with the environment (e.g., hazardous waste streams, toxic releases, etc.)?
5. How does the system interact with personnel (e.g., the need for personal protective equipment)?
6. What sources of energy does the system use, and how is this energy supplied to the system?
7. What are the maintenance requirements of the system?
8. How does the system interact with other systems within the plant?

The above list is illustrative only and must be tailored to individual system designs.

Proper application of the hazard evaluation procedures also requires a sound knowledge of the type of hazards involved within the system. The design engineer must develop a checklist summarizing the types of hazards that warrant further evaluation within the system. This checklist should take the following hazards into account:

- Toxic chemicals
- Fire
- Explosion
- Runaway chemical reactions
- Temperature extremes
- Radiation
- Equipment/instrumentation malfunctions

- System moving parts
- Electrical hazards
- Hazardous noise and vibration
- Mechanical hazards
- Biological hazards
- Environmental pollution
- Pressure hazards

After the system has been defined, the hazard evaluation procedures can be used to identify different types of hazards within the system components and to propose possible solutions to eliminate the hazards. These procedures are extremely useful in identifying system modes and failures that can contribute to the occurrence of accidents, and should be an integral part of different phases of process development from conceptual design to installation, operation, and maintenance.[7,8] The following hazard evaluation procedures are useful in the preliminary and detailed stages of the design process: relative ranking techniques (DOW and MOND hazard indices), preliminary hazard analysis (PHA), "what if" analysis, failure modes effects criticality analysis (FMECA), hazard and operability study (HAZOP), fault tree analysis (FTA), and event tree analysis (ETA).[10] Cause consequence analysis (CCA) combines features of ETA and FTA.[1]

Relative Ranking Techniques: DOW and MOND Hazard Indices

This method provides a quick and simple way of estimating risks in process plants. The procedure assigns penalties for those processes or operations that can contribute to an accident and assigns credits to safety features of the plant that can mitigate the effects of an accident. The penalties and credits are combined into an index that is an indication of the relative ranking of the plant risk.[23] Although both DOW and MOND methods can be used to evaluate risks associated with different processing units, the MOND method considers material toxicity in addition to reactivity and flammability.

Preliminary Hazard Analysis (PHA)

A preliminary hazard analysis (PHA) is a general, qualitative study that yields a rough assessment of potential hazards and means of their rectification within a system. It is called "preliminary" because it usually is refined through additional studies. Part of the U.S. Military Standard System Safety Program, PHA contains a brief description of potential hazards in system development, operation, or disposal. This method focuses special attention on sources of energy for the system and on hazardous materials that might adversely affect the system or environment.[23]

Resources needed to conduct a PHA include plant design criteria, equipment, and material specifications.

The results of a PHA study can be summarized in the form of a table (see Table 9.2) or a logic diagram (as in Figure 9.3). In either format, potential hazards that pose a high risk are identified, along with their cause and major effects. In addition, for each hazard identified, a preliminary means of control is prescribed in the analysis. Thus a PHA is not performed just to develop a list of possible hazards, but is used to identify those hazardous features of a system that can result in unacceptable risks, and to assist in developing preventive measures in the form of engineering or administrative controls or the use of personal protective equipment.[1,9,17,23]

"What If" Analysis

The main purpose of this method is to identify the hazards associated with a process by asking questions that start with "What if . . . ." This method can be extremely useful if the design

Table 9.2. Summary table for PHA.

Hazard	Cause	Major Effects	Corrective/ Preventive Measures
1.	a1		a1
	a2		a2

team conducting the examination is experienced and knowledgeable about the operation; otherwise, the results are incomplete. The examination usually starts at the point of input and follows according to the flow of the process.[18,23]

"What if" analysis begins with defining the study boundaries. Two types of study boundaries are considered: the consequence category boundary includes public risk, employee risk, and economic risk; the physical boundary addresses the section of the plant that should be considered for analysis.

The second step is to obtain all the information about the process that will be needed for a thorough evaluation. This would include, but not limited to: the process materials used and their physical properties; the chemistry and thermodynamics of the process; a plant layout; and a description of all the equipment used, including controls and instrumentation. The last part of the information-gathering step is the preliminary formation of the "what if" questions.

The third step is to select a review team, which usually is composed of two or three members who have combined experience in the process to be studied, knowledge in the consequence category, and experience in general hazard evaluation. If the team is inexperienced, the results may be incomplete or incorrect.

Once the team has been established, the review is conducted. Typically, the review begins with the process inputs and follows through to the outputs. Each of the "what if" questions is ad-

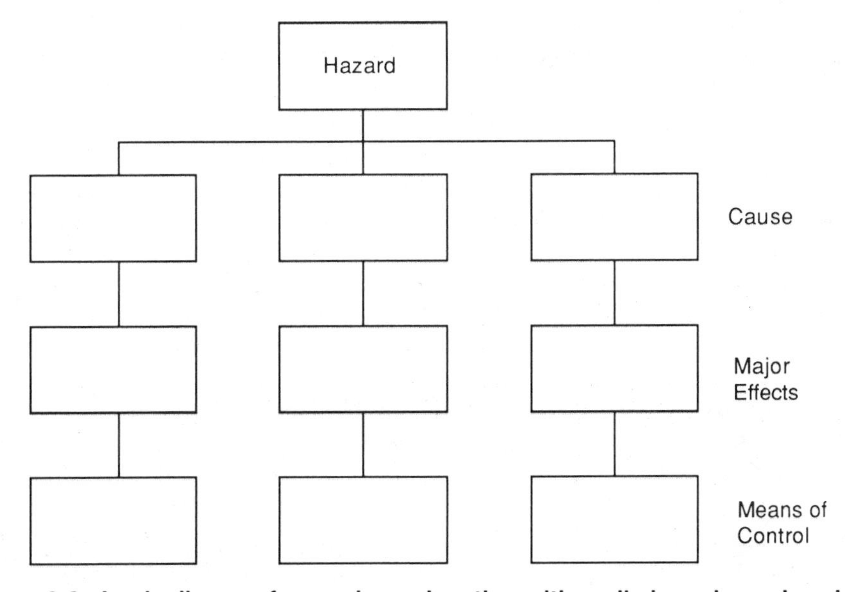

Figure 9.3. Logic diagram for use in conjunction with preliminary hazard analysis.

Table 9.3. "What If" analysis of the ethylene polymerization reactor.

What If . . .	Consequence/Hazard	Recommendation
1. Cooling water pump break down	Runaway reaction/explosion/fire	Stand-by pump/alarm system
2. Too much oxygen fed into reactor	Runaway reaction/explosion/ fire/debris flying	Alarm system feed flow control/ initiator flow control
3. Wrong initiator	None likely	—
4. Valve after reactor gets clogged	Pressure buildup/explosion/fire/ debris flying	Feed flow control/initiator flow control/alarm system
5. Compressor breaks down	None likely	—
6. Trauma to cooling jacket	Runaway reaction/explosion/ fire/debris flying	Temperature alarm/feed flow control

dressed by identifying the hazard and its consequence, and recommending solutions or alternatives to alleviate the risk.[23]

The final step in a "what if" analysis is to report the results in a systematic and easily understood format. An example of a common format can be seen in Table 9.3, which includes the questions, their consequences, and recommendations. Although the reader may not be familiar with the ethylene polymerization process (which will be explained later), Table 9.3 demonstrates the format for presentation of the results of a "what if" analysis.

Failure Modes Effects and Criticality Analysis (FMECA)

Failure modes effects and criticality analysis (FMECA), also known as Failure Modes and Effect Analysis (FMEA), is a systematic method by which equipment and system failures and resulting effects of these failures are determined.[1] FMECA is an inductive analysis; that is, possible events are studied, but not the reasons for their occurrences. FMECA has some disadvantages; human error is not considered, and the study concentrates on system components, not system linkages that often account for system failures.[3] FMECA provides an easily updated systematic reference listing of failure modes and effects that can be used in generating recommendations for equipment design improvement. Generally, this analysis first is performed on a qualitative basis; quantitative data can be applied later to establish a criticality ranking, often expressed as probability of system failures.

Five steps are required for a thorough analysis: First the level of resolution of the study must be determined, followed by development of a format. Then the problem and boundary conditions are defined, and the FMECA table is completed. Finally, the results of the study are reported.[8,23]

In the first step in FMECA, determining a level of resolution, if a system level hazard is to be addressed, equipment in the system must be studied; for a plant level hazard, individual systems within the plant must be examined.

Once the level of resolution of the study has been determined, a format must be developed, and it must be used consistently throughout the study. A minimal format should include each item, with its description, failure modes, effects, and criticality ranking.

The problem and boundary conditions definition includes identifying the plant or systems that are to be analyzed and establishing physical system boundaries. In addition, reference information on the equipment and its function within the system must be obtained. This can be found in piping and instrumentation design drawings as well as literature on individual components or equipment. The final part of problem definition is to provide a consistent criticality ranking definition. In a quantitative study, probabilities often are used for ranking. If the study is being conducted on a qualitative basis, relative scales (Table 9.4) usually are used as a means of rank-

Table 9.4. Suggested criticality rankings based on aerospace hazard classification

Criticality Ranking	Effects on System and Surroundings
I	Negligible effects
II	Marginal effects
IV	Critical effects
IV	Catastrophic effects

Source: Reference 1.

ing. Table 9.4 summarizes hazard classes used in the aerospace industry, which may be used as a relative scale.[1] However, if this type of scale is used, ''negligible, marginal, critical, and catastrophic'' effects should be defined somewhat more clearly. Another more specific criticality ranking scale is suggested by the *Guidelines for Hazard Evaluation Procedures*, published by the Batelle Columbus Division for the American Institute of Chemical Engineers.[23] This scale is summarized in Table 9.5.

The FMECA table should be concise, complete, and well organized. This table should include equipment identification relating to a system drawing or location in order to prevent confusion when similar equipment is utilized in different locations. The table must include *all* failure modes for each piece of equipment, and the effects of each failure along with the associated criticality ranking.

The final step in conducting a FMECA is to report the results. If the table that was prepared is complete, it may be sufficient. Often, however, a report of suggested design changes or alterations also should be included. Table 9.6 shows a sample chart than can be used in FMECA.

Hazard and Operability Study (HAZOP)

The purpose of a hazard and operability study (HAZOP) is to identify problems associated with potential hazards and deviations of plant operation from the design specifications. It is carried out by a multidisciplinary team following a structure that includes a series of guide words. The

Table 9.5. Suggested scale for criticality ranking for qualitative FMECA.

Criticality Ranking	Effects on System and Surroundings
1	None
2	Minor system upset Minor hazard to facilities Minor hazard to personnel Process shutdown not necessary
3	Major system upset Major hazard to facilities Major hazard to personnel Orderly process shutdown necessary
4	Immediate hazard to facilities Immediate hazard to personnel Emergency shutdown necessary

Source: Reference 23.

Table 9.6. Sample chart that can be used in FMECA analysis.

		Effects on			
Equipment	Failure Mode	Other Systems	System	Relative Ranking	Remarks

results of this study are dependent upon the quality of information on the process or plant and the experience of the team members.

A thorough HAZOP study can be accomplished in five steps, the first step being to define the scope and purpose of the study. The scope includes the specific areas of the process to be studied as well as what type of hazard consequences will be considered. The object of the study is included within the purpose.

The second step in classic HAZOP is to select a team to carry out the study. Ideally this team includes five to seven members from different areas within the operation.[9,23] A team leader is chosen, who should have a good general knowledge of the process being studied as well as experience in conducting HAZOP studies.

Once the team has been formed, information gathering must begin. The quality of the study depends on the source of the information used in conducting it. Suggested materials include piping and instrumentation diagrams, flow diagrams, layouts, and any equipment information that may be available. During data collection, the team leader should determine the sequence of study, or study nodes. Each study node is a specific portion of the design that will be studied individually. The leader also should compose a list of guide words such as those summarized in Table 9.7.

The team then will carry out a review of the process, examining each study node individually and applying all guide words to each of its components. The flow diagram in Figure 9.4 suggests a typical sequence to follow when carrying out this study. Each member of the team should contribute equally to the hazard analysis and the final tabulated report.

Table 9.7. Guide words for HAZOP.

General parameters	no more part of less as well as other than reverse
Time parameters	sooner later other than
Position, source parameters	where else other than
Temperature, pressure parameters	higher lower more less

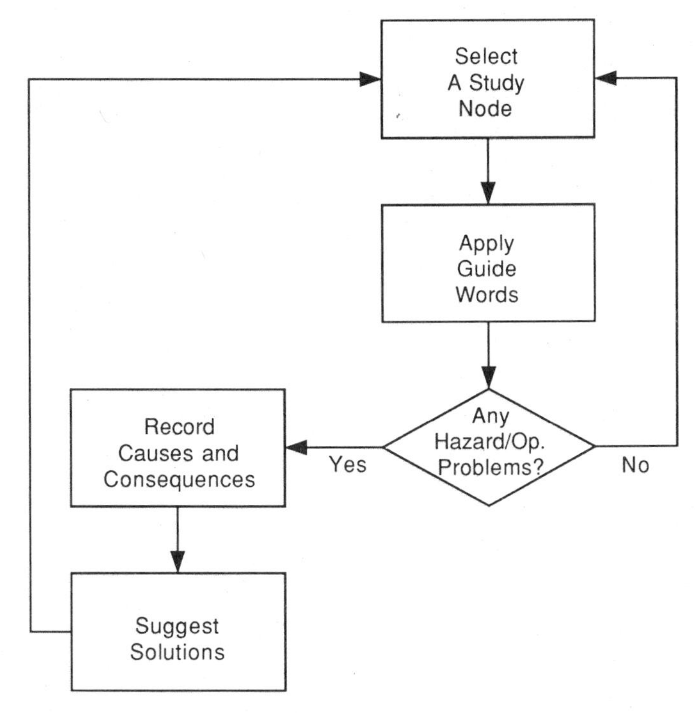

Figure 9.4. Flow diagram for a hazard and operability analysis (HAZOP).

The final report for a HAZOP study should contain all information in tabular format. Each table should include the guide words used, the deviation from the expected operation, the causes of that deviation, any consequences, and suggested actions to alleviate or eliminate the problem. The main purpose of HAZOP studies is to find problems, not to solve them; so only obvious solutions need be suggested. Each parameter should be addressed in an individual table. These tables may be accompanied by a report that includes the scope of the study and any suggestions or general recommendations.

Fault Tree Analysis (FTA)

Fault tree analysis (FTA) was developed in 1961 by H. A. Watson of Bell Telephone Laboratories.[1] This method of hazard evaluation visually demonstrates the interrelationship among equipment failure, human error, and environmental factors that can result in the occurrence of an accident.[10] FTA is a "backward" analysis; a system hazard, or top event, is the starting point, and the study traces backward to find the possible causes of the hazard. Analysis is restricted to the identification of system elements and events that led to the specified failure or accident. FTA employs Boolean logic; this requires that any statement, condition, act, or process be described as only one of the two possible states, such as on/off, fully open/not fully open, and so on. FTA can be computerized, and probabilities of events occurring can be calculated using minimum cut sets. A minimum cut set is the most direct path to a top event; a lower number of cut sets indicates a lower probability of the event occurring.

Three steps are required to conduct a fault tree analysis thoroughly and accurately. First, the undesired event, or top event, is defined. In the second step investigators develop a thorough understanding of the system to be analyzed. This can be accomplished by studying design drawings, equipment specifications, the literature, and operation procedures, as well as any other source information that may be available. The third step is construction of the fault tree. The symbols used in fault tree analysis are displayed in Figure 9.5. The fault tree will begin with the

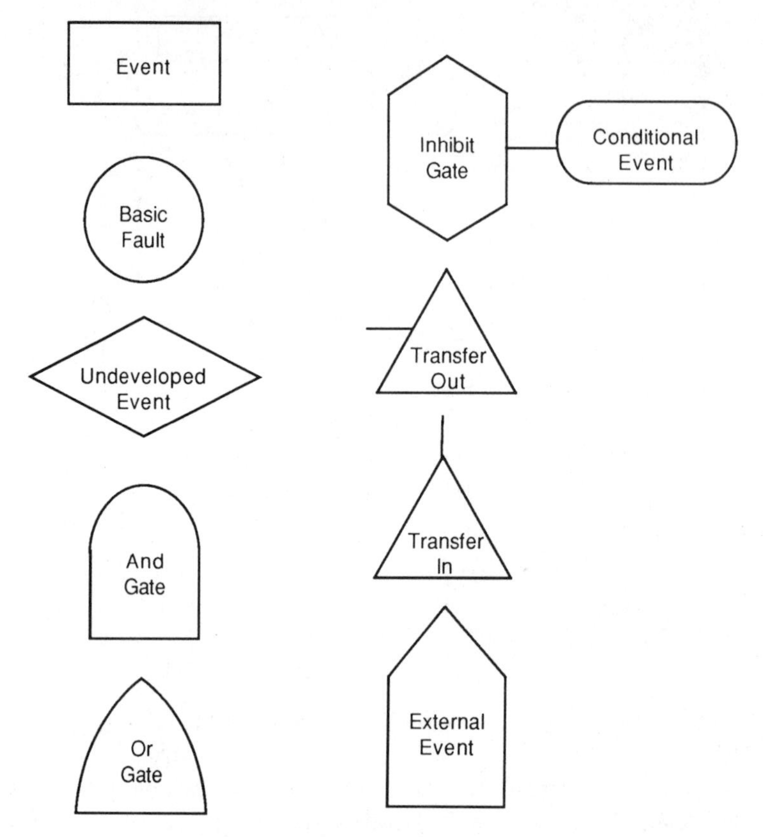

Figure 9.5. Fault tree analysis symbols.

top event, and will address any possible equipment failure, human error, or environmental factors that could result in the top event. "AND" gates are used when the existence of all conditions indicated will contribute to the top event; "OR" gates indicate that any one of the conditions indicated leads to the top event. Undeveloped events are occurrences that are not further addressed, either for lack of necessary information or for other reasons (e.g., the particular event goes beyond the scope of the study). Basic faults are the primary cause of the top event. Basic faults represent a malfunction of equipment that occurs in the environment in which the equipment was intended to operate. Each branch of the fault tree eventually should end up in either a basic fault or perhaps an undeveloped event. Triangles are used for transfer of the fault tree to another location or another page. Figure 9.6 illustrates a fault tree analysis applied to a flammable storage area.

Event Tree Analysis (ETA)

Event tree analysis (ETA) is a forward analysis beginning with an initiating event and proceeding forward to find possible consequences resulting from that event.[1,5] The course of events is determined by the success or failure of various safety functions as the accident progresses.

A complete event tree analysis can be accomplished in four steps: First, the initiating event is identified, and then relevant safety functions are determined. Next, the event tree is constructed, and then the resulting accident event sequences are described.[13,16,23]

The initiating event should be a system or equipment failure, a human error, or a process upset. The process upset can be due to numerous factors, including environmental causes.

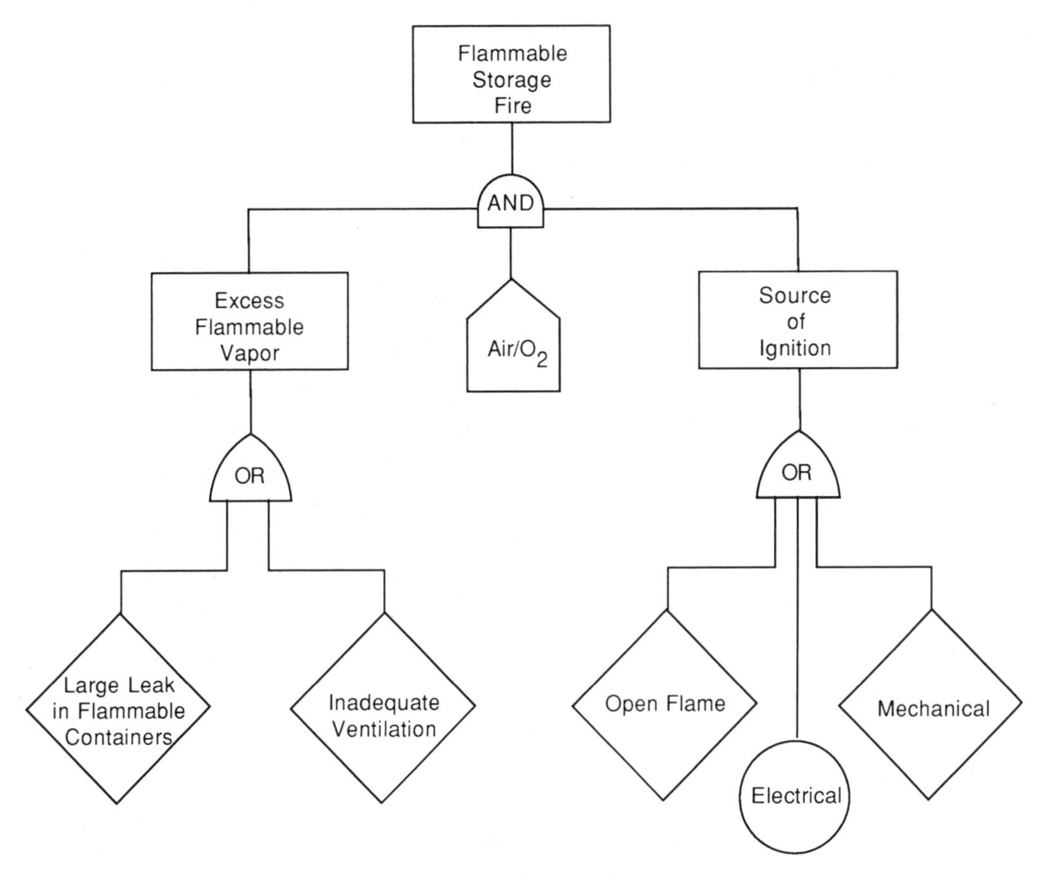

Figure 9.6. Fault tree analysis for a flammable storage fire.

The second step is to identify all the safety functions designed to deal with the initiating event. These safety functions should include any automatically responding safety systems, such as automatic shutdown. Alarms and warning systems, operator actions, and containment methods or barriers also must be considered.

The construction of the event tree begins by placement of the initiating event on the left side of the diagram, and placement of each of the safety functions being considered at the top of the page. Because the event tree will display a chronological development of accidents, the placement of the safety functions should be accurate in this respect. The event tree will show branches at a safety function if and only if the success or failure of that function affects the course of the accident. If this is not the case, there will be no branching at that safety function. Upward branching indicates success of the safety function, and downward branching indicates failure. Figure 9.7 shows a representative event tree structure with the letters representing safety functions. The series of letters at the end of each branch indicate all the safety functions that have failed in that particular path. The end of each branch should contain a notation as to the condition of the system, such as safe, unsafe, unstable, and so on.

The final step in this analysis is to describe each of the sequences contributing to an accident. An accurate description of the expected outcome for each branch must be supplied in the final report. Although the results of this study are qualitative, they can be quantified using minimal cut sets and probabilistic data.

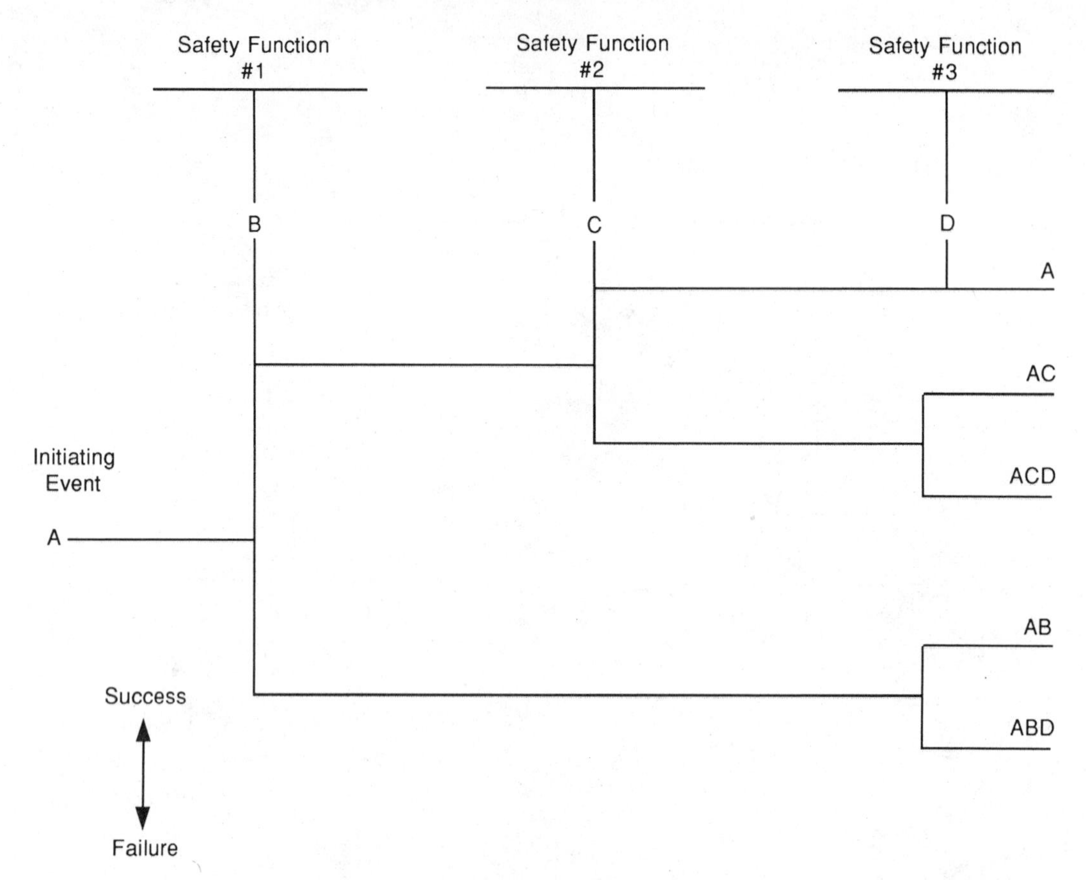

Figure 9.7. Typical event tree structure.

Cause Consequence Analysis (CCA)

Cause consequence analysis (CCA), invented by RISO Laboratories in Denmark,[1] combines the forward thinking features of event tree analysis with the reverse thinking features of fault tree analysis. The result of this analysis is a cause consequence diagram that displays the relationships between accident sequences and their basic causes.

Four basic steps are involved in the completion of cause consequence analysis. The first step is to select an event for analysis. The event should be either a top event, as in FTA, or an initiating event, as in ETA. If a top event is chosen, the second step is to develop a fault tree for that event; otherwise, an event tree diagram should be developed for the chosen initiating event. The third step is either to complete a fault tree for each resulting accident on the event tree formed in step 2, or to complete an event tree for each of the safety functions found in the fault tree developed in step 2. Essentially what is being done is a simultaneous fault tree and event tree analysis, which combines top–down and bottom–up studies. Once steps 2 and 3 are completed, the information obtained must be assembled into one coherent flow diagram. For this purpose, the fault tree symbols are used, and the event tree format is modified in order to easily fit into the fault tree. The symbols used for the event tree portion of the study can be found in Figure 9.8. Using these symbols for consequences and branch points, the event tree diagram can be easily incorporated into the fault tree diagram. Although the results of this study are qualitative, they are quite accurate, and minimum cut sets can be used to quantify the results.

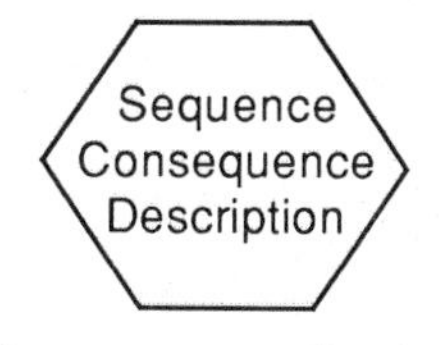

Consequence Symbol

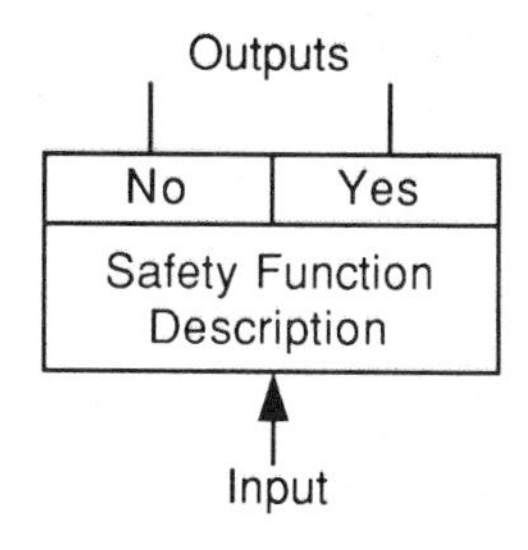

Branch-Point Symbol

Figure 9.8. Symbols used in cause consequence analysis.

Application of Hazard Evaluation Techniques to Preliminary Design of Metal Organic Chemical Vapor Deposition (MOCVD)

Description of the MOCVD Process

Metal organic chemical vapor deposition (MOCVD) is a process whereby optoelectronic and microelectronic photovoltaic thin films are produced. These films are used in semiconductors, photocathodes, and laser diodes, as well as other applications where photovoltaic cells are needed. Thin film technology is based almost entirely upon deposition from the gas phase; chemical deposition is favored for epitaxial growth on single-crystal substrates. In epitaxial growth, the formed crystals follow the crystal pattern of the substrate, or base crystal; this permits the formation of a product that is consistent and predictable in performance.

A basic understanding of the MOCVD process is necessary for one to be aware of the hazards involved in it. The reactants typically are alkyls of group III metals and hydrides of group V elements. The alkyls are stored in stainless steel bubblers in the liquid phase, and these bubblers are kept in carefully controlled refrigerated baths in order to maintain a stable vapor pressure. The gaseous hydride sources often are contained at or near room temperature in dilute mixtures with hydrogen.

Dilute vapors of these reactants are transported near room temperature to a common manifold. The alkyls and hydrides remain separated until their introduction into a reactor containing a heated susceptor, where pyrolysis and deposition occur. The carrier gas normally is purified hydrogen, and the storage and flow systems typically are assembled from stainless steel tubing. Figure 9.9 shows a simplified schematic diagram of the gas-handling system used in MOCVD.[10,13]

Hazards of Gases Used in MOCVD and Effects on Workers and Public Safety

The physical properties and hazardous characteristics of the gases used in MOCVD are critical factors in the design of this process. These properties are summarized in Tables 9.8 and 9.9. Fire and explosion hazards are based upon flash points, the flash point being defined as the lowest temperature at which the air/vapor mixture formed above the surface of a liquid will ignite in the presence of a source of ignition. Liquids with flash points below 100°F are classified as

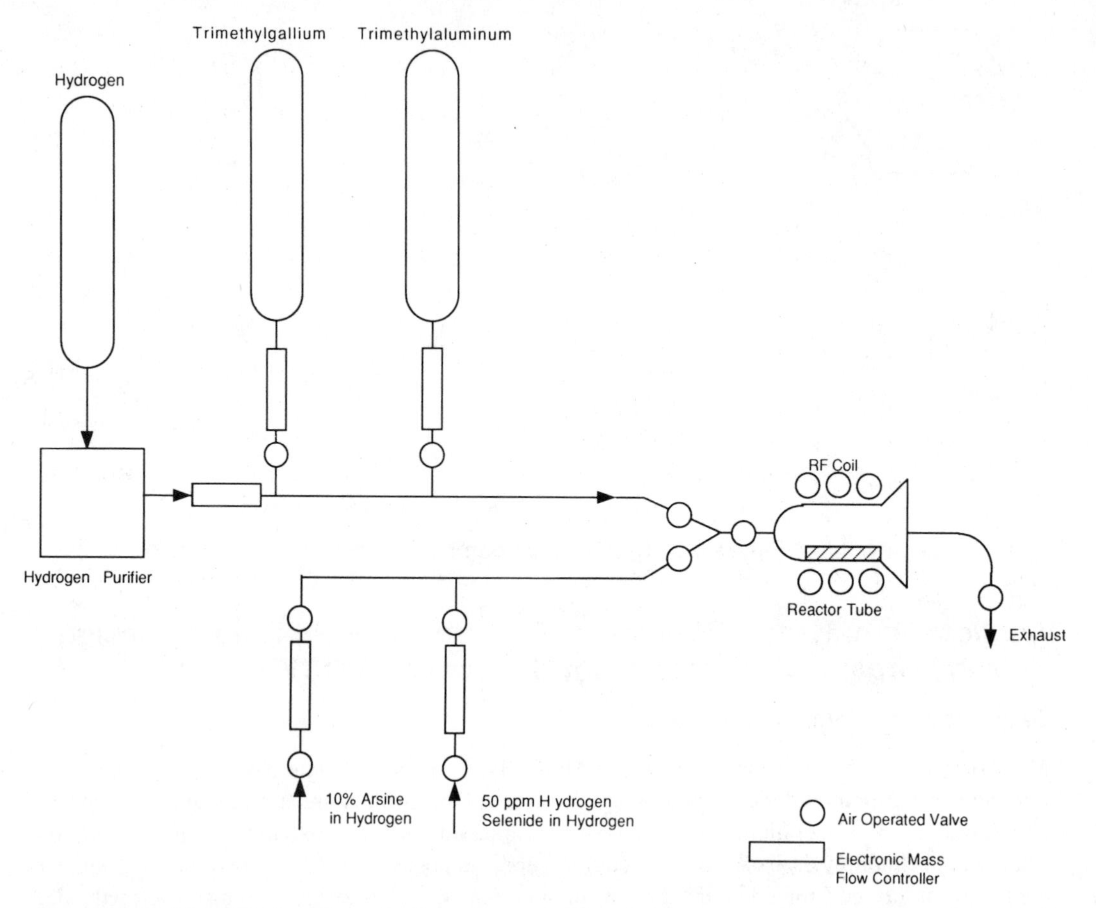

Figure 9.9. Simplified schematic flow diagram of metal organic chemical vapor deposition (MOCVD).

Table 9.8. Physical properties of source gases and doping agents.

		Molecular Weight	Density,* g/cm³	Melting Point, °C	Boiling Point, °C
Trimethylaluminum	$(CH_3)_3Al$	72.07	.752	15.4	126.0
Triethylaluminum	$(C_2H_5)_3Al$	114.0	.837	−52.5	194.0
Trimethylgallium	$(CH_3)_3Ga$	156.9	1.06	−19.0	55.7
Triethylgallium	$(C_2H_5)_3Ga$	156.9	1.06	82.3	142.6
Trimethylindium	$(CH_3)_3In$	114.8	1.57	88.0	134.0
Triethylindium	$(C_2H_5)_3In$	202.0	1.26	32.0	184.0
Hydrogen selenide	H_2Se	81.0	2.121	−64.0	−41.4
Arsine	AsH_3	78.0	2.661	−166.0	−55.0
Triethylantimony	$(C_2H_5)_3Sb$	208.9	1.32	−29.0	160.0
Diethylberyllium	$(C_2H_5)_2Be$	67.1	2.30	12.0	194.0
Tetramethylgermanium	$(CH_3)_4Ge$	132.7	1.01	−88.0	43.6
Dimethylmercury	$(CH_3)_2Hg$	230.7	3.07	—	96.0

*Densities for liquid state.

flammable liquids because they pose a severe fire hazard at room temperature. Liquids with flash points between 100°F and 200°F are classified as combustible liquids, and they present less of a fire hazard than flammable liquids.[15]

In Table 9.9, fire hazards of source gases are classified as moderate or dangerous. Here moderate refers to flash points between 100°F and 200°F, and dangerous refers to flash points below 100°F.

As can be seen in Table 9.9, the source gases used in the MOCVD process also pose severe health hazards, one measure of the degree of health hazard posed by a chemical being its toxicity. Although all chemicals are toxic if the dose administered is high enough, toxic chemicals are divided into three groups: extremely toxic, highly toxic, and toxic. Most toxicity data are obtained by conducting experiments on laboratory animals. The lethal dose that is fatal to half of the test animals when administered orally is called the lethal dose 50, or LD_{50}. An extremely toxic chemical, by definition, has an oral LD_{50} of less than 1 mg/kg (milligrams of chemical per kilogram of body weight). A highly toxic chemical, on the other hand, is defined as one that has an LD_{50} greater than 1 mg/kg but less than 50 mg/kg. A toxic chemical has an LD_{50} between 50 mg/kg and 500 mg/kg.[26]

A typical hydride gas source used in MOCVD is arsine, which is extremely toxic and is dangerously reactive under certain conditions. The primary hazard posed by arsine is exposure through inhalation, and this gas is a carcinogen that also can affect red blood cells, the gastrointestinal system, and the central nervous system. The permissible exposure limit (PEL) set by OSHA for this gas is 0.05 part per million (ppm). In addition, arsine poses severe fire/explosion hazards in the presence of an ignition source or oxidizers such as chlorine and nitric acid. The decomposition product of this gas also can pose severe health hazards.

In addition, arsine is flammable when exposed to flame, and moderately explosive when exposed to chlorine, nitric acid, or open flame. When this gas is heated to decomposition, it emits highly toxic fumes. Personnel who have a potential for exposure to arsine must wear protective gear and a self-contained breathing apparatus. The primary threat posed by arsine to the surrounding community is the accidental release of large quantities of gas.

Alkyls used in the MOCVD process are all pyrophoric; that is, they explode upon contact with oxygen in air. Because of this high reactivity, they must be stored under an inert atmosphere. They also pose dangerous fire and explosion hazards. These compounds also are irritating to mucous membranes and the skin; personnel must wear the appropriate PPE when working with them. Toxicity is not a primary concern for the surrounding community because the gases generally are stored as dilute mixtures with hydrogen. However, the disaster hazard is high because of the highly flammable and explosive nature of the compounds. Employee awareness and safety procedures are essential in the safe handling of alkyls.

Table 9.9. Hazardous properties of source gases.

	Health Hazard	Fire and Explosion
Arsine	Extremely toxic	Moderate
Hydrogen selenide	Extremely toxic	Dangerous
*Trimethylaluminum	Highly toxic	Dangerous
*Triethylaluminum		
*Triethylgallium	Toxic	Dangerous
*Trimethylindium	Toxic	Dangerous
*Triethylantimony	Toxic	Dangerous
*Dimethylmercury	Toxic	Dangerous

*Pyrophoric gases.

Some other chemicals involved indirectly in MOCVD also pose physical and health hazards. Acetone, methanol, and chloroform are used in cleaning the substrates prior to deposition. Chloroform, for example, is a carcinogen, and methanol and acetone present a severe fire hazard. Personnel exposed to these substances should wear the appropriate protective gear and work in well-ventilated areas.

Application of PHA to the MOCVD Process

The greatest risk to personnel and the surrounding community in the MOCVD process is the toxic and pyrophoric properties of the reactant gases. Table 9.10 is the result of concentrating on these two characteristics in the first step of a preliminary hazard evaluation (PHA). In order to complete a thorough PHA on this process, similar studies would include hazards due to the operating environment, operations, facility, and safety equipment. The study of each area would result in a table similar to that completed for toxic gas release and fire explosion hazards.[22]

Table 9.10. Application of preliminary hazard analysis to the design of metal organic chemical vapor deposition.

Hazard	Cause	Major Effects	Corrective/Preventative Measures
Toxic gas release	Leak in storage cylinder	Potential for injury fatalities from large release	• provide warning system • minimize on-site storage • develop procedure for tank inspection and maintenance • develop purge system to remove gas to another tank • develop emergency response system
Toxic gas release	Reactor heater failure	Potential for injury, fatalities from large release	• provide temperature control inside reactor with automatic shutdown of gas flow to the reactor • design collection system to remove and purify/recycle or discard unreacted gases • design control system to detect excess gases in exhaust and shut down gas flow
Toxic gas release	Pressure buildup due to high temperatures in storage cylinders due to refrigeration system failure	Potential for injury fatalities from large release	• provide control system to detect extreme temperature variations and activate backup cooling system • backup cooling system
Toxic gas release	Trauma to storage tanks	Potential for injury, fatalities from large release	• locate gas storage away from unnecessary plant traffic • warning system for personnel in the area such as signs, lights, etc. • training of employees in the area
Toxic gas release	Leakage in process lines	Potential for injury, fatalities from large release	• provide accurate gas monitoring system on-site • provide adequate protective gear/equipment • provide emergency response system • design control system to detect leakage and divert flow to a secondary system

Table 9.10. *(Continued)*

Hazard	Cause	Major Effects	Corrective/Preventative Measures
Toxic gas release	Damage to reactor tube due to high temperature	Potential for injury fatalities from large release	• provide a warning system for temperature fluctuation, divert flow to temporary storage tank
Toxic gas release	Compressor failure	Potential for injury, fatalities from large release	• provide spare compressor with automatic switch-off control • develop emergency response system
Toxic gas release	Reactor outlet becomes plugged	Potential for injury, fatalities from large release	• provide relief valve on reactor with outlet to a temporary storage tank
Explosion, fire	Spark	Potential for fatalities due to toxic release and fire Potential for injuries, fatalities due to flying debris	• design process to maintain spark-inducing equipment at a safe distance from gas-handling equipment • ground possible static producing equipment • develop emergency fire response system • train personnel
Explosion, fire	Reaction with oxygen, water, oxidizing agents	Potential for fatalities due to toxic release and fire Potential for injuries, fatalities due to flying debris	• maintain inert carrier gases or vacuum in all phases • warning system to monitor presence of water or air • design plumbing (water pipes) a safe distance from gas-handling equipment • routine maintenance and inspection to maintain safe operation of all parts of system
Explosion, fire	Overheat in reactor tube	Potential for fatalities due to toxic release and fire Potential for injuries, fatalities due to flying debris	• provide warning system for temperature fluctuation, evacuate reaction tube, shut off input valves, activate cooling system • design control system to detect overheat and disconnect heater
Explosion, fire	Leak in reactor	Potential for fatalities due to toxic release and fire Potential for injuries, fatalities due to flying debris	• connect all reactor intel and outlet lines to a common inert tank; activate the connecting line in emergency • Provide automatic unit shut down as a result of above activation

Application of FTA to the MOCVD Process

Basically a fault tree is a logic diagram used to analyze an undesired event*, where causes that can lead to the undesired event are cataloged and broken down further. This analysis is continued to determine all of the events and combinations of events that can lead to the undesired event.

The first step in the construction of a fault tree is selection of the undesired event. A fault tree can be constructed for virtually any event that can occur within a system, but because only one event is analyzed in a single fault tree, the undesired event usually is an accident or potential accident that is sufficiently important to warrant the study. The undesired event may be a catastrophe such as the release of toxic gases into the community, or it may be an accident with less serious results, such as one that produces a disabling injury.

*All causes that can lead to the undesired event.

Next, it is necessary to reason backward from the undesired event by asking "How could this happen?" In answering this question, primary causes and how they interact to produce the undesired event are identified. the same question then is asked for each of the primary causes, which, in turn, are broken down into events that lead to them, and so on. This logic process is continued until all potential causes have been identified.

Throughout the process, a tree diagram is used to record the events as they are identified. The undesired event is shown at the top of the tree; the primary causes are shown immediately below the undesired event; the events that lead to the primary causes are shown at the next lower level; and so on. The tree branches are terminated when all events that could eventually lead to the undesired event are shown.

The top event for the MOCVD process is selected to be the release of toxic gases from the reactor tube during normal operation. Before the analysis can begin, the existing event, unallowed events, system boundary (reactor), and equipment configuration must be defined, as shown in Figure 9.10. The causes for the top event can be overheat damage to the reactor tube, trauma to the reactor tube, or a pressure buildup within the reactor. Figure 9.11 summarizes the sequence of events that could lead to the top event.

Application of FMECA to the MOCVD Process

The major objective of this method is to identify the location of failures within the system and the impact of such failures. In the usual procedure, each item used in the system is listed on an FMECA chart. Such items include people, equipment, materials, machine parts, and environmental elements necessary for system operations. The exact manner or mode in which each item can fail then is determined.

Table 9.11 summarizes the results of application of FMECA to the MOCVD process. It should be noted that the criticality rankings utilized are somewhat subjective and depend to a large degree

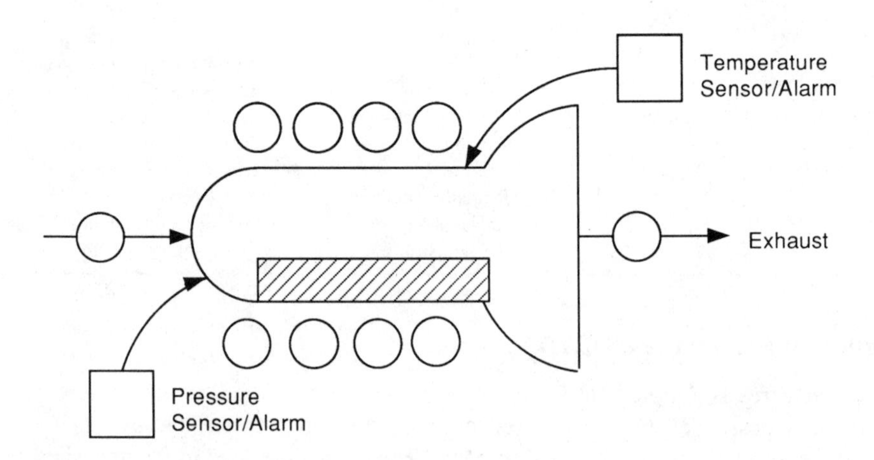

TOP Event – toxic gas release from reactor tube during normal operation
Existing Event – leak in reactor tube
Unallowed Events – power failures, natural disasters
Physical Bounds – as shown
Equipment Configuration – inlet value open, exhaust value open, heater on
Level of Resolution – equipment shown in figure.

Figure 9.10. Preliminary steps for fault tree analysis of the MOCVD process.

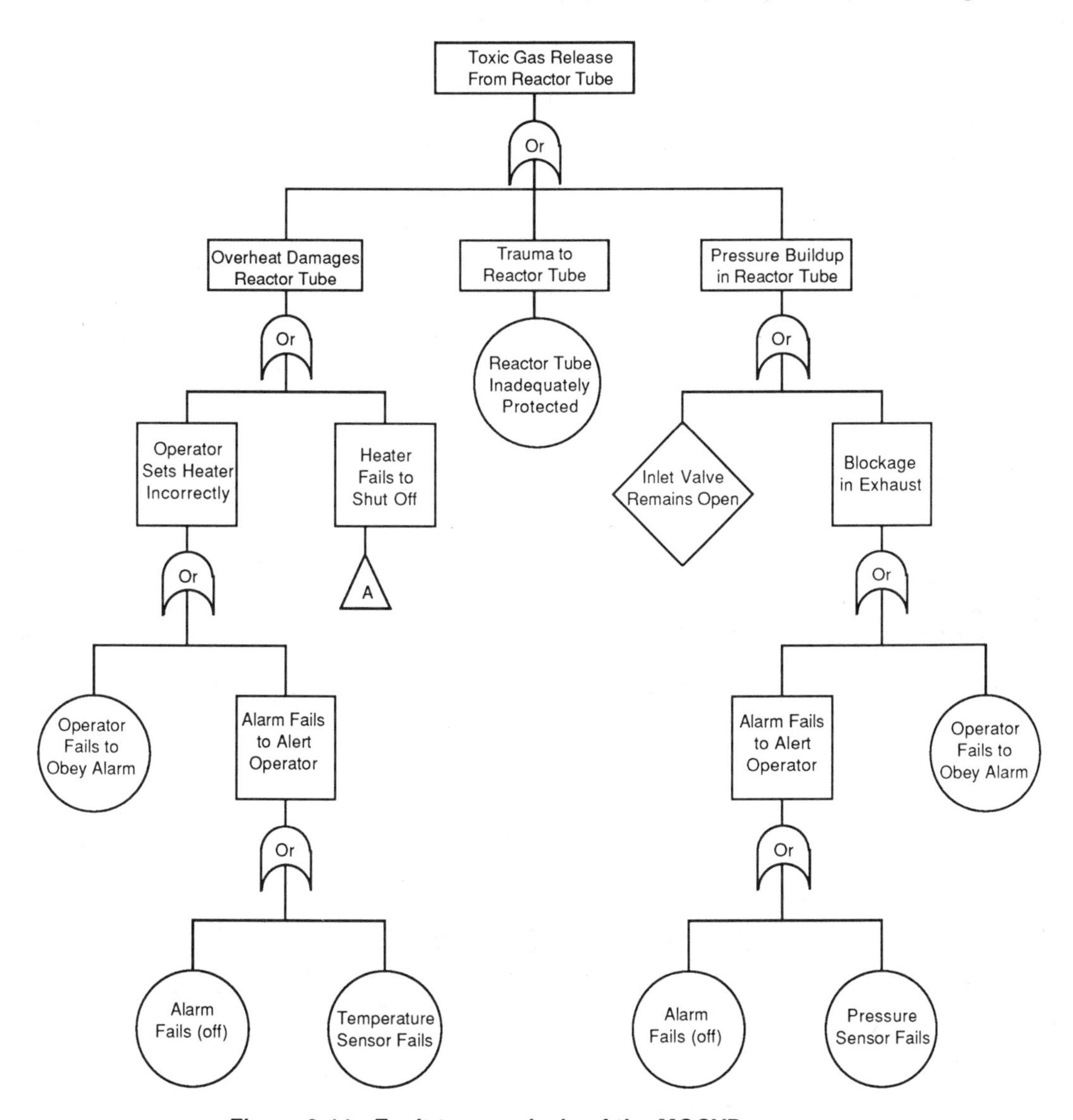

Figure 9.11. Fault tree analysis of the MOCVD process.

on the judgment and experience of the design engineer. This, however, can became an important factor for using and prioritizing hazard mitigation in process design.

Application of HAZOP Studies to the MOCVD Process

As mentioned earlier, the major objective of a HAZOP study is to determine the possibilities of a system's deviating from its design intent. The major goal of this method is to identify design problems rather than to solve them.

Table 9.12 presents a HAZOP study on two parameters of the MOCVD process: flow of tri-methylaluminum (TMAl) and temperature of the TMAl storage tank. Although both of these areas are quite important, it must be emphasized that many more parameters, such as the reactor tube, should be considered in a similar manner. The guide words are chosen from Table 9.7; more specific guide words may be considered for different types of processes.

Table 9.11. Results of failure modes effects and criticality analysis (FMECA) of MOCVD process.

Item	ID	Failure Mode	Effects	Crticality Ranking
Reactor tube	A	Rupture	Release of pyrophoric gas causing fire and release of toxic gases	III IV
Air-operated valve on storage cylinder	B	Rupture	Release of pyrophoric gas; release of toxic gas	III IV
		Failure to close	Excess gas in reactor tube that can cause increase in pressure and rupture of reactor tube	IV
Air-operated exhaust valve	C	Failure to open	Pressure buildup in reactor; possible rupture, release of pyrophoric and toxic gases	IV
		Rupture	Release of pyrophoric and toxic gases	IV
Control on reactor heater	D	Sensor fails inadequate response control system fails	Reactor overheating beyond design specification	II
Pump from storage to reactor	E	Malfunction electrical mechanical	Overheating of reactor tube	II
Transfer line fittings	F	Loose, not properly installed	Leakage of pyrophoric and toxic gases	II
Refrigeration equipment	G	Failure to operate	Increase in vapor pressure; cylinder rupture	IV

Table 9.12. Application of hazard and operability studies (HAZOP) to the MOCVD process parameter flow of trimethylaluminum (TMAl).

Guide Word	Deviation	Consequences	Causes	Recommended Action
no	no flow of TMA1	TMA1 not released to reaction chamber; no reaction occurs; possible unreacted toxic gas in exhaust	faulty valve	pressure-regulated automatic shutdown (at storage tank)
			faulty mass flow controller	pressure-regulated automatic shutdown (at storage tank)
			empty storage tank	monitor tank levels regularly, automatic level, pressure control
more	more flow of TMA1	possible rupture to reaction chamber or valves; release of toxic, pyrophoric gas	faulty mass flow controller	pressure-regulated automatic shutdown; (before reactor)
part of	normal flow of lower concentration TMA1	possible unreacted toxic gas released in exhaust	operator error in flow rate ratios of gases	
			faulty mass flow controller at storage tank for TMA1 or other gases	monitor exhaust with shutdown if unreacted gases detected

Table 9.12. *(Continued)*

Guide Word	Deviation	Consequences	Causes	Recommended Action
less	less flow of TMA1	unreacted toxic gas may be released with exhaust	faulty mass flow controller operator error in choice of flow rates	pressure regulated automatic shutdown warning to operator if flow rates of all gases not in a ratio within limits
higher	higher temperature in storage tank	increased pressure; possibility of Leakage of toxic, pyrophoric gas	faulty cooling unit	backup cooling unit activated by temperature sensor
lower	lower temperature in storage tank	decreased pressure may result in lack of flow of gas resulting in unreacted toxic gases in exhaust	faulty cooling system faulty thermostat	backup cooling unit activated by temperature sensor regular monitoring and maintenance of all aspects of cooling unit

Application of ETA to the MOCVD Process

Event tree analysis focuses on the accident outcomes that may result following an equipment failure within the system. A forward thinking process, ETA starts with an initiating event and analyzes the consequences in terms of both success and failure.

Figure 9.12 represents an event tree analysis for a reactor heater failure in the MOCVD process.

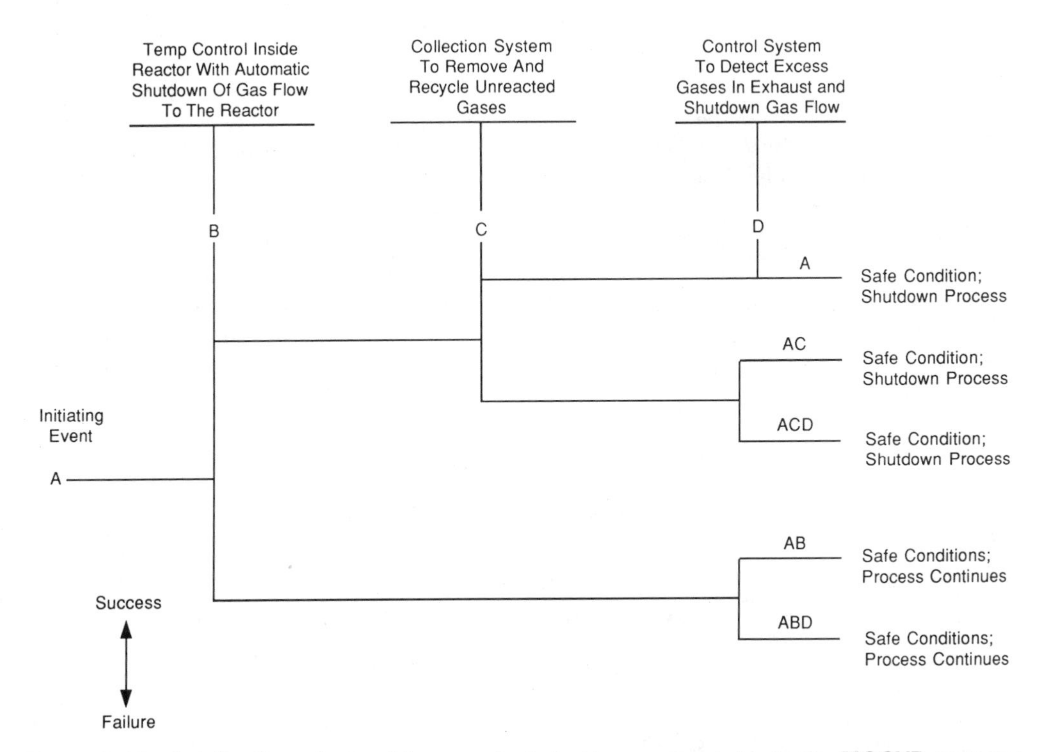

Figure 9.12. Application of event tree analysis to the reactor tube in the MOCVD process.

This method demonstrates the possible outcomes of various equipment failure combinations, and the information can be used to determine if unsafe conditions exist. The event tree is a straightforward and concise means for analyzing initiating effects that could lead to unacceptable risks.

Discussion of Results for the MOCVD Process

Each method utilized to analyze the MOCVD process results in important design information. HAZOP and PHA both include recommended corrective actions in the final report. PHA begins with a hazard and then explores cause and effects; HAZOP, on the other hand, begins with a particular equipment malfunction and studies the cause. FTA begins with a top event and traces it down to all possible basic faults that cause the undesired event, whereas ETA starts with an initiating event and explores the effects of different combinations of failure modes. The final report from an FMECA is in tabular form, and studies one individual failure mode and its effects and criticality on the system. Because of the broad scope of results obtained from different types of analysis, the most beneficial hazard evaluation of the design process should include several methods of evaluation.[2,23]

These procedures are invaluable tools in the safe design and operation of any potentially hazardous process.

Application of Hazard Evaluation Techniques to the Preliminary Design of an Ethylene Production Plant

Ethylene Plant Process Description

Ethylene is the major feedstock used by the petrochemical industry for the production of a variety of synthetic polymers, and is produced by the steam cracking of hydrocarbons such as ethane, propane, naphtha, and gas oil. The steam pyrolysis of hydrocarbons produces ethylene along with a wide range of by-products such as hydrogen, methane, propylene, and butadiene; and because ethylene purity is a critical factor in polymerization units, the mixture of gases obtained from steam cracking must be purified. Therefore, ethylene plants are comprised of a pyrolysis section in which the feedstock is cracked in pyrolysis furnaces to produce ethylene and other gases, and a purification section in which the pyrolysis products are separated and recovered.[6,24,25]

Ethylene production involves high temperatures (1500°F) in the pyrolysis section and cryogenic temperatures in the purification section. The feedstocks, products, and by-products of pyrolysis are flammable and pose severe fire hazards. Benzene, which is produced in small amounts as a by-product, is a known carcinogen. Table 9.13 summarizes some of the properties of ethane (feedstock) and product gases.

Table 9.13. Hazardous properties of materials in ethylene production.

	Toxicity	Fire Hazard	Explosion Hazard
Ethane	Low	Very dangerous	Moderate
Hydrogen	None	Dangerous	Dangerous
Acetylene	Moderate	Very dangerous	Moderate
Methane	Low	Very dangerous	Dangerous
Ethylene	Low	Very dangerous	Moderate
Carbon dioxide	Low	None	None

Source: Reference 26.

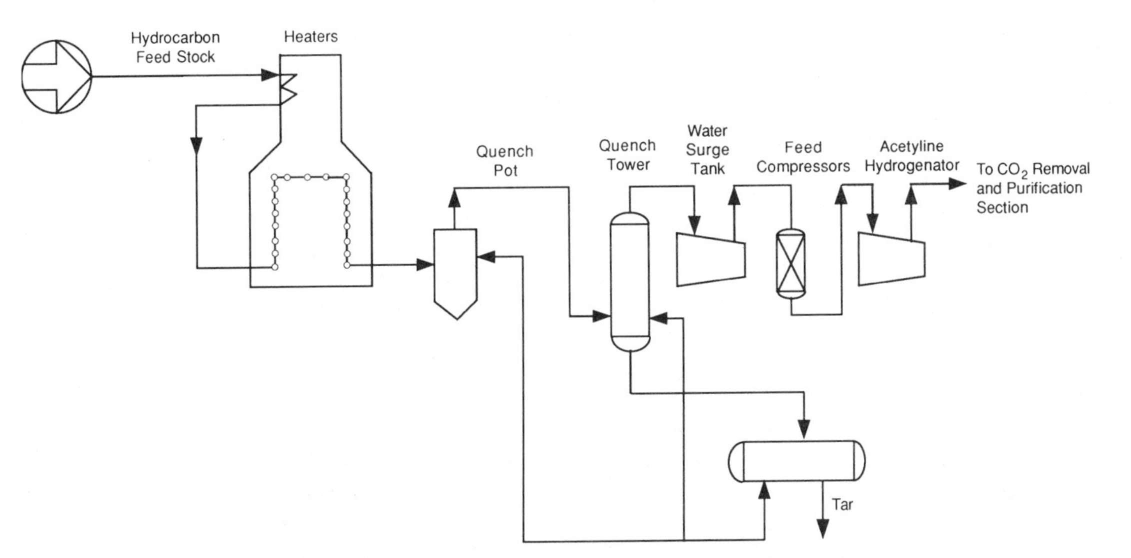

Figure 9.13. Pyrolysis and waste heat recovery section of an ethylene production plant.

Figure 9.13 shows a simplified schematic diagram of the pyrolysis and waste heat recovery section of an ethylene plant.

Prior to entering the pyrolysis reactors, the feedstock is mixed with steam, which reduces the hydrocarbon partial pressure, acts as a heat tansfer medium, and reduces coke laydown inside the reactor tubes. The tubular reactors are heated to reaction temperatures (1,100–1,700°F) by means of direct-fired heaters. The flow of steam and feedstock to the reactors can be adjusted to provide an optimum residence time, which is a function of the feedstock used.

After completion of the cracking reactions in the tubular reactors, the gaseous mixture flows to a quench tower, where the gas temperature is lowered enough to stop the cracking reactions. Oil or water can be used as the cooling medium. Transfer line heat exchangers can be used to recover the heat contained in the product gas and use this energy to produce high pressure steam.

The cooled gaseous products are dried using molecular sieves and compressed to about 500 psig by a multistage compressor. The compressed gas then is sent to an acetylene converter, where acetylene is selectively hydrogenated to ethane. The gaseous mixture then flows to the purification section of the plant, where each component of the gas is recovered by means of cryogenic distillation.

Application of PHA to the Pyrolysis and Waste Heat Recovery Section

Table 9.14 summarizes the results of application of a preliminary hazard analysis to the pyrolysis section of an ethylene plant. The major hazard to both personnel and plant is the fire or explosion hazard of the gases used or produced in the process.

Application of FTA to the Pyrolysis Furnace of an Ethylene Plant

Figure 9.14 demonstrates the initial steps for a fault tree analysis; the top event, bounds, configurations, and unallowed events are specified, and the level of resolution is shown. Once all the limits have been determined, the fault tree is constructed, as in Figure 9.15. Note that every branch of the fault tree ends in a basic fault or cause leading to the top event.

Table 9.14. Example of application of preliminary hazard analysis to an ethylene plant.

Hazard	Cause	Major Effects	Corrective/Preventative Measures
Damage to feed reactor tubes	Feed compressor failure (no endothermic reactions in reactor	Capital loss, downtime Damage to the furnace coils due to high temperature	• provide spare compressor with automatic switch-off control • develop emergency response system
Explosion, fire	Pressure build up in the reactor due to plug in transfer lines	Potential for explosion/fire	• provide pressure relief valve on the reactor tubes • provide warning system for pressure fluctuations (high pressure alarm) • provide auxiliary lines with automatic switch off.
Explosion, fire	Violent reaction of H_2 to acetylene converter with air in presence of ignition source	Potential for injuries, and fatalities due to fire or explosion	• provide warning system (hydrogen analyzer) • eliminate all sources of ignition near hydrogen gas storage area • develop emergency fire response • automatically shut off the H_2 feed • Provide fire fighting equipment
Flammable gas release	ethane storage tank ruptures	Potential for injuries, and fatalities to to fire or explosion	• provide warning control system (pressure control) • minimize on-site storage • develop procedure for tank inspection • develop emergency response system • gas monitoring system
Flammable gas release	CH_4 Storage tank (line) leak/ rupture (fuel for the furnace)	Potential for injuries, and fatalities due to fire or explosion	• provide warning system • minimize on-site storage • develop procedure for tank inspection • develop emergency response system • gas monitoring system
Flammable gas release	Radiant tube rupture in the furnace	Potential for injuries, and fatalities due to fire	• reactor materials of construction • monitor design vs. operating reactor temperature • temperature control instrument
Employee exposure to benzene (carcinogen)	Leak in knock-out pots or during handling benzene	chronic health hazard	• install warning signs in the area • provide appropriate PPE • develop safety procedures for handling and clean up • monitor concentration of benzene in area to meet TLV requirements
Fire/explosion in acetylene converter	runaway reaction (exothermic)	fatality injury loss of capital	• install temperature control on converter • install pressure relief on reactor responding to temperature control
flammable atmosphere	Leak in transfer lines	fire/explosion	• install combustible gas meter in sensitive areas • provide adequate fire fighting equipment • emergency shutdown • educate and train personnel on emergency procedures

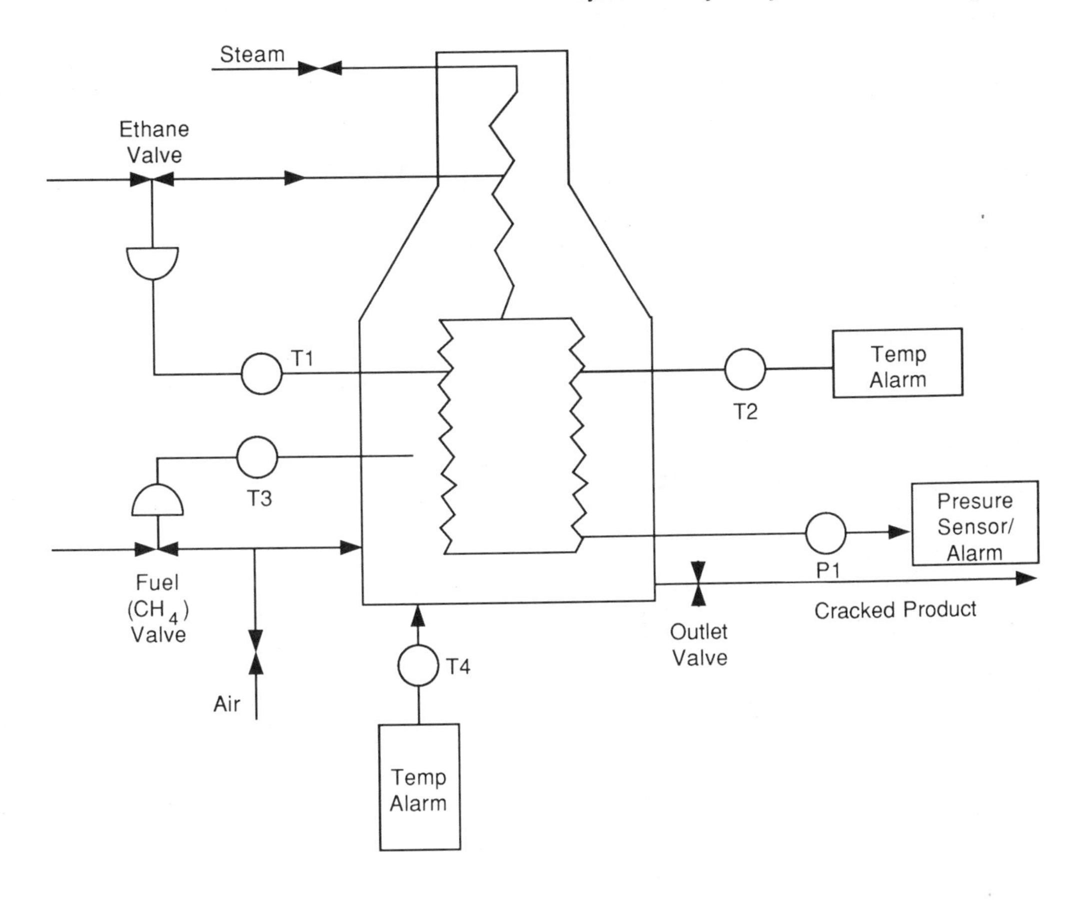

Top Event – Fire Explosion in the Furnace
Existing Event – Abnormal temperature and radiant tube rupture
Unallowed Events – Electric power failures
Physical Bounds – As shown in the figure
Equipment Configurations – Inlet feed values open, outlet value open
Level of Resolution – Equipment shown in figure.

Figure 9.14. Fault tree analysis preliminary steps for an ethylene production plant.

As can be noted from Figure 9.13, steam and ethane are mixed prior to entering the reactor tubes where pyrolysis reactions take place. All feed and product lines must be equipped with appropriate control devices to ensure safe operation.[20]

Discussion of Results

Fault tree analysis results in a flow chart that breaks down a top event into all possible basic causes. Although this method is more structured than preliminary hazard analysis, it addresses only one individual event at a time. To use FTA for a complete hazard analysis, all possible top events must be identified and investigated; this would be extremely time-consuming and perhaps unnecessary in a preliminary design. As can be noted from the previous analysis of an ethylene plant, one of the major disadvantages of FTA is lack of recommendations for preventive and corrective measures. However, FTA has the advantage of pinpointing the sequence of events that

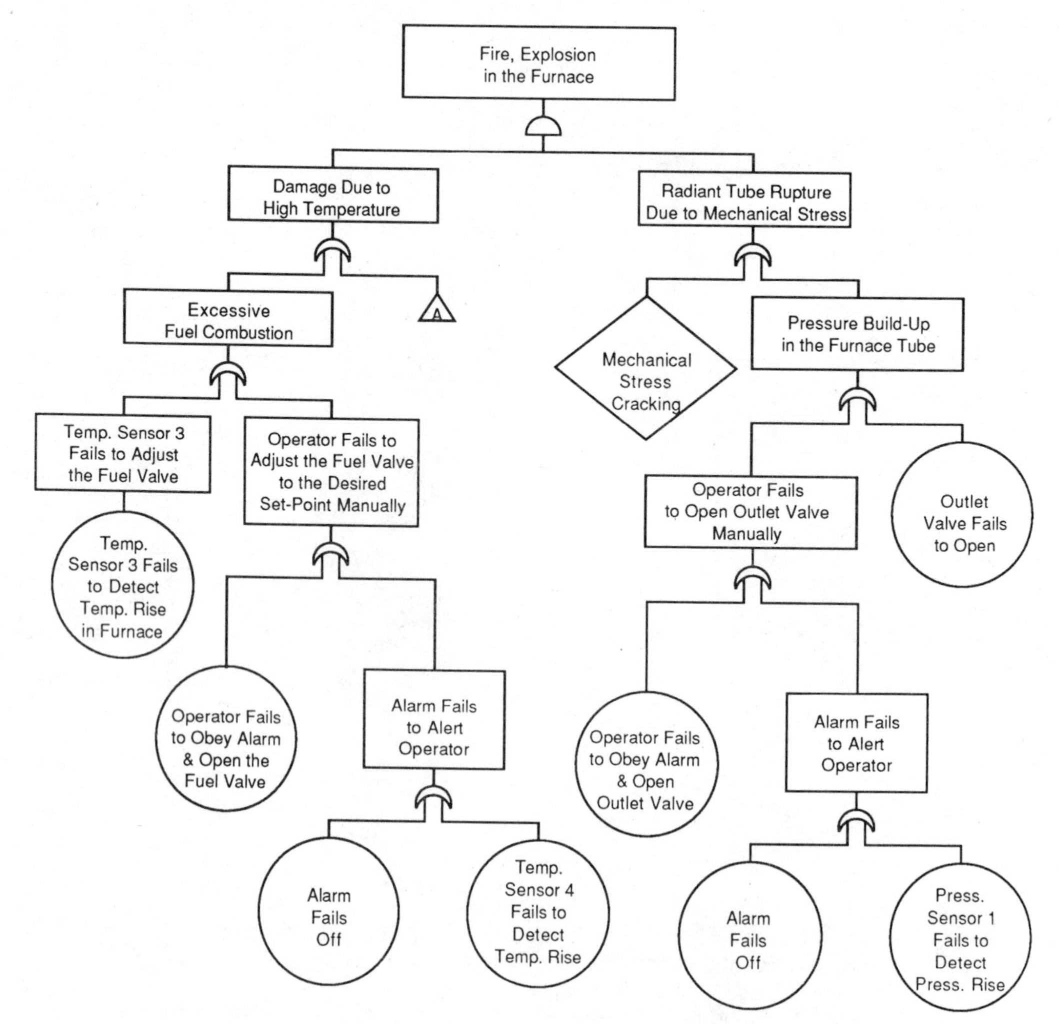

Figure 9.15. Application of fault tree analysis to an ethylene plant preliminary design.

could lead to an undesired top event. Once these causes have been identified, an experienced design team can recommend solutions in the form of design alternatives and/or instrumentation. In recommending solutions, the probability, severity, and economics of each case must be taken into account. For example, the problem of temperature control failure in the reactor tubes as a result of disruption in ethane flow can be solved by installing flow controls on the lines. Although flow controls on feed and steam lines are installed for the purpose of controlling the residence time in the reactor and product distribution, the flow controllers also contribute to temperature control in the reactor. These interactions are important, and their effects must be taken into account as a conceptual design develops into a flow diagram and finally into a piping and instrumentation diagram.

The previous analysis also indicates that PHA not only has the capability of identifying major hazards in the process, but also recommends corrective measures at the very early stages of design. This is extremely important in the development of new technologies and in feasibility studies. The overall economic picture of a process can change drastically as a result of instru-

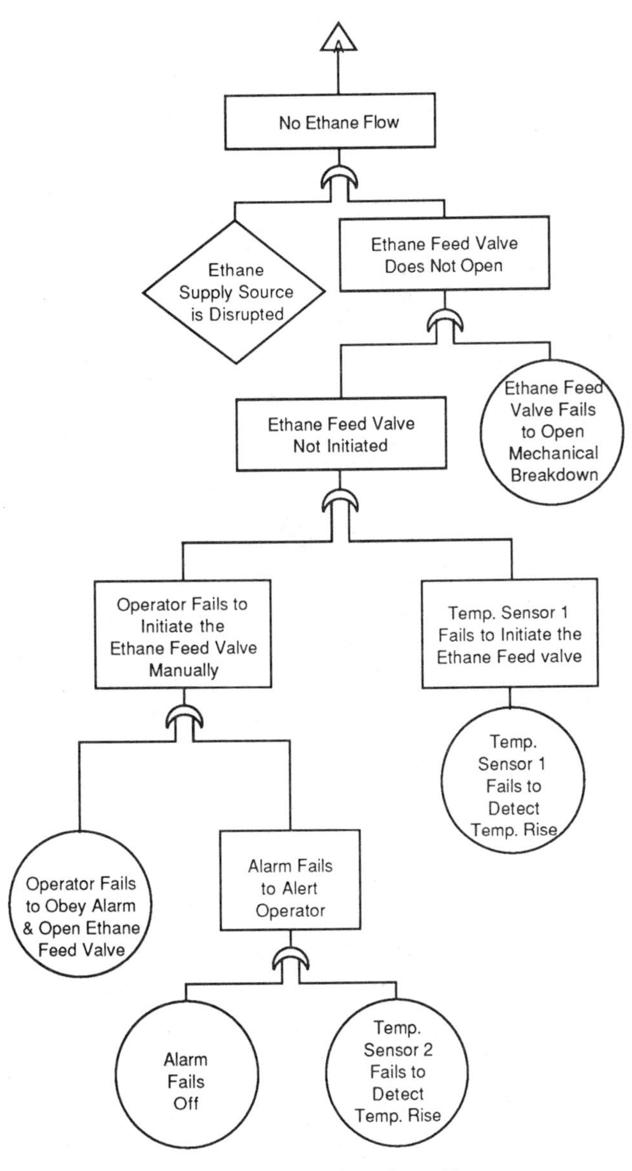

Figure 9.15. ***(Continued)***

mentation and/or procedures to render an acceptable risk or to bring the plant into compliance with regulations.

Application of Hazard Evaluation Procedures to the Preliminary Design of an Alkylation Process

Alkylation Process Description

Alkylation, used widely in the petroleum industry to produce high-octane gasoline, involves the reaction between a low molecular weight olefin and an isoparaffin in the presence of an acid catalyst. For example, propylene and isobutane can react according to the following reaction to produce gasoline:

$$\underset{\text{Propylene}}{CH_3{-}CH{=}CH_2} + \underset{\text{Isobutane}}{CH_3{-}\underset{|}{\overset{}{CH}}(CH_3){-}CH_3} \rightarrow \underset{\text{2,3-Dimethylpentane}}{CH_3{-}CH(CH_3){-}CH(CH_3){-}CH_2{-}CH_3}$$

Although the reactions taking place in an alkylation reactor are numerous and relatively complex, the major reactions always involve combination of a low molecular weight olefin and an isoparaffin, as demonstrated above.

The acid catalysts used in the alkylation process include mainly hydrofluoric acid (HF) and sulfuric acid (H_2SO_4). Most alkylation processes in operation today use HF as the catalyst for operating temperature flexibility. As the HF process developed by the Phillips Petroleum Company is among the most widely used alkylation processes, this process description will focus on the Phillips process.

Figure 9.16 shows a simplified schematic flow diagram of the Phillips alkylation process. In this process isobutane is catalytically alkylated with olefins that exist in refinery off gases, in the presence of liquid hydrofluoric acid. Recycle isobutane from the fractionating tower is mixed with olefin and isobutane feed before entering the alkylation reactor. The feed must be passed through dryers to remove any moisture prior to its entry into the alkylation reactor.

The alkylation reactor/settler, designed exclusively by the Phillips Petroleum Company (see Figure 9.17), has no moving parts and contains a moving bed of hydrofluoric acid that provides high dispersion of feed into the catalyst and almost instantaneous conversion of feed to alkylate product.

The operating conditions in the reactor are mild; the temperature can range from 25°C to 45°C, and the operating pressure is selected in such a way as to maintain fluids in their liquid states.

After completion of the alkylation reactions, the mixture of catalyst, alkylate, and unreacted

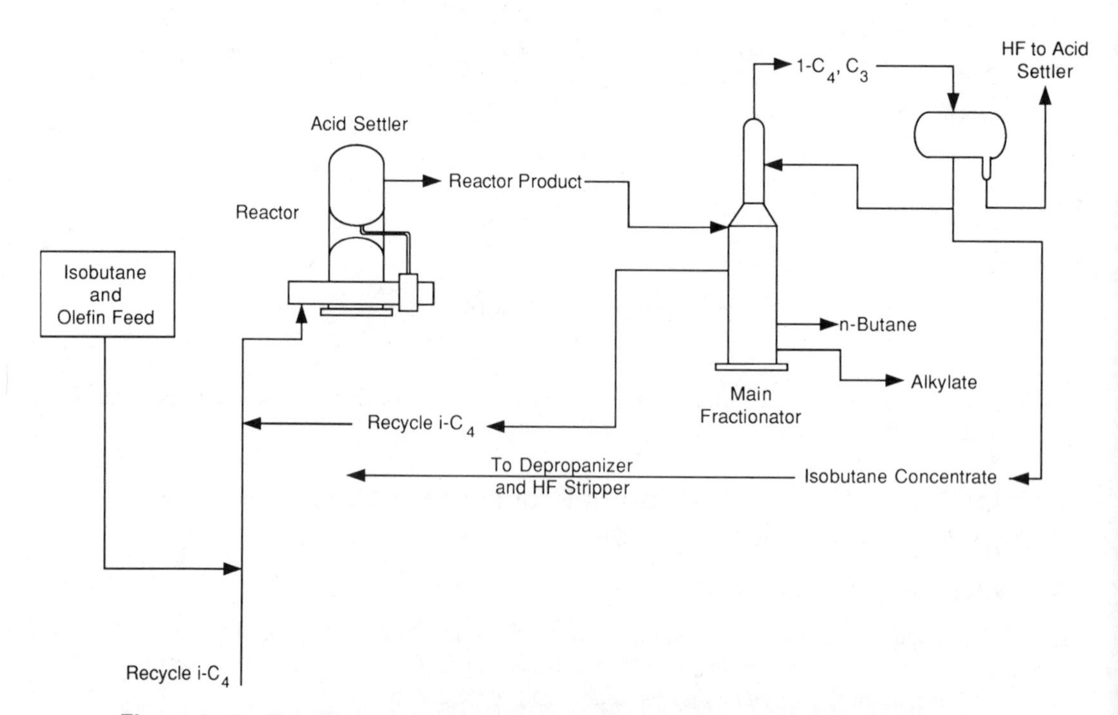

Figure 9.16. Simplified flow diagram of a hydrofluoric acid alkylation process.

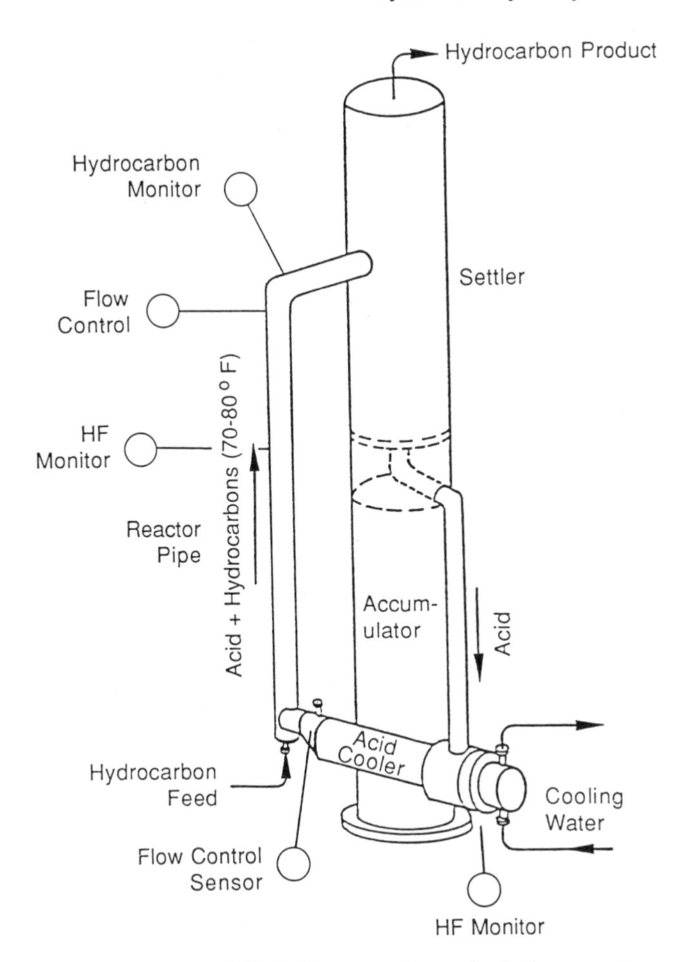

Figure 9.17. Simplified diagram of an alkylation reactor.

reactants flows upward to an acid settler (see Figure 9.17), where acid hydrocarbon phase separation takes place. The hydrocarbon phase has a lower density than hydrofluoric acid, and because the two phases are virtually immiscible, the acid settles at the bottom of the settler.[6]

The hydrocarbon phase, which contains the alkylate product, is withdrawn from the top of the settler and is sent to the main fractionator, where the alkylate is separated from hydrocarbon gases. The acid is withdrawn from the bottom of the settler and sent through a water-cooled heat exchanger before being recycled back to the reactor.

Application of ETA to the Alkylation Reactor

The major hazards in the alkylation process are the flammability of gaseous and liquid hydrocarbons and the health hazards posed by hydrofluoric acid. Table 9.15 summarizes some of the hazardous properties of materials used in the process. Hydrofluoric acid, which has a TLV of 3 ppm, can cause severe health hazards ranging from skin burns to death as a result of overexposure. Because of its hazardous properties, HF requires the implementation of special handling and storage procedures as well as the use of proper protective equipment.

Figure 9.18 shows an ETA analysis focusing on the possibility of an acid leak from the lines carrying the acid into and out of the water-cooled acid cooler. (It should be noted that ETA provides an opportunity for study of the combination of scenarios resulting from an undesired

Table 9.15. Hazardous properties of materials commonly used in alkylation.

	Toxicity	TLV*	Fire Hazard	Explosion Hazard
Butene	Low	—	Very dangerous	Moderate
Propene	Low	—	Very dangerous	Moderate
Butane	Moderate	800 ppm	Very dangerous	Dangerous
Propane	Moderate	1,000 ppm	Very dangerous	Very dangerous
Hydrofluoric acid	Dangerous	3 ppm	—	—

*TLV = threshold limit value.
Source: Reference 26.

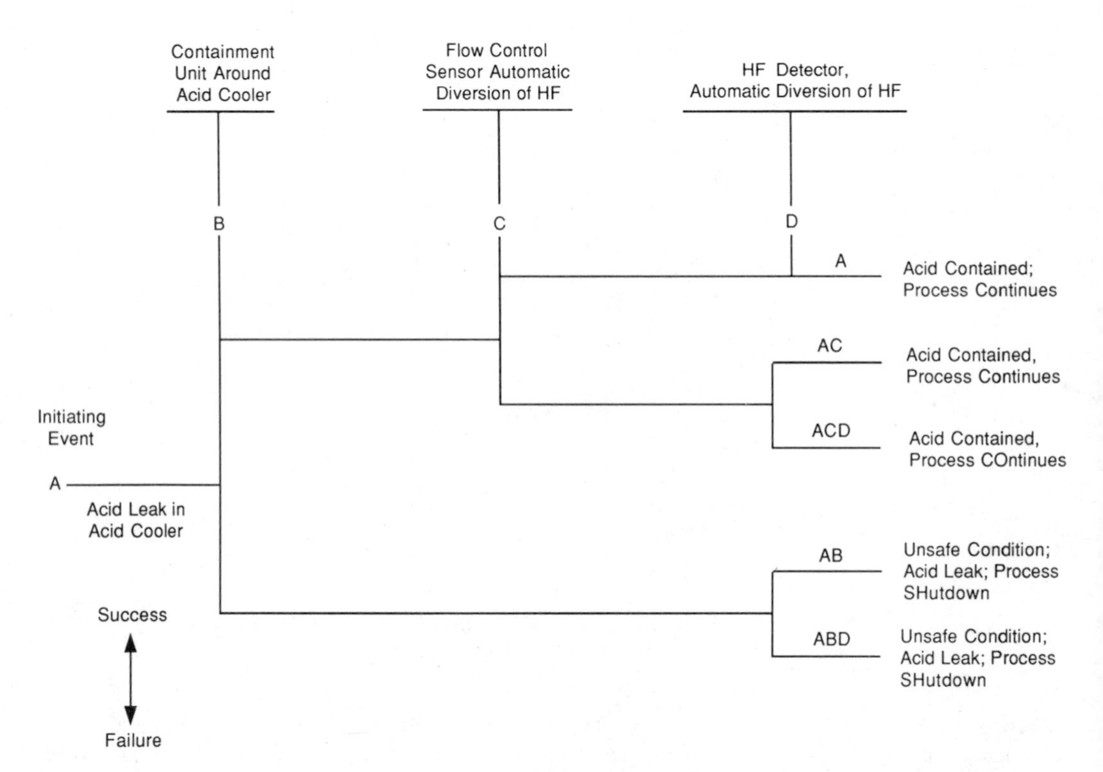

Figure 9.18. Event tree analysis for an acid leak around the acid cooler of an alkylation reactor.

initiating event.) Figure 9.19 shows a similar analysis for the reactor tube failure. Safety issues such as these can be identified and dealt with properly even in the very preliminary and early stages of design.

Application of CCA to the Acid Containment Unit of the Alkylation Reactor

Figure 9.20 demonstrates the application of cause consequence analysis (CCA). The starting point for this analysis is the event tree in Figure 9.19. Here the acid containment unit has been isolated for study, and the consequences and reasons for the success or failure of the containment unit are highlighted. It should be noted that cause consequence analysis combines the backward thinking scheme of fault tree analysis for top events with the forward thinking scheme of event tree analysis for initiating events. Although cause consequence analysis can become very complex

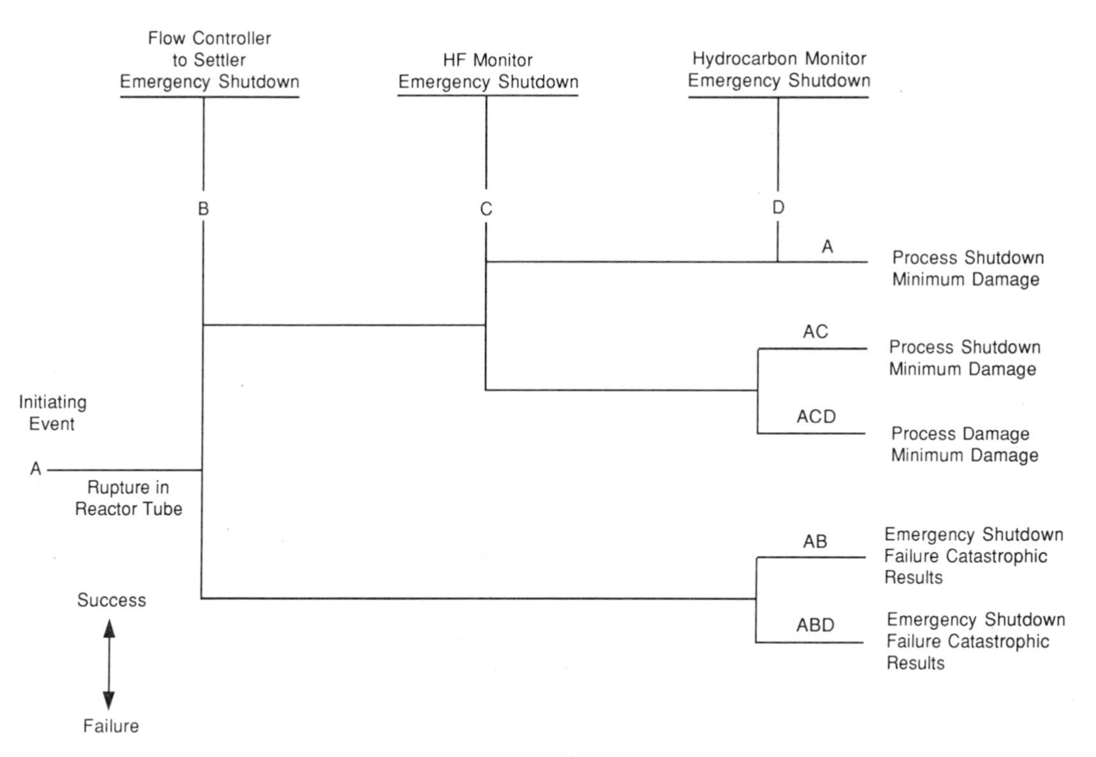

Figure 9.19. Event tree analysis for an acid leak due to reactor tube failure.

and involved in a short period of time, it is a valuable tool in the identification of possible accident scenarios and their consequences.

Discussion of Results

As the above analysis performed on the alkylation process shows, cause consequence analysis can become rather complicated. The main advantage of this method, however, lies in the fact that it combines both the forward thinking pattern of ETA and the reverse thinking pattern of an FTA analysis. The best way to carry out a cause consequence analysis is first to complete an event tree and a fault tree analysis and then to combine the components of each into a cause consequence analysis. Although CCA offers the advantages of both ETA and FTA, the latter two analyses are somewhat more readable and easier to utilize in the design process than CCA.

Application of Hazard Evaluation Procedures to the Preliminary Design of a High Pressure Low Density Polyethylene Plant

Process Description

Polyethylene, one of the most widely used thermoplastics, is obtained by the polymerization of ethylene:

$$-\overset{|}{\underset{}{C}}=\overset{|}{\underset{}{C}}- \rightarrow [-\overset{|}{\underset{|}{C}}-\overset{|}{\underset{|}{C}}-\overset{|}{\underset{|}{C}}-\overset{|}{\underset{|}{C}}-]_n$$

The polymerization process can be carried out at either low or high pressure, depending on the intended use of the final product. Polyethylene obtained by the low-pressure process has a higher

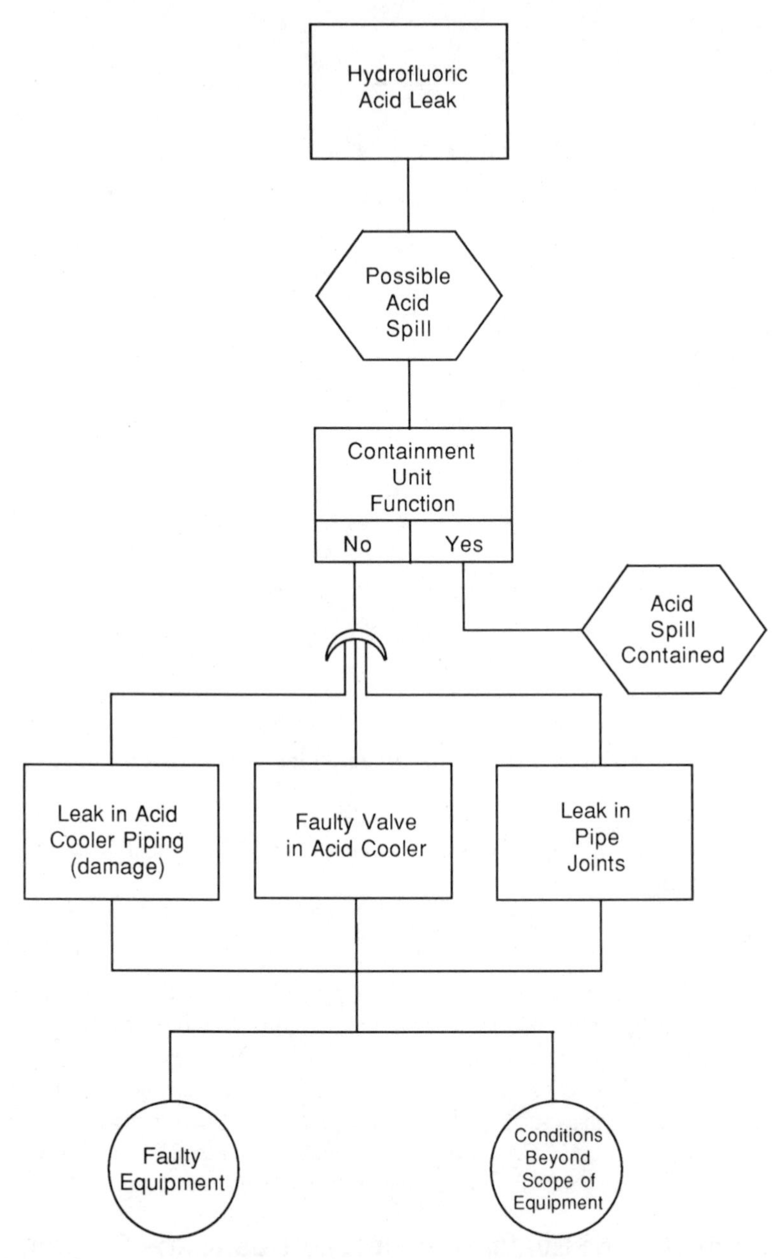

Figure 9.20. Cause consequence analysis for an acid leak in the alkylation process.

degree of crystallinity and as a result, a higher density as compared to polyethylene obtained via the high pressure process. Polyethylene has found widespread use in packaging, construction, agriculture, household items, and the rubber industry.[12,14]

Figure 9.21 shows a schematic simplified flow diagram for a high pressure low density polyethylene plant. Polymerization grade ethylene (99.5% pure) is compressed to about 1,500 atmospheres using a multistage reciprocating compressor (makeup compressor), and is heated to a temperature of about 350°F. The molecular weight of the polymer is directly proportional to the reactor pressure. After the addition of small amounts of an initiator, which acts as a catalyst for

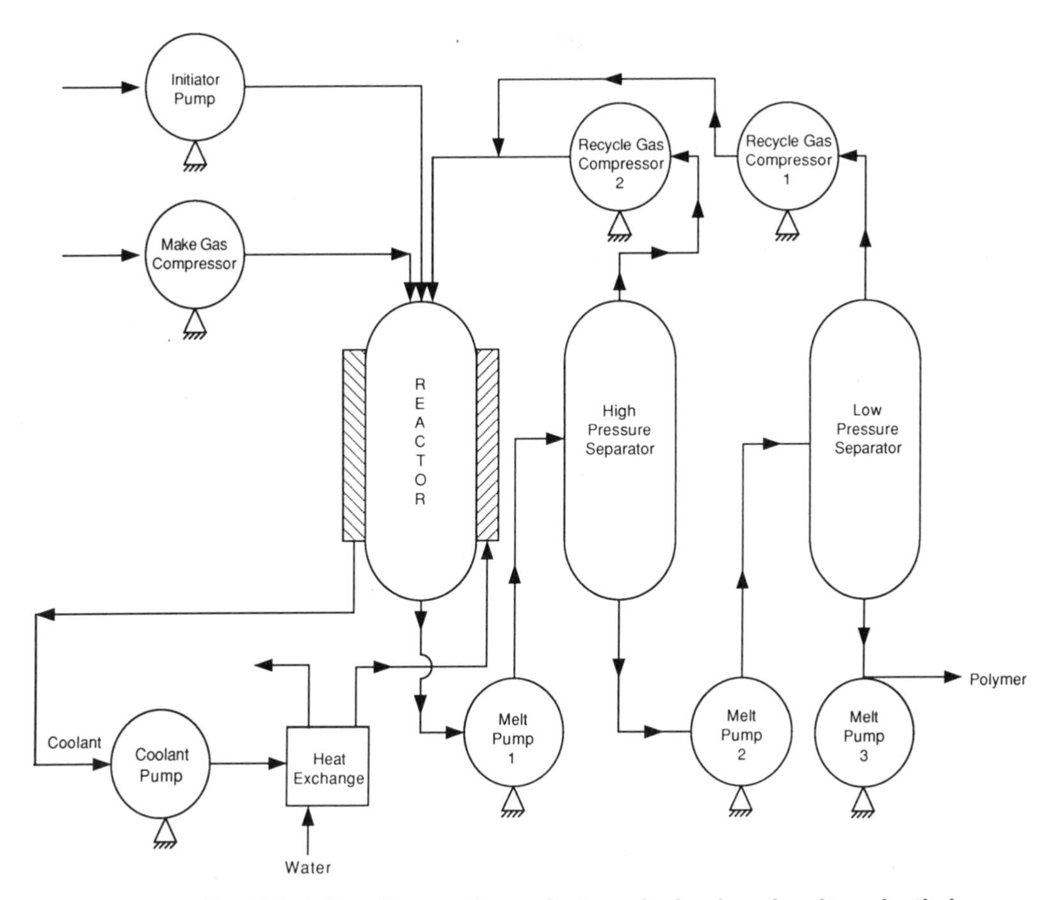

Figure 9.21. Simplified flow diagram for a plant producing low density polyethylene.

the polymerization reactions, the mixture of ethylene and initiator enters the polymerization reactor, a tube with a high length-to-diameter ratio, to facilitate the removal of heat generated by polymerization. Because of the large exothermic heat of polymerization and the difficulty of removing this heat, the conversion of ethylene per pass is kept in the 15 to 25% range. Some reactors use a cooling jacket with a suitable heat transfer fluid to remove the heat of polymerization. It is critical to remove the heat generated within the reactor, or there is the possibility of a runaway reaction (a self-accelerating reaction that, if not controlled, can lead to a disaster). If the heat generated within the polymerization reactor were not removed, it would raise the reactor temperature, thereby increasing the rate of the polymerization reactions and generating more heat, which would increase the reactor temperature further.

The mixture of polymer and unreacted ethylene is removed from the reactor by means of a melt pump and sent to a high pressure separator for recovery of the bulk of the unreacted ethylene. The ethylene recovered from the high pressure separator is fed to the suction port of the recycle gas compressor and is fed back to the polymerization reactor. The polymer, which contains some unreacted ethylene, is pumped out of the high pressure separator using a melt, and is sent to a low pressure separator for recovery of the remainder of the unreacted ethylene. The overhead from the low pressure separator is sent to a recycle gas compressor and is fed back to the reactor. The polymer is pumped out of the low pressure separator by a melt pump, and is sent through water tanks to a pelletizer. The pelletized product is devolatilized and purged to remove any residual monomer before it is sent to packaging.

Table 9.16. "What If" analysis for high pressure low density polyethylene process (simplified diagram).

What If	Consequence/Hazard	Recommendations
coolant pump to reactor fails	Runaway condition in reactor explosion/fatality	• Provide accurate temp. monitoring in reactor/ • backup pump/high temp. alarm • relieve reactor pressure in reactor through automatic control to stop reactions • automatic shut off of ethylene flow
Coolant temp. to Jacket is high	Eventual runaway condition in reactor	• Provide adequate temp. control on coolant line • Use heat exchanger flow control to adjust inlet temp.
There is runaway condition in reactor	Explosion/fire/fatality	• Provide adequate temp. control on coolant line • Use heat exchanger flow control to adjust inlet temp. • Install rupture disk/relief valve to relieve pressure to stop reactions • Emergency shut down procedure
Recycle gas compressor 1 or 2 fails	None likely	• Spare compressor or shutdown Procedure
The melt pump fails	High level in reactor more polymerization, runaway eventually, exceed design pressure	• Provide level and flow control schemes to activate spare pump or shut the flow of monomer • Shut down procedure if no spare pump
Leak at suction or discharge of compressors	Fire/explosion	• Use monitoring devices to ensure no flammable gas is released
Ethylene leaks out of process lines	Fire/explosion	• Provide adequate flammable gas monitoring devices
Monomer/initiator ratio out of control	Eventual runaway reaction, fire, explosion	• Provide flow control on the initiator and monomer lines

Application of "What If" Analysis to the Preliminary Design of a Low Density Polyethylene Plant

Table 9.16 summarizes the results of a preliminary "what if" analysis for a high pressure low density polyethylene production. This information is compiled in tabular form, and recommendations are provided. The success of a "what if" analysis depends upon the analyst's familiarity with the process and experience with this type of hazard evaluation. For example, the possibility of a runaway reaction may not seem obvious to someone unfamiliar with the polymerization process.[19]

Application of HAZOP to the Preliminary Design of a Low Density Polyethylene Plant

As mentioned earlier, HAZOP focuses on how certain parameters of a plant an deviate from their design value. Although during the course of a HAZOP study the solution to certain problems might become evident, it should be noted that the main objective is the identification of design problems.

Table 9.17 is an example of a HAZOP study for two parameters in a high pressure low density

Table 9.17. HAZOP study of a polyethylene plant.

Guide Word	Deviation	Consequences	Causes	Recommended Action
Parameter: reactor temperature				
Higher	Higher reaction temperature	Runaway reaction in reactor	Coolant pump to reactor fails	Provide temperature control Provide high temperature sensor/alarm Provide pressure relief valve with automatic feed from temperature control system Provide spare coolant pump
			coolant temperature high	Use heat exchanger temp. control to adjust inlet cooler temperature
Lower	Lower reactor temperature	Poor or no reaction; Poor quality product	coolant temperature low	Provide temperature monitoring in reactor Use heat exchanger to adjust inlet coolant temperature
Parameter: How rate of ethylene and polyethylene and initiator				
No (polymer)	No flow	Level buildup in reactor	Melt pump 1 fails	Provide level control in reactor with automatic flow through a spare pump
Less (ethylene)	Less flow	System upset Product quality System shutdown	Make up or recycle compressor failure	Provide a spare compressor with automatic switch off from the failed compressor
More (initiator)	More flow	More polymerization Possibility of runaway conditions Product quality off specification	Initiator pump malfunction	Provide adequate flow controls on both initiator and monomer lines to maintain the desired initiator to monomer ratio
Less (initiator)	Less flow	Less polymerization Reactor temperature imbalance Affects downstream equipment such as heat exchangers	Make up and recycle gas compressor	Provide flow controllers on ethylene and initiator lines

polyethylene process. Because the reaction temperature and the flow rate of ethylene are important for both the quality and the quantity of the product as well as safety considerations, these two areas are good candidates for study. A complete HAZOP also will contain studies of temperature in the preheat and cooling stages, the flow rate of the initiator, and the integrity of all equipment.[21]

Discussion of Results

As can be seen in the case studies of the polyethylene plant, both "what if" and HAZOP provide valuable insight for the safe design of a process at the preliminary stages. However, although both techniques provide recommendations for a safe design, HAZOP is somewhat more organized than "what if" in terms of guidelines. The analyst must only choose all desired parameters and then apply the appropriate guide words. "What if" analysis, on the other hand, does not follow any specific guidelines, and its success depends largely on the experience of the design team conducting the study.

Application of Hazard Evaluation Procedures to the Batch Process of Industrial and Military Explosive Production

Many people regard explosives as chemical products used solely by the military. Although chemical explosives have been used extensively by the military for destructive purposes, many engineering projects such as the construction of dams, roads, and underground mining would have been impossible without the use of explosives.

The commercial production and sale of explosives are a twentieth century phenomenon, but explosive mixtures such as black powder were known to the Chinese many centuries ago. The discovery of nitroglycerin, nitrocellulose, and dynamites in the mid-nineteenth century formed the cornerstone for the discovery and use of a variety of more sophisticated and powerful explosives.

Explosives are defined as unstable chemical compounds that can decompose as a result of thermal or mechanical shock. The decomposition reactions normally are accompanied by the generation of large quantities of heat and gases, and when confined the gases can exert a large amount of force on the walls of their container.

Chemical explosives can be classified as detonating and deflagrating. The detonating explosives are further classified as primary and secondary explosives. Examples of primary detonating explosives are mercury fulminate and lead azide; TNT and picric acid are secondary explosives. Black powder and nitrocotton are deflagrating or low explosives.

Primary detonating explosives are extremely unstable compounds that can decompose violently by a thermal or mechanical shock. These compounds are extremely dangerous, and extreme care and caution should be exercised in the their production, handling, and use. They usually are used in small quantities to initiate detonation in secondary explosives, which are relatively more stable compounds.

Properties of Explosive Compounds

Explosive compounds intended for commercial or military use must satisfy certain requirements, which are determined by standardized tests, and include sensitivity, stability, and brisance.

The sensitivity of an explosive to mechanical shock is determined by finding the height from which a standard weight must be dropped to detonate the explosive compound. The brisance is related to the explosive force associated with the detonation of the compound and is measured by the degree to which an explosive compound can shatter sand in a confined space. Although the primary property of interest in the production of any explosive compound is its unstability, it should satisfy stability safety requirements under normal conditions. Table 9.18 summarizes the properties and products of decomposition of some of the more common explosive compounds.[25]

Table 9.18. Properties of some of the more commonly used explosives.

Identity	Formula	Decomposition Products	Heat Released, cal/kg	Explosion Temp., °C	Pressure, kg/cm^2	Velocity of Wave, (m/s)
TNT	$C_7H_5(NO_2)_3$	H_2, CO, C, N_2	656	2,200	8,386	6,800
Trinitroglycerin	$C_3H_5(NO_3)_3$	H_2O, O_2, N_2	384	1,100	5,100	4,100
Picric acid	$C_6H_2(OH)(NO_2)_3$	CO, H_2O, H_2, N_2	847	2,717	9,960	7,000
Ammonium nitrate	NH_4NO_3	H_2, O_2, N_2	384	1,100	5,100	4,100
Nitrocellulose	$C_{24}H_{29}O_9(NO_3)_{11}$	CO, CO_2, H_2O, N_2	1,250	2,800	10,000	6,100
Gun powder	2 KNO_3 + 3C + S	N_2, CO_2, K_2S	501	2,090	2,970	NA

Source: Reference 25.

Manufacture of Explosives

Chemical explosives are manufactured by a variety of processes, depending on their intended use. Because of the unstable nature of these compounds, most of the manufacturing processes pose both physical and health hazards. In the following section, the manufacture of nitrocellulose will be discussed, and preliminary hazard analysis (PHA) and "what if" analysis will be applied to a preliminary design of this process.

Manufacture of Nitrocellulose

Although the explosive properties of nitrated cotton have been known for a long time, a major problem with the use and commercialization of this compound has been its inherent instability and rapidity of explosion. In recent years, with the discovery of stabilizing compounds and techniques to prolong its storage life, nitrocellulose has found widespread use as a military propellant.

The empirical formula for cellulose is $[C_6H_7O_2(OH)_3]_n$. It is a rather complex molecule with an average molecular weight in the neighborhood of 300,000 and a wide range of molecular weight distribution. The fundamental cellulose molecule contains three hydroxy groups, which can be esterified with nitric acid according to the following reaction:

$$C_6H_7O_2(OH)_3 + 3HNO_3 + H_2SO_4 \rightarrow C_6H_7O_2[ONO_2]_3 + 3H_2O + H_2SO_4$$

In addition to nitrate esters, some sulfate esters are formed, as a result of the presence of sulfuric acid. The sulfate esters are extremely unstable compounds that could generate a dangerous acid condition in the powder storage area if not properly removed. It is critically important to prevent an acidic condition in the finished nitrocellulose product, as that would greatly catalyze and enhance the decomposition reactions and could lead to disastrous explosions. Normally a stabilizer—which is a compound with basic properties, such as diphenylamine—is added to the finished product to neutralize any excess sulfuric or nitric acid.

The Process

Figure 9.22 shows a simplified schematic diagram for a nitrocellulose manufacturing process.

Cellulose-containing compounds such as wood, cotton liners, or pulp are boiled in a caustic solution in process vessels called kiers. The product of kiers flows into another vessel, where bleaching is accomplished by compounds such as NaOCl or $Ca(OCl)_2$. After bleaching, the cellulose is dried in dryers at about 105°C. The product of the dryer is fluffed and weighed, and then is fed into a nitrator where the proper amounts of nitric and sulfuric acid are added to the charge to carry out the nitration reactions. One nitrator charge normally contains 32 lb of cellulose, which is mixed and agitated with 1,500 lb of acid at 30°C for approximately 30 minutes.

After completion of the nitration reactions, the nitrator charge is dropped into a centrifuge, where separation between nitrated cellulose and the mixed acid takes place. The spend acid is partially recovered for reuse in the nitrator, and the nitrocellulose is washed with boiling water and again washed in a beater.

In order to neutralize the residual ester sulfates and acids, the nitrocellulose product is washed with boiling water and sodium carbonate solution. The water then is removed by a centrifuge, and the product is stored.

Hazard Analysis

Tables 9.19 and 9.20 summarize the results of PHA and "what if" analysis of the nitrocellulose manufacturing process. As can be noted from these tables, hazard evaluation procedures can identify and mitigate the process hazards by recommending safety procedures and/or instrumentation.

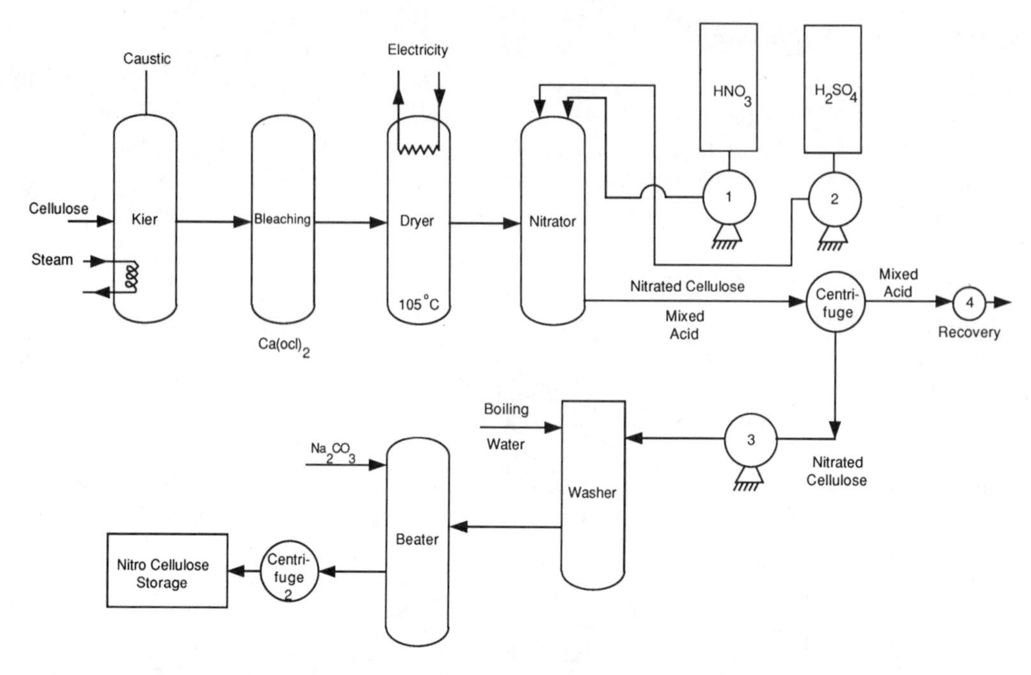

Figure 9.22. Nitrocellulose production.

Table 9.19. Preliminary hazard analysis for nitrocellulose production.

Hazard	Cause(s)	Major Effects	Preventive Measures
Explosion in storage area	Acid condition in finished product	Explosion fatality fire	• Install sulfate and acid analyzing instrument after centrifuge 2. Divert flow to temporary storage with subsequent flow to centrifuge 2 inlet.
Explosion in nitrator as a result of excess acid and high temperature	Malfunction of pumps 1 or 2; malfunction of nitrator heater	Explosion fatality, fire	• Install adequate flow controllers on sulfuric and nitric acid lines. Install temperature control on nitrator with a high temperature alarm.
Explosion in or down-stream of the nitrator	Dryer heater malfunction resulting in high temperatures	Fire, fatality	• Install temperature control on dryer with automatic shut off and high temperature alarm.
Nitrated cellulose and mixed acid spill	Centrifuge 1 or 2 malfunction	Employee exposure to hazardous substances	• Develop procedures for use of personal protective equipment. • Develop emergency and spill cleanup procedures. • Install instrumentation for emergency shut off.
Nitrated cellulose spill	Malfunction of centrifuge 2	Possible explosion, fatality, fire	• Use appropriate PPE. • Develop emergency and spill response procedures. • Install instruments for automatic shutoff.
Acid spill from storage tanks	Leak in storage or acid lines	Employee exposure to corrosives	• Require appropriate PPE to match potential hazards. • Develop procedures for storage tanks and lines inspection.

Table 9.19. *(Continued)*

Hazard	Cause(s)	Major Effects	Preventive Measures
			• Install acid monitoring devices in area, compare with PEL or TLV. • Develop procedures for general safe acid handling.
Caustic spill while transferred to kiers	Pump malfunction or employee mishandling	Employee exposure to corrosives (health hazard)	• If pump used, install instrumentation to stop flow in case of pump malfunction or leak. • Develop caustic handling Standard Operating Procedure (SOP). • Develop emergency and spill response procedures. • Determine appropriate PPE to be used.
Nitrated cellulose spill	Pump 3 malfunction or mishandling from centrifuge	Explosion, fire, fatality	• Install instrumentation to detect, alert and correct malfunction. • Develop SOPs for handling and spill response.

Table 9.20. "What If" analysis for nitrocellulose production.

What If	Consequence/Hazard	Recommendation
There is excess sulfate and acid in finished product?	Explosion, fire, fatality	• Analyze for excess sulfate and acid prior to storage. • Flow controllers on acid lines or proper acid handling procedures. • Adequate PPE. • Emergency response procedures.
Pump 1 or 2 malfunction or acid handled improperly?	Acid condition in product; employee exposure to health hazards	• Install flow controllers on acid lines or develop adequate acid handling and charging procedures. • Determine appropriate PPE to be used by employees. • Develop emergency and spill response procedures.
Centrifuge 2 malfunctions?	Nitrocellulose spill/explosion, fire, fatality	• Install appropriate instruments to shut off equipment in case of a malfunction. • Determine appropriate PPE for personnel near this centrifuge while operating. • Develop emergency and spill response procedures in case of a malfunction and spill.
The acid storage tanks leak?	Personnel exposure to health hazards	• Develop procedures for acid storage tank inspection. • Install acid monitoring instruments. • Develop spill clean up procedures.
The caustic line leaks or pump to kiers malfunction?	Caustic spill, personnel exposure to a corrosive health hazard.	• Install instrumentation to detect, alert caustic leaks and stop the flow of caustic to kiers. • If caustic handled manually, determine appropriate PPE and develop handling procedures. • Develop emergency and spill response procedures.

Instructor's Guidelines
Importance of Incorporation of System Safety Topics into Senior-Level Design Projects

Thousands of accidents occur throughout the United States every day. Most are caused by the failure of people, equipment, supplies, or surroundings to behave or react as expected. Statistics published by the National Safety Council reveal that 98% of all industrial accidents are caused by unsafe conditions and unsafe acts, with natural disasters responsible for only 2% of the accidents (see Chapter 8). Lack of attention to safety topics at the design stage of a process would undoubtedly create inherently unsafe conditions in the process, which could lead to disastrous accidents.

As the nation's technology is mobilized in response to ever increasing social demands to improve the quality of life through advancing technology, new dimensions must be added to the role of the design engineer to accomplish this task in an occupationally and environmentally safe manner.

Public reaction to unsafe design and operation of industrial process has resulted in strict occupational and environmental laws in recent years. However, the considerable public pressure on industry to ensure that public interests are incorporated into the design, operation, and waste disposal aspects of industrial processes has not been transmitted proportionately to the nation's engineering schools, which are responsible for the training of future engineers, and, in part, for the continuing education of practicing engineers.

Traditionally, American engineering schools have done a superb job of training engineering students in the fundamental laws of nature governing their fields and in the application of those laws to engineering problems; but they perhaps have been less successful in conveying to engineering students the importance of occupational and environmental safety in the design process and the criticality of the legal and moral responsibilities of the engineer toward society.

It is not uncommon for senior-level graduating engineers to think only in terms of the technical aspects of the profession with little or no emphasis on the real issues of safety and environment in the design process. These graduates get a rude awakening when they join industrial firms and find that safety and environmental issues are real problems that must be dealt with, and that they have little or no training in these areas. Although some companies have made a commitment to train their young engineers in occupational safety and environmental problems, not all new engineers are lucky enough to work for those companies; many are employed by companies with rather poor safety records, and with no academic training in these areas, they end up learning about safety and environmental problems through trial and error and unnecessary, senseless accidents.

Also some engineering schools do not have the right attitude toward safety issues. Some, in fact, lack the right systems for compliance with occupational safety and environmental laws. For example, it is utterly ridiculous to present any safety or environmental training to graduating engineers where food and beverages are allowed in chemical laboratories and where for lack of a proper disposal system, dangerous chemicals are poured down laboratory sinks.

It is time for the engineering schools to take a more responsible position in regard to occupational safety and environmental health problems. Unless safety is regarded as a science and is incorporated into the engineering curriculum in a systematic manner, regard for safety can easily turn into lip service with no meaningful results.

The senior-level design project offers the perfect medium for an introduction to system safety. Hazard evaluation procedures should be discussed, and the students can apply one or more of these methods to portions of their design project. This assignment will give the students an appreciation of the time and effort that go into such analyses. In addition, students should recognize the cost-saving benefits of identifying, reducing, or eliminating hazards at the design stage rather than after implementation. Students in design courses have enough core knowledge of processes to understand, at least fundamentally, the repercussions of system safety. For these reasons, hazard evaluation procedures and system safety topics should be addressed in all senior-level design projects.

Educational Objectives

Incorporation of system safety topics into senior-level design courses can accomplish several goals. The graduating engineer realizes (maybe for the first time) that in order to make a process both economical and operable, safety and environmental issues must go hand in hand with the technical aspects of the project. The student also can develop an appreciation of both legal and moral responsibilities of the engineering profession. The incorporation of system safety topics into preliminary design projects also can give the student the minimum tools required to apply the scientific laws of nature to the design and operation of hazardous technologies in an occupationally and environmentally safe manner.

The system safety approach provides a thorough, systematic means for addressing workplace hazards. Because of the complexity of production, construction, and processes, informal hazard evaluations are no longer sufficient; a systematic approach is more economical and results in more complete analyses. A thorough assessment of the risks inherent in a process can minimize losses due to downtime and worker injury. In addition, public intolerance of dangerous failures and accidents is becoming a force to contend with. Stricter legislative requirements are most likely to be met by utilization of systematic hazard evaluation approach.

The system safety approach begins with defining the system to be studied, allowing for human, equipment, and environmental interactions. The scope and depth of the study are determined; the prime factors in this determination include the projected cost of analysis, schedule deadlines, and available human resources. Safety hazards are then identified; these hazards can be the result of equipment failure, human error, environmental conditions, or any combination of those factors.[10] After identification of the hazards, decisions must be made as to whether or not the related risks are acceptable. If the risks are not acceptable, plans to reduce or eliminate them are identified and analyzed for efficiency, cost, and workability.

System Safety as an Integral part of the Design Process

The techniques of system safety can be applied to the design and risk assessment of any process or operation. These procedures are independent of the nature or type of process or operation and can easily be applied to minimize risks.

The first risk assessment effort must be made at the design stage. Risk reduction is far less costly before equipment is in operation. Once a process is operating, hazard abatement may be carried out through safety devices or isolation of the hazard; however, addressing the hazard at the design stage can make it possible to entirely eliminate a risk before a plant is built. Identifying and presenting hazards at the design stage saves time and money by reducing the need for equipment modification, downtime, and litigation costs. The design engineer for any process must recognize his or her moral and legal responsibility, and his or her obligation to incorporate systematic hazard evaluation into the design process.

For the reasons mentioned above, it is extremely important to familiarize students of all engineering disciplines with system safety techniques and to incorporate these procedures into their design projects. It should be noted that incorporation of safety topics in engineering curricula is also a requirement for accreditation by the Accreditation Board for Engineering and Technology (ABET). Apparently, the senior-level design project is the ideal place for the incorporation and application of system safety topics. As pointed out earlier, hazard evaluation procedures have enough flexibility in their application that they can be easily tailored to fulfill the design and

Table 9.21. Application of preliminary hazard analysis to a highly toxic hazard material (HTHM) storage tank.

Hazard	Consequence	Cause	Corrective/Prevention Measures
Release of toxic gas	Fatalities injuries	Rupture in storage tank	• Tank materials of construction
Release of huge amount of toxic gas	High release of gas into the community, fatality	Fire in tank farm; Explosion of storage tank	• Community preparedness • Fire prevention techniques and equipment in tank farm
Release of toxic gas	High release of gas into the community	Collapse of tank foundation due to earthquake	• Down wind siting • Improve structural design of foundation
Release of toxic gas	High release of gas into the community	Rupture in main transfer line	• Gas analyzer with automatic diversion of flow • Minimize piping
Release of toxic gas	High release of gas into the community		• Plant layout
Release of toxic gas	High release of gas into the community	Small leak in line or from tank	• Toxic gas analyzer • Personnel PPE?

safety requirements of all engineering disciplines. For example, the release of a toxic gas from a highly toxic hazard material storage tank can pose a severe environmental problem; and system safety techniques can easily be applied to design a storage tank that will minimize the risk of toxic gas release.

Table 9.21 summarizes the application of PHA to a highly toxic hazard material (HTHM) storage facility. This analysis identifies the hazard, its cause, and its consequences. Also, possible preventive and corrective measures are suggested.[27]

References

1. Henley, E. J., and H. Kamamoto, *Reliability Engineering and Risk Assessment*, 1st Ed., Prentice-Hall, Englewood Cliffs, N.J. (1981).
2. *Hazard Classification Systems*, Hazardous Materials Inturmation Center, Inter/Face Associates, Inc., Middletown, Conn. (1986).
3. Firenze, R. J., *The Process of Hazard Control*, 1st Ed., Kendall/Hurt Publishing Co., Dubuque, Iowa (1978).
4. Hammer, W., *Occupational Safety Management and Engineering*, 3rd Ed., Prentice-Hall, Englewood Cliffs, N.J. (1983)
5. Green, A. E., *Safety Systems Reliability*, John Wiley & Sons, New York (1983).
6. Meyers, R. A., *Handbook of Petroleum Refining Processes*, McGraw-Hill Book Co., New York (1986).
7. Hanes, N. B., and A. M. Rossignol, *Comprehensive Occupational Safety and Health Engineering Academic Program Development Strategy*, U.S. Dept. of Health and Human Services (1984).
8. Roland, H. E., and B. Moriarty, *System Safety Engineering and Management*, 1st Ed., John Wiley & Sons, New York (1983).
9. Page, G. A., Hazard evaluation procedures, American Society of Safety Engineers, Professional Development Conference and Exposition, Las Vegas, Nev. (1988).
10. Kavianian, H. R., C. A. Wentz, R. W. Peters, and L. E. Martino, Total concepts in safety systems management for hazardous materials handling and design of hazardous processes, Annual Loss Prevention Symp., AlChE Spring 1989 National Meeting, Houston, Tex. (Apr. 1989).
11 McElroy, F. E., *Accident Prevention Manual*, 8th Ed., National Safety Council, Chicago (1983).
12. Moore, G. R., *Properties and Processing of Polymers*, 1st Ed., Prentice-Hall, Englewood Cliffs, N.J. (1984).
13. Kavianian, H. R., G. V. Brown, and J. K. Rao, *Safety Systems Management for Design of Hazardous Technologies*, ASSE (1988).

14. Rosen, S. L., *Fundamental Principles of Polymeric Materials*, 1st Ed., John Wiley & Sons, New York (1982).
15. NFPA, *Fire Protection Handbook*, 5th Ed., NFPA, Quincy, Mass. (1976).
16. Kavianian, H. R., L. J. McIntyre, E. Shanahan, and L. Tami, Application of Hazard Evaluation Procedures to the Design of a Hazardous Industry, School of Engineering, California State University, Long Beach, Calif. (1988).
17. Kavianian, H. R., R. Orr, R. Arbuckle, and A. Edwards, Hazard Analysis and Safety Management of a Radioactive Gas Handling Process, School of Engineering, California State University, Long Beach, Calif. (1988).
18. Dollah-Kanan, M., Z. H. Mustaffa, and Z. Abidin, Safety System Management for Design of Hazardous Technologies, California State University, Long Beach, Calif. (1988).
19. Norsham, K., M. Kamaluddin, and D. N. Nguyen, Preliminary Hazard Evaluation of Polyethylene Plant, California State University, Long Beach, Calif. (1988).
20. Abdulmalik, M., E. Firoszabadi, H. Kavianian, and S. Panahshashi, Safety and Hazard Assessment for Preliminary Design of Ethylene Plant, California State University, Long Beach, Calif. (1988).
21. Renshaw, L., B. Brand, and A. Roubanis, Hazard Evaluation of a Low Density Polyethyene Reactor, California State University, Long Beach, Calif. (1988).
22. Jaafar, S., F. Rajab, and N. Saaidin, Preliminary Hazard Evaluation for Hydrogen Plant, California State University, Long Beach, Calif. (1988).
23. *Guidelines for Hazard Evaluation Procedures*, Batelle Columbus Division, The Center for Chemical Process Safety, American Institute of Chemical Engineers, New York (1985).
24. Kniel, L., O. Winter, and K. Stork, *Ethylene, Keystone to the Petrochemical Industry*, Marcel Dekker, New York (1980).
25. Shreve, R. N., and J. A. Brink, Chemical Process Industries, 46h Ed., McGraw-Hill, New York (1977).
26. Plunkett, E. R., *Handbook of Industrial Toxicology*, 3rd Ed., Chemical Publishing Co., New York (1987).
27. Little, A. D., and R. Levine, *Guidelines for Safe Storage and Handling of High Toxic Hazard Materials*, Center for Chemical Process Safety, American Institute of Chemical Engineers, New York (1988).
28. Slote, L. *Handbook of Occupational Safety and Health*, John Wiley & Sons, New York (1987).
29. Kavianian, H. R., J. K. Rao, and G. V. Brown, Toxic gas hazard management in microelectronics and optoelectronics industries, *Proceedings* HAZMACON 88, Anaheim, Calif. (Apr. 5–7, 1988).
30. Henley, E. J., and H. Kumamoto, *Reliability Engineering and Risk Assessment*, 1st Ed., Prentice-Hall, Englewood Cliffs, N.J. (1981).
31. Paustenbach, D. J., Should engineering schools address environmental and occupational health issues?, *J. of Professional Issues in Engineering*, Vol. 113, No. 2, ASCE (Apr. 1987).
32. Talty, J. T., and J. B. Walters, Integration of safety and health into business and engineering school curriculum, *Professional Safety, Proceedings ASSE*, pp. 26–32 (Sept. 1987).
33. Kavianian, H. R., and J. K. Rao, Should engineering schools address environmental and occupational health issues?, Discussion paper, *J. of Professional Issues in Engineering* (Nov. 12, 1987).
34. Cowan, P. A., and J. K. Rao, Disaster abatement and control in the chemical process industry through comprehensive safety systems engineering and emergency management, Lessons from Flixborough and Bhopal, *Proceedings* World Conference on Chemical Accidents, Institute Supervisor di Sainta, Rome (July 7–10, 1987).

Selected Bibliography

35. *Safety in Academic Chemistry Laboratories*, American Chemical Society Committee on Chemical Safety, Washington, D.C. (1985).
36. Talty, John, NIOSH project officer, *Engineering Control of Occupational Safety and Health Hazards*, U.S. Dept. of Health and Human Services (Jan. 1984).
37. Ferry, T. S. *Safety Program Administration for Engineers and Managers*, Charles C. Thomas, Springfield, Ill. (1984).
38. Maust, L., Gr. Costrini, M. Emanuel, M. Givens, C. Zmudzinski, and J. Coleman, MOCVD of III-V compound epitaxial layers, *Semiconductor International* (Nov. 1986).

39. *Occupational Exposure Limits for Airborne Toxic Substances*, International Labour Office, Geneva (1980).
40. Bretherick, L., *Handbook of Reactive Chemical Hazards*, Butterworth's, London and Boston (1979).
41. Crone, H., *Chemicals and Society*, Cambridge University Press (1986).
42. O'Riordan, T., and W. Sewell, *Project Appraisal and Policy Review*, John Wiley & Sons, New York (1981).
43. Mendeloff, J., *Regulating Safety: An Economic and Political Analysis of Occupational Safety and Health Policy*, MIT Press, Cambridge, Mass. (1979).
44. Brown, D., *Systems Analysis and Design for Safety*, Prentice-Hall, Englewood Cliffs, N.J. (1976).
45. Sax, N., *Dangerous Properties of Industrial Materials*, Van Nostrand Reinhold, New York (1979).
46. Material Safety Data, Alta Products, Danvers, Mass. (Mar. 1987).
47. Novhi, A., and R. Stirn, Heteroepitaxial Growth of Cdrx Mnx Te on Ga As by Metal Organic Chemical Vapor Deposition, Jet Propulsion Laboratory, Pasadena, Calif. (Sept. 1987).
48. Woltson, R. G., and S. M. Vernon, Safety in thin film semiconductor deposition, SERI Photovoltaics Safety Conference (Jan. 1986).
49. Sax, N. I., and R. J. Lewis, *Hazardous Chemicals Desk Reference*, Van Nostrand Reinhold, New York (1987).

Exercises

1. A flammable liquid storage contains one hundred 55-gallon drums of gasoline with a flash point of −40°F. Assuming that all electrical devices are of the approved type, construct a fault tree diagram depicting all possible scenarios that can lead to a fire in the storage.

2. In a hazardous waste site the cleanup crew uses level A protection with self-contained breathing apparatus (SCBA). Perform a preliminary hazard analysis (PHA) to identify the failure modes for the SCBA and recommend procedures, equipment, and/or instrumentation for preventive measures.

3. As part of a feasibility study for commercialization of the metal organic chemical vapor deposition (MOCVD) process, it is necessary to determine the system failure modes and their effects on personnel, community, and equipment safety. Perform a hazard and operability (HAZOP) analysis using the reactor temperature as the selected parameter.

4. A high pressure low density polyethylene plant is operating under 1500 atmospheres and 300°F. Perform fault tree and event tree analyses for the possibility of a runaway reaction in the polymerization reactor. Combine the fault and event tree analyses into a cause consequence diagram.

5. You are the process engineer in charge of operation of an HF alkylation unit in a petroleum refinery. The acid catalyst is stored in a 5,000-gallon tank from which it is pumped to the unit for use as a makeup catalyst. Perform a "what if" analysis on the acid storage tank, concentrating on the different ways that an acid leak can occur. Discuss the occupational safety and environmental health effects of such a leak, and recommend procedures, equipment, and/or instrumentation to minimize risks.

6 In a chemical laboratory located on the fourth floor of a heavily populated building, several bottles of chemicals are stored in wooden cabinets. The laboratory, which has dimensions of 25 × 25 × 15 ft, is used by 18 students at a time to perform their chemistry experiments. During a recent inspection it was discovered that nitric acid (an oxidizer and corrosive), methyl ethyl ketone (an organic peroxide), and a petroleum derivative with a flash point of −25°F are stored together in one of the cabinets. Perform a preliminary hazard analysis to identify the possible accidents that can occur. Discuss the effects of each accident, and recommend procedures, equipment, and/or instrumentation to minimize risks.

7. In an industrial operation two workers are responsible for running parts through a caustic tank that generates corrosive vapors at concentrations much above the TLV. Perform a preliminary hazard analysis to identify the physical and health hazards associated with this operation. Your analysis should result in recommendations for engineering controls, to reduce the concentration of corrosive vapors to below the TLV value, and for appropriate personal protective equipment for worker safety.

8. In an ethylene plant, ethane is used as a feedstock in pyrolysis reactors located inside a furnace operating at 1500°F. Perform a preliminary hazard analysis and a failure mode effects and criticality analysis analysis (FMECA) on the reactor tubes and furnace assembly. Your analysis should identify any safety instrumentation devices (such as flow controllers, temperature controllers, etc.) that might be needed for the safe operation of the heaters. Draw a simplified diagram of the heater assembly (with reactors inside), and mark any instrumentation required on the diagram.

Chapter

10

Container and Spill Management

Containers of Hazardous Waste

Accidents can occur during the handling of hazardous waste drums and other containers. The hazards may include detonation, fires, explosions, vapor generation, and physical injury resulting from moving heavy containers by hand and working around stacked drums, heavy equipment, and deteriorated drums.

The identification of practices and procedures for the safe handling of drums and other hazardous waste containers is helpful in setting up a waste container handling program.[1]

Inspection of Containers

The appropriate procedures for handling drums and other containers depend on their contents. Thus, prior to any handling, workers should inspect containers visually to gain as much information as possible about their contents. Visual inspection also can help personnel to determine the integrity of a container prior to handling it. Generally, containers should be inspected for any obvious corrosion, swelling, or punctures. If a container's integrity is in question, either its contents should be pumped to another container, or the container should be placed into an overpack. The inspector should look for:

- Symbols, words, or other marks on the containers indicating that its contents are hazardous (e.g., explosive, corrosive, toxic, and flammable).
- Symbols, words, or other marks on a container indicating that it contains discarded laboratory chemicals, reagents, or other potentially dangerous materials in small-volume individual containers.
- The drum type, which is also a good indication of the associated hazards of the drum's contents, as summarized in Table 10.1.
- Signs of deterioration, such as corrosion, rust, and leaks.
- Signs that the container is under pressure, such as swelling and bulging.
- The type of the drumhead, as listed in Table 10.2.

Conditions in the immediate vicinity of containers may provide information about their contents and associated hazards. Monitoring should be conducted around containers using instruments such as a gamma radiation survey instrument, an organic vapor monitor, or a combustible gas meter. The results of this survey can be used to classify containers into preliminary hazard categories; for example, leaking or deteriorated, bulging, explosive or shock-sensitive, containing small-volume individual containers of laboratory wastes or other dangerous materials.

As a precautionary measure, personnel should assume that unlabeled drums contain hazardous materials until their contents are characterized. Also, they should bear in mind that drums frequently are mislabeled—particularly drums that are being reused. Thus, a drum's label may not accurately describe its contents.

Table 10.1. Special Container and Drum Types and their Associated Hazards.

Drum or Container Type	Potential Hazard
Polyethylene or PVC-lined drums	Usually they contain strong acids or bases; if the lining is punctured, strong acids or bases can quickly corrode the steel, resulting in a significant leak or spill.
Exotic metal[a]	Usually an extremely dangerous material is contained in these very expensive drums.
Single-walled drums used as a pressure vessel	Reactive, flammable, or explosive substances may be contained in drums that have fittings for both product filling and placement of an inert gas, such as nitrogen.
Laboratory packs	They contain expired chemicals and process samples from laboratories, hospitals, and similar institutions. Individual containers within the lab pack often are packed in absorbent material. They may contain incompatible, shock-sensitive, highly volatile, or highly corrosive materials, radioisotopes, or very toxic exotic chemicals. Laboratory packs may be an ignition source for fires at hazardous waste sites.

[a]Aluminum, nickel, stainless steel, or another unusual metal.
Source: Reference 1.

Table 10.2. Information on Drum Contents Provided by the Drumhead Type.

Drumhead type	Information
Removable lid	Designed for solid materials
Bung[a]	Designed for liquids
Contains liner	May contain a highly corrosive or otherwise hazardous material

[a]A bung is a stopper or plug in the lid hole for emptying and filling the drum.
Source: Reference 1.

Handling Drums

The purpose of safety measures in the handling of drums is to (1) respond to any obvious problems that might impair worker safety (e.g., leakage or the presence of explosive substances), (2) prepare drums for sampling, and (3) organize drums into different areas to facilitate their characterization and remedial action.

Drum handling has the potential for causing accidents; so drums should be handled only if it is necessary. Prior to handling, all personnel should be trained on the hazards of handling and instructed to minimize handling as much as possible. In all phases of handling, personnel should be alert to new information about potential hazards, and should respond to these hazards before continuing with routine handling operations. Overpack drums (larger drums in which leaking or damaged drums are placed for storage or shipment) and an adequate volume of absorbent should be kept near areas where minor spills may occur. Where major spills are possible, a containment berm should be constructed, which should be high enough to contain the entire volume of the liquid in the drums. If drum contents do spill, personnel trained in spill response should be called to isolate, contain, and clean up the spill.

The equipment that can be used to move drums includes: (1) a drum grappler attached to a hydraulic excavator; (2) a small front-end loader, which can be either loaded manually or equipped with a bucket sling; (3) a rough-terrain forklift; (4) a roller conveyer equipped with solid rollers; and (5) drum carts designed specifically for drum handling. The drum grappler provides maximum safety for drum handling; it keeps the operator away from the drums so that there is less likelihood of injury if a drum detonates or ruptures. Figure 10.1 shows drum handling with a

Figure 10.1. Handling drums with a forklift. (Source: Argonne National Laboratory)

forklift, and Figure 10.2 shows a drum grappler. In case of an explosion, grappler claws help protect the operator by partially defecting the force of the explosion.

Manual Drum Handling

Although handling drums with mechanical equipment is preferable, drums sometimes must be moved manually. This section discusses the dangers and safety rules for manual drum handling.[1,2]

When drums are moved manually, special care should be exercised in following safety rules. Potential injuries that may result if drums are not handled safely include strained backs, smashed fingers, hernias, broken legs or feet, and severe cuts on the hands.

Workers should consult the following checklist before moving any drum by hand:

- Check your workspace to make sure you have enough room.
- Plan your route before you move the drum.
- Check the rolling surface for tripping hazards.
- Check the bottom chime (i.e., the rim) to make sure you will get an even roll.
- Check the top chime for any burrs or slivers.
- Check the bung to ensure that it is tight and will not leak.
- Check the top for any collected water or grease that might cause your hands to slip and lose control.
- If you plan to use a pallet, check its condition. Do not use a broken pallet.

Drums usually are ''broken'' loose before movement. The methods used for safe breaking of drums are pushing, pulling, and a combination of pushing and pulling.[3]

When the pushing technique is used, the following procedure is recommended.

- Place your hands on the near chime at shoulder width.

Figure 10.2. Handling drums with a grappler. (Source: Argonne National Laboratory)

- Move your shoulders low and close to the drum.
- Slowly push the drum forward with your legs until you feel it reach its balance point.

Pulling is necessary when drums are in a cluster. The following steps are recommended for pulling a drum:

- Grip the far chime with one hand and the near chime with the other.
- Brace your foot at an angle across the bottom chime while your hands and feet form a straight line.
- Before you pull, check the position of your fingers for possible pinch points.
- Pull back and let the weight of your body tip the drum.

The pushing and pulling method is used when a drum is against a wall. Workers pushing and pulling drums use the following procedure:

- Place your hands at the near drum position with the hands apart at shoulder width.
- Brace the drum with your foot to prevent it from sliding, and shift your weight to the rear foot.
- Pull and drag the drum a few inches to the left and then to the right.

Rolling Drums. The rolling of drums can be very dangerous if not done properly. The steps undertaken depend on the direction of roll. For example, a worker should observe the following steps when rolling drums to the left:

- Place your left hand high on the chime and your right hand low.
- Use both hands to carefully roll the drum.

- As your right hand reaches the top position, quickly switch the left hand to the top position.
- Lift your hands and place them into position. Do not slide your hands because they may be cut or burned.
- Keep your feet separated for better equilibrium.
- Turn your body slightly away from the drum, but do not get too far away. Keep one leg next to the drum (nearly touching it) for extra stability. (*Note*: When rolling drums, *never* cross your arms or legs because you may lose control of the drum.)

Once a drum has reached its final destination, it must be positioned. Positioning a drum is the reverse of breaking the drum.[2,3]

Palletizing Drums. When placing drums on a pallet, workers should use the following procedure:

- Position the drum close to the pallet.
- Make sure the pallet is in good condition.
- Break the drum loose using the pull technique.
- Keep your shoulder low and your hands and feet in a straight line, and use the weight of your body to roll the drum until half the bottom chime is over the pallet.
- Use the weight of your body to counterbalance the drum as you set it down.
- Push it the rest of the way with your legs. Keep your shoulders low and close to drum.[3]

General Handling Procedures for All Drums

The following procedures can be used to maximize worker safety during drum handling and movement:[1]

- Train personnel in proper lifting and moving techniques to prevent back injuries.
- Make sure the vehicle selected has a sufficient rated-load capacity to handle the anticipated loads, and make sure the vehicle can operate smoothly on the available road surface.
- Air-condition the cabs of vehicles to increase operator efficiency.
- Protect the operator with heavy splash shields.
- Supply operators with appropriate respiratory protective equipment when needed.
- Have overpack drums ready before any attempt is made to move drums.
- Before moving anything, determine the most appropriate sequence in which the various drums and other containers should be moved. For example, small containers may have to be moved first to permit heavy equipment to enter the area to move the drums.
- Exercise extreme caution in handling drums that are not intact or tightly sealed.
- Ensure that operators have a clear view of the roadway when carrying drums. Where necessary, have ground workers available to guide the operator's motion.

Drum Handling Procedures for Specific Hazardous Situations

Drums Containing Shock-Sensitive or Explosive Waste. If a drum is suspected to contain explosive or shock-sensitive waste as determined by visual inspection, it should not be handled until the appropriate authorities have been notified.

Bulging Drums. If a drum is critically overpressurized, as evidenced by bulging or swelling, it should be isolated with a barricade or steel demolition net until the pressure can be relieved remotely. If it is not possible to set up a barricade, a tarpaulin may be used to cover the drum, provided that the cloth is positioned remotely using long poles or rods. However, it should be emphasized that the mere weight of the tarpaulin or a change in position of the drum could cause rupture. Slow venting using a bung wrench and plastic cover over the drum has worked for less critical situations, but this should be attempted only by experienced personnel using proper protective equipment and extreme caution. The following points should be kept in mind in dealing with bulging drums:[1,2]

- Pressurized drums are extremely dangerous. These drums should not be moved unless it is absolutely necessary.
- If a pressurized drum has to be moved, handle the drum with a grappler unit constructed for explosive containment. Move the bulged drum only as far as is necessary. Exercise extreme caution when working with or adjacent to potentially pressurized drums.

Lab Packs. Laboratory packs (i.e., drums containing individual containers of laboratory materials normally surrounded by cushioning absorbent material) have the potential for creating fires at hazardous waste sites. Such containers should be considered to hold explosive or shock-sensitive wastes until otherwise characterized. If handling is required, personnel should take the following precautions:[1]

- Prior to handling or transporting lab packs, make sure that all nonessential personnel have moved a safe distance away.
- Whenever possible, use a grappler unit constructed for explosive containment.
- Maintain continuous communication with the site safety officer until handling operations are complete.
- If lab packs need to be opened, place an explosion-proof shield between the worker and the drum.
- Ensure that proper PPE is used, and that it is compatible with the hazards posed.
- If crystalline material is noted at the neck of any bottle, handle it as a shock-sensitive waste acid or other similar material.
- Palletize and secure the repacked drums prior to transport.

Drums Containing PCBs. Before 1977, polychlorinated biphenyls (PCBs) were manufactured commercially in the United States and marketed under the trade name Aroclor®. The industrial manufacture of PCBs was stopped in the United States in 1977 after it was discovered that PCBs accumulate and persist in the environment and cause toxic effects in animal testing. PCBs are considered to have relatively low acute toxicities. No information is available for humans, but for experimental animals the acute oral LD_{50} is greater than 750 mg/kg in all species tested, which is not extremely toxic. In humans exposed to PCBs, the skin and liver are the main sites affected. The EPA has regulations regarding the use, storage, and marking of PCBs and PCB-containing items (40 CFR Part 761). In order to comply with the law and protect worker safety, the following points should be considered in the handling of PCBs:

- All PCBs and PCB items should be marked according to specifications described in 40 CFR 761.40.
- EPA recommends that all PCB items be disposed of by incineration.
- Containers for storing PCBs must be checked for leaks at least once a month.
- PCBs must be packaged in metal drums meeting DOT specifications.
- Transformers containing PCBs must be palletized and tightly sealed to prevent leakage. A large sheet of polyethylene must be placed under the transformer and extended at least one quarter of the way up the side of the pallet. Enough vermiculite must be provided on the sheet to absorb PCB leakage.

- Head space must be provided in PCB liquid drums so that the drums will not be full at 130°F (54.4°C).
- Containers must be securely closed and constructed to prevent leakage caused by changes in temperature, humidity, and altitude during transportation and in-transit handling.
- Metal containers having corrosion or dents at the chime or a seam, soldered, or welded area are potential hazards.
- In handling containers of PCBs, adequate PPE must be used to prevent worker exposure.
- A spill of 10 lb or more of PCBs must be reported to EPA.

Leaking, Open, and Deteriorated Drums. If a drum containing a liquid cannot be moved without rupture, its contents should be immediately transferred to a sound drum using a pump appropriate for transferring that liquid. Immediately place the empty drum in an overpack container.

If the drum has a small puncture or leak, it may be possible to use wooden plugs or stoppers to temporarily stop the leak.[1,2]

Opening Drums

Drums usually are opened and sampled in place during a site investigation. The factors discussed below enhance the efficiency and safety of drum-opening operation.

Preferably, drum opening should be separated from the drum staging and drum removal areas to prevent a chain reaction in case of fire or explosion. Drums that contain reactive, explosive, or toxic materials should be placed in a drum bunker, which should be built in an isolated area surrounded by sandbags or concrete. The ground should be sloped so spills can flow into a centrally located sump. Alternatively, the drum can be placed in a spill pan that has a volume adequate to collect the contents of the drum. The pan should have a drain for recovering the wastes. A plexiglass shield should be installed between the worker and the drum.

Sensing probes of any monitoring equipment should be placed near the drum with indicating meters behind the plexiglass. Drums must be moved from the staging to the opening area one at a time, using a grappler unit.

At some sites where the workspace is limited, the drum staging and opening areas can be combined. This should be done using extreme caution and with safety rules in mind. Drums should be staged in rows of two or in groups of four with sufficient distance between them to allow for entry of sampling or emergency equipment. A distance of one foot or more should be provided between individual drums in a row to minimize the possibility of a chain reaction in case of fire or explosion.[2]

There are three basic techniques used for drum opening: manual opening with nonsparking bung wrenches, drum deheading, and remote drum opening. The choice of opening device largely depends on the contents of the drum and the number of drums to be opened. Remote control opening devices always should be considered, as they maximize worker safety. Manual opening devices such as bung wrenches and drum deheaders should be used only when the drum contents are not reactive or explosive. In such cases, appropriate PPE should be used.[2,3]

Bung wrenches can be used to remove the bung plug located on the side or head of the drum. These plugs are threaded plugs of various designs, and a number of universal bung wrenches are available for bung removal. These wrenches should be made of nonsparking materials such as bronze and aluminum to eliminate any ignition source that could cause fire or explosion if drum contents are flammable. Manual drum opening with bung wrenches should not be performed unless the drum is structurally sound, and the drum contents are not reactive or explosive. When opening drums with bung wrenches, workers should keep the following points in mind:[1,2]

- Appropriate PPE that matches the specific hazard must be used.

- Before drums with bungs on the head are opened, they should be placed in an upright position.
- Drums with bungs on the side should be placed on their side with the bung plugs up before being opened.
- The worker should use a slow wrenching motion and a steady pull across the drum.
- If there is any evidence of a chemical reaction, pressure buildup, or release of potentially toxic fumes, personnel should immediately leave the opening area and arrange for remote opening devices.

Drum deheading is an inefficient, relatively unsafe, and time-consuming process. It may be desirable for certain situations where the number of drums to be opened is small, and the drum contents are not hazardous. A problem workers frequently encounter when using a drum deheader is that it cannot cut those parts of the drum head that have been dented, although a drum dekinker can be used to straighten dented chimes for deheading.[2]

As remote drum opening is the safest available technique for drum opening, it always should be considered. Two basic tools originally developed by EPA and modified by several contractor groups are available:

1. A remote bung remover, which is an air-operated wrench that uses compressed air. A specially devised mounting bracket and a nonsparking bung socket are used to spin the bung from the top or side of the drum.
2. A hydraulic drum plunger, which forces a penetrator into the drum and seals the resulting holes. A sample can be drawn through the hollow stem of the penetrating device, and the device is left in the hole to provide sealing.

The following checklist is helpful for the safe opening of drums:

- Use equipment that is compatible with the drum contents and construction.
- If a supplied-air respiratory protection system is used, place a bank of air cylinders outside the work area to provide breathing air in case of an accidental disruption of supplied air, as in a compressor failure.
- Protect personnel by keeping them at a safe distance from the drum opening area. If personnel work near the drums, place an explosion-resistant shield between the personnel and the drums to protect them in case of detonation. Locate controls for drum opening, monitoring, and fire suppression equipment on the personnel side of the explosion-resistant plastic shield.
- If possible, monitor continuously during opening. Place sensors of monitoring equipment, such as colorimetric tubes, radiation survey instruments, explosion meters, organic vapor analyzers, and oxygen meters as close as possible to the source of contaminants.
- Exercise great care in using remote-controlled devices for opening drums, including pneumatically operated impact wrenches to remove drum bungs, hydraulically or pneumatically operated drum piercers, and backhoes equipped with bronze spikes for penetrating drum tops in large-scale operations. Remember the safety tips listed below when opening drums with these devices.
- Do not use picks, chisels, or firearms to open drums.
- Hang or balance the drum-opening equipment to minimize worker exertion.
- If the drum shows signs of swelling or bulging, perform all steps slowly. Relieve excess pressure prior to opening, from a remote location if possible, using such devices as a pneumatic impact wrench or a hydraulic penetration device. If pressure must be relieved manually, place a barrier such as explosion-resistant plastic sheeting between the worker and the bung to deflect any gas, liquid, or solid that may be expelled as the bung is loosened.
- In opening exotic metal drums and polyethylene- or PVC-lined drums, exercise extreme caution, as these drums usually contain extremely hazardous materials.

- Do not open or sample individual containers within laboratory packs.
- Reseal open bungs and drill openings as soon as possible with new bungs or plugs to avoid explosion and vapor generation. If an open drum cannot be resealed, place it in an overpack drum. Plug any openings in pressurized drums with pressure-venting caps set to a 5-psi release to allow venting of vapor pressure.
- Decontaminate equipment after each use to avoid mixing incompatible wastes.[1,2]

Container Sampling

Container sampling can be one of the most hazardous activities that workers undertake because it often involves direct contact with unfamiliar wastes. Hazardous material sampling has the potential for exposing workers to both acute and chronic health hazards, so sampling procedures must be developed to minimize worker exposure to hazardous chemicals. The methods developed must be appropriate for a wide range of chemicals and situations because of the unknown nature of many hazardous wastes. The option of using disposable sampling equipment always must be considered because decontamination if the field may be impractical. An appropriate sampling method must satisfy the following requirements:[2]

- Safety: Before a sampling method can be developed, a risk evaluation of worker exposure must be conducted. Based on the outcome of this evaluation, appropriate methods along with suitable PPE can be selected to minimize the risks.
- Practicality: The sampling method must be simple, pragmatic, and proven.
- Representativeness: The objective of any sample analysis program is to obtain hazard information for a large population based on a few samples. Therefore, any successful sampling method must gather representative samples.
- Economics: The costs of equipment, labor, and operational maintenance need to be analyzed in relation to overall benefit. Instrument durability, disposable equipment, cost of decontamination, and degree of precision and accuracy also should be considered.

The basic objective of any sampling program is to produce a set of samples representative of the source under investigation and suitable for subsequent analysis. Specifically, the objective of sampling hazardous wastes is to acquire information that will assist investigators in identifying and quantifying compounds present. Of utmost importance is representativeness: the sample needs to be chosen so that is possesses the same qualities or properties as the material under consideration, based upon the analytical techniques used.

Sample size, also an important criterion, must be carefully chosen with respect to the properties of the entire container and the limitations of the analytical procedure. For example, although the entire contents of an intact 55-gallon drum certainly can be considered a representative sample of the drum material, it is an impractical sample because of its bulk. Alternatively, too small a sample size can be just as limiting, as representativeness and analytical volume requirements might be jeopardized.

A third criterion is maintenance of sample integrity: the sample must retain the original properties of the container through collection, transport, and delivery to the analytical laboratory. Degradation or alteration of the sample through exposure to air, excess heat or cold, microorganisms, or contaminants must be avoided.

Homogenous materials in containers present no special problem in efforts to obtain a representative sample. Because there is no change of composition throughout the material in the container, any sample size obtained can be considered to be representative of the entire material. Examples of homogeneous materials are well-mixed liquids, gases, and pure compounds.

Heterogenous materials can be divided into those that show discrete changes of quality, such as reactive solids, and those that show continuous quality changes throughout the material, such as layers of liquids and solids. Heterogenous materials do present special problems in obtaining a representative sample. Special techniques such as random mixing and composite sampling must be utilized for these materials.[1,2]

Sampling Equipment

The proper sampling equipment should allow representative samples to be easily taken, while avoiding hazardous spills and splashes. There is a variety of sampling equipment available for taking samples from drums and other containers of hazardous materials. The E-Z sampler is a chemically resistant polypropylene sampler, with a 125-ml borosilicate glass collecting bottle that screws into the lower end of the sampler. There is also a bomb sampler, made of corrosion-resistant stainless steel and consisting of a cylindrical reservoir chamber, a weighted plunger that seals the chamber at the bottom, and a cable attachment for suspending the apparatus and activating the sampling device. Also available is a dipper sampling device, which provides for easy sampling from large tanks and open drums.[4]

Sampling Plan

Before any sampling activities are begun, it is imperative that a work or sampling plan be developed to identify the purpose and goals of the program and the equipment, methodologies, and logistics to be used during the actual sampling. This plan should be developed when it becomes evident that a field investigation is necessary.[1]

Personnel should consult the following checklist when developing a sampling plan:

- Obtain historic background information about the wastes.
- Determine which drums should be sampled.
- Select the appropriate sampling device(s) and container(s).
- Develop a sampling plan that specifies the number, volume, and locations of samples to be taken.
- Develop standard procedures for opening drums, sampling, and sample packaging and transportation.
- Use appropriate PPE during sampling, decontamination, and packaging of the sample.

When manually sampling from a drum, workers should use the following techniques:

- Sample only after the drum-opening operations have been completed.
- Do not lean over other drums to reach the drum being sampled.
- Cover drum tops with plastic sheeting or other suitable uncontaminated material to avoid excessive contact.
- Never stand on drums because it is extremely dangerous to do so. Use mobile steps or a platform to safely sample the elevated drums.
- Obtain samples with either glass rods or pumps. Do not use contaminated items because they may contaminate the sample and may not be compatible with the waste in the drum.

Cleanup of Spills

Spills may occur during container handling or as a result of leakage from damaged drums or other containers. These spills have the potential for causing environmental damage and health problems. For example, if a hazardous substance leaks into a drain, it can contaminate the local water supply, causing harmful health effects. In addition, a spilled substance is flammable, explosive, or reactive, it can create physical hazards.[1,3]

Workers should follow these steps when controlling and containing small spills:

- Identify the spilled substance.
- Evaluate the hazards that may be encountered during control, containment, and cleanup of the spill.
- Establish communication with the appropriate personnel.
- Ensure that people are protected.
- Implement the appropriate control procedures to stop the flow of the spill.
- Contain and clean up the spilled substance.

Identification of the spilled substance and its characteristics is a critical factor in determining the most effective cleanup procedures. Container labels, shipping papers, and MSDSs can help provide valuable information on the identity of the hazardous substance. The time to read labels and MSDSs is not during an emergency, however. Workers should consult these sources of information before working with hazardous substances so that they are familiar with potential hazards as well as recommended spill and emergency procedures.

The next step is to evaluate the hazards posed by the spilled substance. The evaluation should include the identification of all potential health, physical, and environmental hazards, such as vapors or the presence of electrical, thermal, or mechanical energy sources that could act as sources of ignition. This evaluation should be conducted from a safe distance upwind of the spill. If the hazardous substance is flammable, all sources of ignition must be identified and removed.

If a spill occurs inside a building and is not released into the outside environment, the appropriate person within the organization must be contacted. This person should be identified in the organization's spill prevention, control, and countermeasures (SPCC) plan. The SPCC plan also should provide the following information:

- A description of the steps that facility personnel should take to clean up the spill.
- A list of those individuals designated to respond to spill emergencies.
- A list of available cleanup equipment.
- The names, addresses, and telephone numbers of state and local emergency response agencies that may be called in to assist in spill cleanup.

Emergency cleanup crews should be told what the source of the spill was, whether the source has been stopped, and approximately how much has been spilled. If a reportable quantity of the spill is released into the environment, the National Response Center (telephone no.: 1-800-424-8802) must be notified. Stiff penalties are imposed if the appropriate agency is not notified about a reportable spill. The reportable quantities for spilled substances have been established by EPA and can be found in 40 CFR Parts 117 and 302.

In order to protect personnel, the spill area must be secured; and if any hazards exist, all personnel who are not essential in cleaning the spill must be evacuated. It is equally important that the cleaning crew wear appropriate PPE that matches the hazards posed by the spilled substance. Information about the type of PPE required can be obtained from the MSDS. PPE used in spill cleanup usually includes eye protection, safety shoes, chemical-resistant gloves, body suits, respirators, and personal monitors. However, because of the importance of using the correct respirator in a cleanup situation, the following points should be remembered regarding respirator selection and use:

- There are two broad categories of respirators: air-purifying and atmosphere-supplying.
- Air-purifying respirators control exposure to harmful substances by cleaning the air with mechanical filters or with materials that remove *specific* gases or vapors.

- When air-purifying respirators are used, the respirator must be matched to the specific hazard that it was designed to protect against; it can be used only in areas with normal oxygen concentrations; and the containment level and degree of use cannot exceed the capacity of the respirator.
- Of the two types of atmosphere-supplying respirators, air-line respirators have the advantage of being lightweight and providing breathing air for extended periods of time. However, the hose can limit mobility, pose a tripping hazard, and become tangled or torn.
- SCBA respirators, the other type, have a limited air supply but increase mobility.
- No one should attempt to choose or wear a respirator without adequate training because respirators vary and must be matched to the specific hazard.

The flow of the leaking substance must be stopped or slowed before containment can be effective. This can be accomplished under most circumstances by turning off a leaking valve, shutting down pumps or compressors, or repairing a leaking container or drum. Before repairing a drum, the worker must ensure that spill cleanup and containment equipment is available.

When repairing a drum with a leak around the chime or seams, or repairing a drum with a puncture near its top or bottom, turn the drum so that the leak is upright to stop the flow of the spill, and then attempt to patch the leak.

To repair small punctures, large irregular holes, and small linear cracks, one may use insoluble mastic or putty, chemical patches, wooden plugs, or rubber balls. The insoluble mastic and chemical patch can be used for small punctures above the flow line that are not subject to pressure. Care should be exercised to ensure that the patching material is compatible with the drum contents. If the patch will be subjected to pressure, a soft wooden plug is desirable.

Larger irregular holes should be patched with a gasket-backed plate held in place with a compression washer, nut, and bolt. Small linear cracks can be patched by driving mastic or cloth into the crack with a wedge. In some cases, appropriate heavy-duty tape can be used as a temporary seal when the patch area is dry enough to provide an effective seal.

Once the leak has been patched or plugged, the drum must be covered with a plastic bag to prevent leakage after being placed inside an overpack drum. The worker should invert the overpack drum and slowly lower it over the patched drum while carefully avoiding pinch points. Then the two drums should be gently tipped over and righted, with the patched drum now inverted safely inside the overpack drum and positioned so that pressure is not on the patched leak. Finally, the overpack drum should be sealed and labeled appropriately.

If the leak occurs on the side of the drum, a slightly different technique for overpacking is more effective: After patching the leak, keep the drum on its side with the patched leak upward. Next, lift one end of the drum slightly and cover the drum with a plastic bag. Before lowering the drum, place a wooden block under the middle of the drum so that it is resting slightly off the ground. Then slip the overpack drum over the end of the patched drum and gently lift the end of the patched drum while sliding it into the overpack drum. Right the two drums, seal the overpack, and apply the appropriate label.

Finally, it is necessary to contain the spilled substance within the immediate area, preventing its flow off-site or into nearby water supplies and following appropriate cleanup procedures.

Methods for the containment of hazardous materials are numerous and to a large degree involve the instincts and innovativeness of the first responders. Some of the methods of containment include dikes, dams, or other barriers, which afford direct containment, plug drainage ditches, and protect natural topography, ponds, and streams.[1]

Another instinctive reaction of first responders to a spill should be to prevent mixing when more than one material is spilled. There are few events in which a first responder cannot pause a few seconds to make a significant contribution with a shovel or a bale of straw. Three or four shovelfuls of dirt or sand often can restrain the flow of a large volume of spilled chemicals, if placed well and in a timely manner.

The dams and dikes referred to in this discussion can restrict or stop the flow of spilled hazardous chemicals. They are not things of beauty or the product of a particular design, but are trial-and-error mounds of dirt placed by eyeball leveling and common sense; but if effective, they ultimately may be worth millions of dollars.

In order to contain spills, dikes are formed with chemically inert sorbents, such as diatomaceous earth, vermiculite, or amorphous silicate. If those substances are not available, earth, sand bags, or other readily accessible materials can be used.[3] Once dikes are formed to hold the spill within a limited area, neutralization or sorption techniques can be used to further contain it.

Neutralization is the elimination of the hazardous properties of spilled acids and bases by chemical action. Sulfuric, hydrochloric, and nitric acids can be neutralized by adding sodium bicarbonate to them. Sodium, ammonium, and potassium hydroxides can be neutralized with weak acids such as acetic acid. *Caution*: Before an attempt is made to contain a spill by neutralization, the *exact* chemistry of the spilled material must be known. Using the wrong neutralizer or the wrong concentration of neutralizer can produce very hazardous reactions and create a lethal situation.

Sorption in spill cleanup is the process of absorbing the spill on or within the pores of a sorbent. Some sorbents are generally nonreactive and can be used for various materials, including acids and bases. Examples of sorbents are straw, clay, vermiculite, activated carbon, and foamed plastic. Before using a sorbent on a spilled material, one should make sure that the sorbent is compatible with the chemical, and that no adverse chemical reaction will take place. Sorbents should be applied to a spill carefully, from the outer edges to the center, to reduce the possibility of its spreading.

When applying a sorbent to a flammable liquid, one should take the following special precautions:

- Use nonsparking tools to avoid starting a fire.
- Ensure that the appropriate firefighting equipment is available in case of an emergency.
- Use caution when cleaning up and disposing of the materials, as the sorbent probably will not change the hazardous properties of the spilled substance.
- Follow the disposal procedures recommended for the specific hazardous substances.

The ventilation in the area must be adequate before an attempt is made to apply a sorbent to a substance that gives off hazardous vapors.

All wastes must be disposed of in accordance with state and federal requirements. Precautions should be taken when disposing of waste material to ensure that the environment will not be polluted, and additional health hazards will not be created.

Hazardous Waste Site Emergencies

The nature of work at hazardous waste sites is such that emergencies are continually possible, no matter how infrequently they actually occur. Emergencies happen quickly and unexpectedly and require an immediate response. At a hazardous waste site, an emergency may be as limited as a worker experiencing heat stress or as far-reaching as an explosion that spreads toxic fumes throughout a community. Any on-site hazard can precipitate an emergency. Chemicals, biological agents, radiation, or physical hazards may act alone or in concert to create explosions, fires, spills, or other dangerous and harmful situations, as shown in Table 10.3.

Site emergencies are characterized by their potential for complexity, because of numerous unidentified toxic chemicals whose effects may be synergistic. Hazards may potentiate one another. For example, a flammable spill can feed an existing fire. Rescue personnel attempting to remove injured workers may themselves become victims. This variability means that advanced planning, including anticipation of different emergency scenarios and thorough preparation for contingencies, is essential to protect worker and community health and safety.[1]

Table 10.3. Emergencies at Hazardous Waste Sites.

Fire
Explosion
Release of toxic vapors
Reaction of incompatible chemicals
Collapse of containers
Discovery of radioactive materials
Minor worker accidents from slips or falls
Medical problems such as heat stress, heat stroke, or chemical exposure
Personal protective equipment failure such as air source failure, tearing of PPE
Physical injuries from hot or sharp objects, or vehicle accidents
Electrical burns, shock, or electrocution

Source: Reference 1.

Contingency Plan Requirements

When an emergency occurs, decisive action is required. Rapidly made choices may have far-reaching, long-term consequences, and delays of even minutes can create life-threatening situations. Personnel must know how to immediately rescue or respond, and emergency equipment must be on hand and in good working order; so planning is essential to the effective handling of emergencies. A contingency plan should be developed.

The contingency plan, which is a written document that sets forth policies and procedures for responding to site emergencies, should incorporate the following:

- Personnel considerations: roles, lines of authority, training, and communication.
- Site mapping, evacuation routes, and decontamination stations.
- Medical assistance and first aid.
- Appropriate equipment and emergency procedures.
- Documentation and reporting.

The contingency plan should be designed as a discrete section of the site safety plan and should be rehearsed regularly. It should be compatible and integrated with the pollution response, disaster, fire, and emergency plans of local, state, and federal agencies. Periodically it should be reviewed to incorporate new or changing site conditions or information.

The plan should specify emergency response roles for on-site workers and off-site response personnel as well as procedures for others who may be on-site, such as contractors, other agency representatives, and visitors. (Some emergency personnel and their responsibilities are summarized in Table 10.4.) Although personnel deployment is determined on a site-by-site basis, pertinent general guidelines and recommendations are discussed below. In all cases, the organizational structure should show a clear chain of command and must be flexible enough to handle multiple emergencies, such as a rescue and a spill response.

In an emergency situation, the project team leader must assume total control of decision-making on the site. This individual must be identified in the emergency response plan and be backed up by a specified alternate with the authority to resolve all disputes about health and safety requirements and precautions. The project team leader should be authorized to seek and purchase supplies as necessary.

As an immediate, informed response is essential in an emergency, all site personnel and others entering the site, including visitors, contractors, off-site emergency groups, and other agency representatives, must have some emergency training. This training should relate directly to site-specific anticipated situations, and must be reviewed regularly through site-specific drills. Training records must be maintained in a training logbook.

Table 10.4. Some Key Personnel Involved in Emergency Response.

Personnel	Responsibilities
Project team leader	Directs emergency response operations. Serves as liaison with appropriate government officials.
Site safety officer	Recommends that work be stopped if any operations threaten worker or public health or safety. Knows emergency procedures, evacuation routes, and appropriate telephone numbers, including ambulance, medical facility, poison control center, fire department, and police department. Notifies local public emergency official. Provides for emergency medical care on-site.
Rescue team	Rescues any workers whose health or safety is endangered.
Communication personnel	Civil defense organizations and local radio and television stations—provide information to the public during an emergency.

Source: Reference 1.

Off-site emergency personnel such as local firefighters and ambulance crews often are first responders and run a risk of acute hazard exposure equal to that of any on-site worker. These personnel should be informed of ways to recognize and deal effectively with on-site hazards. Persons lacking such information may inadvertently worsen an emergency by an improper action (e.g., spraying water on a water-reactive chemical and causing an explosion). Inadequate knowledge of the on-site emergency chain of command may cause confusion and delays. The project team leader should provide off-site emergency personnel with information about site-specific hazards, appropriate response techniques, site emergency procedures, and decontamination procedures.

Internal emergency communication systems are used to alert workers to danger, convey safety information, and maintain site control. Radios or field telephones often are used when work teams are far from the command post. Alarms or short clear messages can be conveyed by audible or visual signals, which also may back up the telephone system.

Off-site sources must be contacted by telephone for assistance or to inform officials about hazardous conditions that may affect public or environmental safety.

Detailed information about the site is essential for advance planning. For this purpose, a site topographic map is a valuable tool, serving as a graphic record of the locations and types of hazards, a reference source, and a method of documentation. The map can be used for planning and training, and as a basis for developing potential emergency scenarios and alternative response strategies.

In an emergency, the project team leader must know who is on-site and be able to control the entry of personnel into hazardous areas to prevent additional injury and exposure. Only necessary rescue and response personnel should be allowed into the site.

Also, it is vital in an emergency for the project team leader and rescue personnel to rapidly determine where workers are located and who may be injured. A passive locator system, that is, a written record of the location of all personnel on-site at any time, could be used to help locate personnel in an emergency.

No universal recommendation can be given for uniformly safe evacuation distances because of the wide variety of hazardous substances and releases at a site. For example, a "small" chlorine leak may call for an isolation distance of only 140 ft, whereas a "large" leak may require an evacuation distance of one mile or more, depending on climatic conditions.

Safe distances can be determined only by the site safety officer at the time of an emergency, based on site- and incident-specific factors. However, planning potential emergency scenarios will help familiarize personnel with relevant points. Factors that influence safe distances may include:

- Toxicological properties and physical state of the substance.
- Quantity, rate, and method of release.
- Vapor pressure and density of the substance.
- Wind speed and direction, atmospheric stability, height of release, and air temperature.
- Local topography characteristics that may affect the response.

On-site safety and first aid stations can be set up for local emergencies that do not require site evacuation.

In case of site-wide evacuation, off-site stations may be located at the most convenient exit. These stations will provide for emergency needs such as first aid for injured personnel, clean dry clothing, washwater for chemical exposure, and communications. In a site-wide evacuation, they can be used to house evacuation exit equipment, thereby reducing security problems.

A severe emergency, such as a spill, fire, or explosion, may cut workers off from the normal exit. Therefore, alternate routes for evacuating victims and endangered personnel should be established in advance, marked, and kept clear. The following guidelines will help in establishing safe evacuation routes.

- Place the evacuation routes in the predominantly upwind direction.
- Consider the accessibility of potential routes. Take into account obstructions, such as locked gates, trenches, pits, tanks, or drums, and the extra time or equipment needed to maneuver around or through them.
- Develop two or more routes that lead to safe areas and are separate or remote from each other, because one may be blocked by a fire, spill, or vapor cloud.
- Mark evacuation routes and areas that do not offer safe escape or that should not be used in an emergency.
- Consider the mobility constraints of personnel wearing protective clothing and equipment.
- Make escape routes known to all who go on the site

If an incident may threaten the health or safety of the surrounding community, the public must be informed of it and possibly evacuated from the area. Site management should plan for this in coordination with the appropriate local, state, and federal groups, such as civil defense, police, local radio and television stations, municipal transportation systems, and the National Guard.

In medical emergencies, procedures should be developed for decontaminating victims, protecting medical personnel, and disposing of contaminated PPE and wastes.

The decision as to whether to decontaminate a victim is based on the type and severity of the illness or injury and the nature of the contaminant. For some emergency victims, immediate decontamination may be an essential part of lifesaving first aid. For others, decontamination may aggravate the injury or delay lifesaving treatment. If decontamination does not interfere with essential treatment, first wash, rinse, and/or cut off protective clothing and equipment.

If decontamination cannot be done, wrap the victim in blankets, plastic, or rubber to reduce contamination of other personnel. Then alert emergency and off-site medical personnel to potential contamination, and instruct them about specific decontamination procedures if necessary.

Emergency equipment is necessary to rescue and treat victims, to protect response personnel, and to mitigate hazardous conditions on-site. All equipment should be in working order and available when an emergency occurs. Safe and unobstructed access for all firefighting and emergency equipment should be provided at all times.

Procedures should include cleaning and inspection of the PPE, and refilling all empty SCBA tanks and preparing them for emergencies immediately after normal use. PPE should be stocked at levels higher than those normally required so that emergency hazards can be effectively managed (i.e., at a site where level C equipment normally is used, level A and B equipment should be available for emergencies).

Response operations usually begin with a trouble notification and continue all the way through the preparation of equipment and personnel for the next emergency. These emergency response procedures will now be discussed.

Emergency Procedures

In an emergency a site alert should be sounded to notify personnel to stop work activities, reduce background noise to aid communication, and begin emergency procedures.

On-site emergency response personnel should be notified about the emergency, including what happened, where it happened, to whom it happened, when it happened, how it happened, the extent of damage, and what aid is needed.

To ensure an adequate response to the emergency, the type of incident must be identified based upon its cause and the extent of chemical release and transport, as well as the extent of damage to structures, equipment, and terrain.

Rescue and response action should be based on the information available at the time of the emergency. Emergency response or rescue should not be attempted until backup personnel and evacuation routes have been identified. Use of the buddy system, which should be enforced at all times, will allow no one to enter an exclusion zone or hazardous area without a partner. Also, personnel in an exclusion zone should be in the line of sight of or have other communication with the project team leader.

The number, location, and conditions of casualties and victims should be determined as quickly as possible to identify the treatment required and personnel, if any, who are missing.

The potential for fire, explosion, and release of hazardous substances should be quickly assessed, based upon the types of chemicals on-site. All personnel on-site should be located relative to the hazardous areas.

The potential for danger to the off-site population or the environment should be determined.

Equipment and personnel resources must be provided for victim rescue and hazard mitigation, depending upon the number of personnel available for response, the resources available on-site, and the resources available from outside groups and agencies.

Off-site personnel or facilities, such as ambulance, fire department, and police, should be contacted to request their assistance.

The hazardous situation must be brought under control to prevent the spread of the emergency.

Victims should be removed and decontaminated, and established procedures should be used to decontaminate uninjured personnel.

Injuries and damage to victims should be treated, in order to stabilize them before they are moved. The hazardous condition also must be stabilized (e.g., repack or empty filled runoff dikes).

Chemical contamination of the transport vehicle, as well as of ambulance and hospital personnel, can be minimized by adequately protecting the rescuers. If it is not possible to decontaminate the victims before transport, cover them. The transport personnel should be provided with disposable coveralls, disposable gloves, and supplied air, as necessary, for their protection. If appropriate, response personnel should accompany the victims to the medical facility to give advice on decontamination.

Site personnel should be evacuated to a safe distance upwind of the incident. Over time the hazards may diminish and permit personnel to reenter the site, or they may increase and require public evacuation.

Public safety personnel should be informed of any need to evacuate the nearby off-site population. A large-scale public evacuation is the responsibility of local or state government authorities.

Critique of the Contingency Plan

Despite the best efforts of all parties at an emergency site, there always will be room for improvement. Before normal site activities are resumed, personnel should be educated and equipped to handle another site emergency. All equipment and supplies should be inspected and restocked, replaced or repaired as necessary. Equipment should be cleaned and refueled for future use. Finally, all aspects of the contingency plan should be reviewed and revised according to the new site conditions and the lessons learned from the prior emergency response. Following any emergency response there is considerable information available that could be useful in updating the contingency plan. Questions such as the following can be answered:

- Cause: What caused the emergency?
- Prevention: Was it preventable? If so, how?
- Procedures: Were inadequate or incorrect orders given or wrong actions taken? Were they a result of bad judgment, wrong or insufficient information, or poor procedures? Could procedures or training be improved?
- Site profile: How did the incident affect the site profile? How were other site cleanup activities affected?
- Community: How was the safety of the community affected?
- Liability: Who was liable for any damage?

The project team leader should initiate the investigation and documentation of the incident. This is important in all cases, but especially so when the incident has resulted in personal injury, on-site property damage, or damage to the surrounding environment. Such documentation may be useful in avoiding recurrences, as evidence in legal actions, for assessment of liability, and for historical review by government agencies. Documentation can include written transcripts or bound field notebooks. Care should be taken to assure the accuracy of the information by recording it objectively.

A chain-of-custody procedure should be used, with each person making an entry and dating and signing the document. All entries should be made in ink and never be erased or altered, even if the details change or revisions are made. The person making the change or revision should mark a horizontal line through old, inaccurate material and initial and date the change.

Documentation should be as complete as possible, including the following:

- Chronological history of the incident.
- Facts about the incident and when they became available.
- Names and titles of the relevant personnel.
- Action decisions made, and by whom. Orders given to whom, by whom, and when. Actions taken, such as who did what, when, where, and how.
- Types of samples; test results.
- Possible exposure of site personnel.
- History of all injuries or illnesses during the incident or later as a result of the emergency.

References

1. NIOSH, *Occupational Safety and Health Guidance Manual for Hazardous Waste Site Activities*, National Institute for Occupational Safety and Health (1987).
2. Wagner, R., et al., *Drum Handling Manual for Hazardous Waste Sites*, 1st Ed., Noyes Data Corp., Park Ridge, N.J. (1987).
3. ITS, *Drums and Other Small Spills; Control, Containment, and Cleanup; Leaders Guide*, Industrial Training Systems Inc., Marlton, N.J. (1987).
4. *EMED 1987*, Graphic Communications Catalog, Buffalo, N.Y. (1987).

Exercises

1. Discuss the inspection of containers that contain hazardous wastes.
2. Describe why laboratory packs of hazardous waste are potentially more dangerous than a drum of liquid hazardous waste?
3. What precautions should be taken prior to the movement of a hazardous waste drum?
4. Describe the safe handling of bulging drums that contain a hazardous waste.
5. Discuss the regulatory requirements for drums that contain PCB liquids.
6. What is the objective of container sampling, and what are the important criteria for a drum sampling program?
7. Describe the important considerations of a drum sampling plan.
8. Discuss the steps to be taken in controlling and containing small spills.
9. What items should be included in a contingency plan for responding to site emergencies?
10. What information should be determined about how to respond to site emergencies?
11. Why should a site contingency plan be critiqued following an emergency at the site?

Chapter

11

Radiation Hazards

The literature on radiation hazards is immense. The intent of this chapter is to deal with the subject in broad terms; for more detailed and comprehensive information, the reader should consult the references at the end of this chapter.

The electromagnetic spectrum, shown in Figure 11.1, contains radiation wavelengths ranging from cosmic rays (very short wavelengths) to radio frequency radiation (relatively long wavelengths). In order to study the effects of radiation on living organisms it is convenient to divide radiation into nonionizing and ionizing groups.[1]

Ionizing Radiation

Ionizing radiation, in general, is any electromagnetic or particulate radiation capable of producing ions either directly or indirectly as a result of interaction with matter. The biological effects of ionizing radiation represent the effort of the living organism to deal with the excess energy left in it after interaction of its atoms with an ionizing ray or particle. In order to fully understand the nature and effects of radiation exposure, a brief review of the pertinent atomic physics is essential.[1-4]

The Atom

All matter is made up of atoms, and the interaction of radiation with matter takes place on an atomic level; so an understanding of atomic structure is essential to an understanding of radiation interactions and radiation damage in living organisms.

An atom contains a central nucleus made up of positively charged particles the protons, and electrically neutral particles, called neutrons. Almost all of the mass of an atom is contained in the nucleus, which is surrounded by negatively charged particles, the electrons rotating around it in orbits. The number of electrons and the number of protons in any atom are equal, so that the atom, in its normal state, is electrically neutral. The number of protons in any atom is called the atomic number, and the number of protons and neutrons the atomic mass.

Like electrons, protons and neutrons are in constant motion; and as a result of this motion, the protons exert a repulsive force on each other. This repulsive force is overcome by a cohesive force that binds the mutually repellent protons together. The energy required to overcome the repulsive forces among protons is called the binding energy of the nucleus.

Isotopes and Radioactive Decay

In some elements atoms that have the same atomic number can have a different atomic mass; that is the number of protons is the same in all the atoms, but the number of neutrons varies. These different species of the element are called the isotopes of that element. Some isotopes, called stable isotopes, have enough binding energy in their nucleus to hold the nuclear particles together. Sometimes, however, the amount of nuclear binding energy within the nucleus of an isotope is

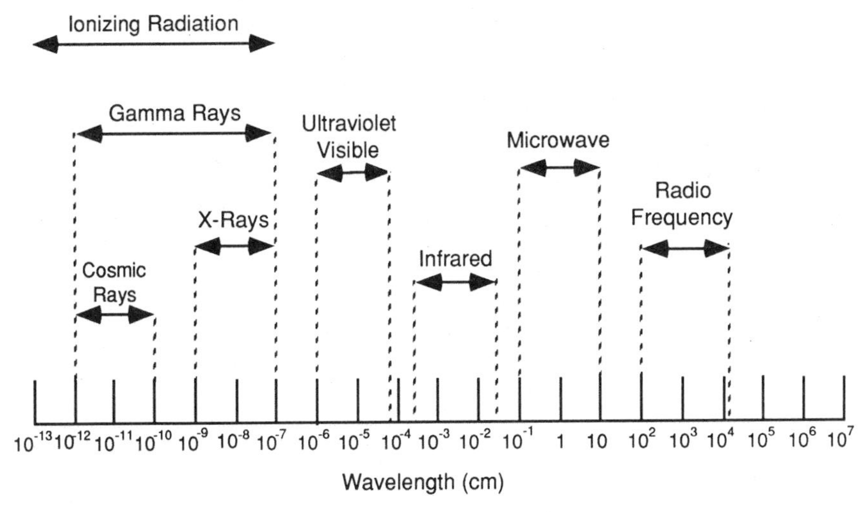

Figure 11.1. The electromagnetic spectrum.

not enough to hold the nuclear particles (protons and neutrons) together, and as a result part of the nucleus is lost, in a process of nuclear disintegration called radioactive decay. Those isotopes of a given element that do not have enough nuclear binding to hold the particles together are unstable species called radioisotopes.

Electrons

The negatively charged electrons move around the nucleus in designated shells or orbits in the form of an electron cloud, and are held in their orbits by the attractive force exerted on them by the positively charged protons in the nucleus The amount of attractive force between two oppositely charged particles increases as the distance between the two particles becomes smaller; so the electrons that are closer to the nucleus are under a relatively large attractive force compared to the electrons in the outer orbits.

The electrons occupying the outermost orbit, which are called the valence electrons, are responsible for chemical reactions. Because of their large distance from the nucleus, the valence electrons are under a relatively small attractive force. With the application of a small amount of energy, these electrons can move even farther away from the nucleus to orbits known as optical orbits. Electrons that occupy the optical orbits are said to be excited. The excitation, however, does not last long, and eventually the excited electrons return to their original orbits. The energy that was supplied to electrons to make them excited in the first place is not lost; it is radiated from the atom.

When radiation involves the optical orbits, optical (ultraviolet) radiation is produced. On the other hand, when radiation involves the inner shell electrons (with more energy), ionizing (X-ray) radiation results (see Figure 11.2).

Ionization and Excitation

The interaction between an ionizing ray and matter could result in either ionization or excitation of the atoms.

The principal way by which an ionizing radiation dissipates its energy in matter is by ejection of one or more orbital electrons. (The removal or orbital electrons from an atom is called ionization.) Because orbital electrons are under the influence of the attractive forces of protons in the nucleus, the ionizing radiation must have an energy level that is equal to or exceeds the energy

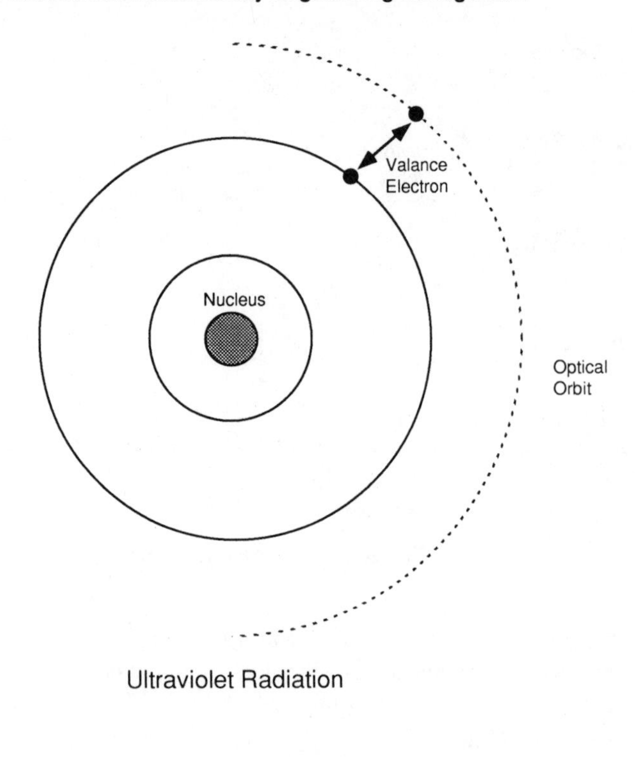

Ultraviolet Radiation

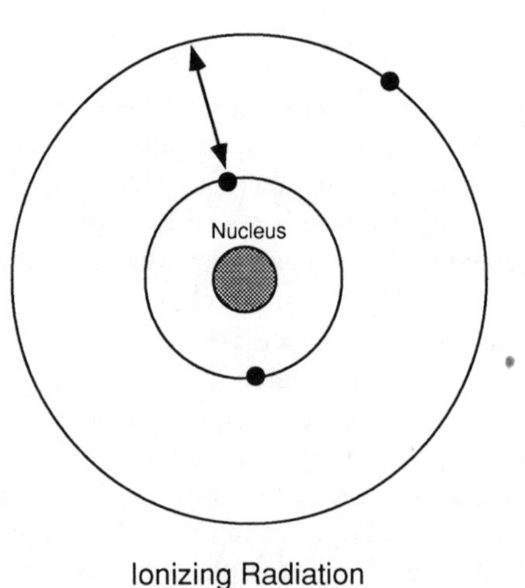

Ionizing Radiation

Figure 11.2. Electron excitement and radiation.

of the nuclear attractive forces. Although atoms are electrically neutral, when one or more electrons are removed from an atom, the atom becomes positively charged and forms an ion pair with the removed electron(s).

Not all interactions between an ionizing ray and matter result in ionization. Sometimes the ionizing ray raises the energy level of certain electrons, moving them to orbits farther away from the nucleus. This process, which is less drastic than ionization, is called excitation. It is estimated that most of the energy of ultraviolet radiation in human tissue is dissipated by excitation.

Types and Origins of Ionizing Radiations

As mentioned earlier, the protons and neutrons within the nucleus of an atom are in constant motion, exerting attractive and repulsive forces on each other. As a result of this energy transfer among particles, the nucleus of some atoms become inherently unstable, releasing a portion of the atomic particles. The atoms that have an unstable nucleus are called radioactive atoms, and the process by which atomic particles are released from the nucleus is called radioactive decay.

Most ionizing radiations are a result of radioactive decay. Table 11.1 summarizes the type and origin of some of the more commonly observed. ionizing radiations.

Alpha rays consist of a stream of positively charged particles traveling at high speed, produced by the radioactive decay of certain heavy elements. These particles have a relatively low penetrative power and a rather high ionization power.

Beta rays can be either negatively charged (negatrons) or positively charged electrons (positrons), produced by radioactive decay. The beta particles are more penetrative than alpha rays and have a moderate power of ionization.

Gamma rays are similar to X rays of shorter wavelength. They are emitted when radioactive transformations take place and usually accompany alpha and beta radiation.

X rays are a result of the ionization of matter by a stream of fast-moving charged particles. When electrons are ejected from the inner orbits, the electrons from outer shells fall to the lower orbits to fill their place. As a result of this energy transformation, X rays are produced.

Neutrons can be emitted spontaneously by certain radioactive materials, and can be produced by bombarding the nucleus of certain isotopes of hydrogen. Although neutrons do not carry any electrical charge, they can initiate radioactive decay in atoms of other elements.

Units of Measurement

Before we discuss the biological effects of ionization radiation, it is essential to review the units used to measure radiation and express radiation exposure data:

Curie (Ci): A measure of the activity of the radioactive material, one curie is equivalent of 3.7×10^{-10} disintegrations per second.

Rad: The rad is a measure of the dose of ionizing radiation to the body in terms of energy absorbed per unit mass of tissue. One rad corresponds to the absorption of 100 ergs per gram of tissue (the erg being the unit for measuring energy in the CGS system).

Table 11.1. Types of ionizing radiation.

Type	Charge	Produced By
Alpha	+	Radioactive decay of heavy atoms
Beta (negatron)	− negative electron	Radioactive decay
Beta (positron)	+ positive electron	Radioactive decay
Gamma ray	0	Radioactive decay
X ray	0	Rearrangement of orbital electrons
Neutron	0	Cyclotron

Source: Reference 2.

Roentgen: A special unit used for measuring exposure to radiation, one roentgen equals 2.58 $\times 10^{-4}$ coulomb per kilogram of air.

Rem: The rem is a measure of the dose of any ionizing radiation to the body tissue in terms of its estimated biological effect relative to a dose of one roentgen. The relation of the rem of other dose units depends on the biological effects under consideration and on the conditions for irradiation. Each of the following is considered to be equivalent to a dose of one rem:

- A dose of one roentgen due to X or gamma radiation.
- A dose of one rad due to X, gamma, or beta radiation.
- A dose of 0.1 rad due to neutrons or high energy protons.
- A dose of 0.05 rad due to particles heavier than protons with sufficient energy to reach the lens of the eye.

Biological Effects of Ionizing Radiation

Overexposure to radiation can produce two types of effects in humans, somatic and genetic.[2,3,5]

The somatic effect is the effect of radiation on organs, tissue, or the whole body. Somatic effects can vary over a wide range, from rapid death due to large exposures to reddening of the skin due to minimal exposure. Some somatic effects of concern, such as cancer, cataracts, and life shortening, may be delayed for long periods. Within the body, cells and tissues respond with varying degrees of sensitivity to radiation. Partial-or whole-body radiation and age also are important factors in determining the somatic effects of overexposure.

Genetic effects of overexposure to radiation are of particular concern because radiation-induced mutations can be carried to subsequent generations be defective genes. Marked adverse effects on the developing germ cells of both the testis and the ovary have been observed. Complete sterility can occur, as a result of the total destruction of sperm cells. Even if some sperm cells survive irradiation, partial or functional sterility may occur in the male if the number of functional sperm produced is decreased substantially.

The Immediately Lethal Effects

The major effect of whole-body exposure to ionizing radiation is shortening of the life of the living organism. The principal factor in determining the length of time that the life is shortened is the dose level of radiation to which the living organism is exposed. Other factors, such as age, sex, type of species,and diet, also play an important role in the length of lifetime shortened.[2]

Doses of ionizing radiation that brings death within approximately 30 days are referred to as immediately lethal and the action of that radiation is classified as acute.

Studies in laboratory animals have revealed that exposure to increasing doses of ionizing radiation over the whole body results in nausea, vomiting, hair loss, loss of appetite, a general but undefined feeling of being unwell, soreness in the throat, diarrhea, and weight loss. As the dose level is increased, in addition to the above signs, some of the animals begin to die. The immediate lethal effect of total body radiation usually is the failure of a vital organ. The order of sensitivity of the vital organ systems to radiation, progressing from the most sensitive to the most resistant, is: the hemopoietic system, the gastrointestinal system, and the central nervous system.[2]

Late Effects of Radiation

Laboratory animals exposed to doses of ionizing radiation that are not immediately lethal may seem to have recovered after a short period of time. As these animals grow older, however, they usually have a higher incidence of tumor development compared to animals of the same group that have not been exposed to radiation. These exposed animals age faster and die earlier than unexposed animals. Life shortening, carcinogenesis, sterility, aging, and cataracts are among the long-term or late effects of ionizing radiation.[2,5,6]

Fertility

Laboratory data on test animals indicate that exposure to ionizing radiation can impair fertility, at least on a temporary basis. Permanent sterility can occur at higher doses of radiation.[1,2] These studies indicate that after a dose of radiation to the gonads, there is a period of continued fertility, followed by a period in which fertility is impaired. If the dose of radiation is high enough, the impairment in fertility might become permanent and result in sterility. Sterility does not necessarily require total loss of sperm; to the contrary, normal, viable sperm may be present. Sterility can be produced, however, if the number of viable sperm is so reduced that the probability of fertilization of an egg becomes unlikely.[2,6]

Some human data are available on the effects of radiation on fertility, mainly from studies conducted on atomic-bomb survivors and on the victims of nuclear reactor accidents.

Studies on atomic-bomb survivors demonstrate that relatively low doses can decrease the production of sperm cells, but effects on spermatogenesis are transient; the sterilizing dose in the male is probably much greater than about 400 to 500 rads—that is, it probably exceeds the mean lethal dose to the whole body.[6] Little is known regarding the delayed effects of radiation on fertility in these exposed populations, nor is there information on the extent of impairment, if any, in the male and female populations exposed.[7-9] Followup studies of Japanese atomic-bomb survivors and of women exposed to fallout have failed to demonstrate any long-term effect on fecundity.[6,15]

Clinical data are also available on male radiotherapy patients and men exposed during criticality accidents at nuclear-reactor installations.[10,11] Careful sperm-count studies after limited partial-body radiation exposure have indicated that if sterility occurs, normal sperm counts can return in about one year after doses of 100 rads and even in three years after exposures in the near-lethal range.[10,11] Acute whole-body exposure has not been shown to cause permanent sterility in males.[11] The sterilizing dose therefore exceeds the lethal whole-body dose for acute radiation. Similarly, sterilization of the human testis has never been shown to result form continuous or fractionated low-dose exposure.[9,10,12,13]

In women, radiotherapy experience has suggested that acute doses of 300 to 400 rads or slightly higher doses given to two or three fractions result in permanent sterility.[2,15] If fractionation is protracted over a two-week period, much larger doses (possibly 1,000–2,000 rads) are required for sterilization, depending on the age of the woman.[11]

Cancer

Cancer induction is considered to be the most important somatic effect of low-dose ionizing radiation. The induction of cancer by radiation is detectable only by statistical means; that is, the cancer of any given person cannot be attributed with certainly to radiation, as opposed to some other cause. In general, the smaller the dose of radiation, the smaller the likelihood is that radiation was the cause.[6]

There are good observational data relative to cancer induction in humans over a range of higher doses, but little direct evidence is available for doses of a few rads. Estimation of the excess risks at these low doses usually involves extrapolation from observations at higher doses on the basis of assumptions about the nature of the dose–response relationship. In considering cancers attributable to radiation exposure, the following comments are pertinent:

- Cancers induced by radiation are indistinguishable from those occurring naturally; hence, their existence can be inferred only on the basis of a statistical excess above the natural incidence.

- Cancer may be induced by radiation in nearly all the tissues of the human body.
- Tissues and organs vary considerably in their sensitivity to the induction of cancer by radiation.
- The natural incidence of cancer varies over several orders of magnitude, depending on type and site of origin of the neoplasm, age, sex, and other factors.
- With respect to the excess risk of cancer from whole-body exposure to radiation, solid tumors now are known to be of greater numerical significance than leukemia. Solid cancers characteristically have long latent periods; they seldom appear before 10 years after radiation exposure and may continue to appear for 30 years or more after radiation exposure. In contrast, the excess risk of leukemia appears within a few years after radiation exposure and largely disappears within 30 years after exposure.
- The major sites of solid cancers induced by whole-body radiation are the breast in women, the thyroid, the lung, and some digestive organs.
- The incidence of radiation-induced human breast and thyroid cancer is such that the total cancer risk is greater for woman than for men. Breast cancer occurs almost exclusively in women, and absolute-risk estimates for thyroid cancer induction by radiation are higher for women than for men (as is the case with the natural incidence). With respect to other cancers, the radiation risks in the two sexes are approximately equal.
- There is now considerable evidence from human studies that age is a major factor in the risk of cancer from exposure to ionizing radiation.
- Various host or environmental factors may interact with radiation to affect the cancer incidence in different tissues. These factors may include hormonal influences, immunologic status, exposure to various oncongenic agents, and nonspecific stimuli to cell proliferation in tissues sensitive to cancer induction by radiation.
- The time elapsing between irradiation and the appearance of a detectable neoplasm is characteristically long, that is years or even decades.
- It is not yet possible to estimate precisely the risk of cancer induction by low-dose radiation because the degree of risk is so low that it cannot be observed directly.
- Despite the difficulties and uncertainties, a clear-cut increase in incidence or mortality with increasing radiation dose has been demonstrated for many types of cancer in human populations, as well as in laboratory animals.

Leukemia

Although it is difficult to directly relate the effects of ionizing radiation to the induction of leukemia, certain types of leukemias are known to be induced by ionizing radiation in humans. Although leukemia is mostly observed after a relatively short period of time following high level radiation over the whole body, it also can result from external radiation given at rather low dose rates over the total body or to large parts of the bone marrow. It is believed that leukemia results from the deposition of radioactive elements such as radium and thorium in the body.[2,6]

Shortening of Life Span

Laboratory animal test data indicate that exposure to ionizing radiation can result in a faster rate of aging and a shortening of the life span.[1,2,5] The amount of shortening of life in test animals appears to be dose-dependent. Although smaller amounts of radiation shorten life to a lesser degree than do large amounts, the response over a range of doses appears to be linear.

Cataracts

Exposure of the eyes to ionizing radiation can result in cataracts. Because a large dose of radiation is required for cataract information, they are rather uncommon late effects of exposure to ionizing radiation.[2]

Table 11.2. Permissible radiation dose in a restricted area.

Part of Body	Rems per Calendar Quarter
Whole body, head and trunk, active blood-forming organs, lens of eyes or gonads	1.25
Hands and forearms, feet and ankles	8.75
Skin of whole body	0.5

Source: Reference 14.

OSHA Requirements

OSHA's regulations on ionizing radiation are contained in 29 CFR 1910.96. OSHA has specific regulations on the exposure of workers to radiation in restricted areas, worker exposure to airborne radioactive materials, precautionary procedures and personal monitoring, and radiation caution signs, labels, and signals.

Exposure of Individuals to Radiation in Restricted Areas

OSHA required employers to ensure that employees who work in a restricted area do not receive, in any three-month period, a dose of ionizing radiation that exceeds the values outlined in Table 11.2. A restricted area, as defined by OSHA, is any area that is controlled by the employer in order to protect employees from overexposure to radiation or radioactive materials.

An employer may legally exceed the doses outlined in Table 11.2 only if one of the conditions set forth in Paragraph b(2) of 29 CFR 1910.96 applies.[14]

Exposure of Individuals to Airborne Radioactive Materials

OSHA has set limits on exposure to airborne radioactive materials, both for employees who work 40 hours per week and for individuals who are under 18 years of age. These limits have been outlined in Tables 1 and 2 of Appendix B in 10 CFR Part 20.

Precautionary Procedures and Personal Monitoring

OSHA requires every employer who handles radioactive material to conduct a survey of his or her work area to ensure compliance.

Employers also are responsible for providing any monitoring devices such as film badges, pocket chambers, pocket dosimeters, or film rings (see Figure 11.3) for those employees who could be overexposed to radiation.

Caution Signs, Labels, and Signals

OSHA requires that all radiation symbols use a magenta or purple color on a yellow background. The conventional radiation symbol prescribed by OSHA is the three-bladed design shown in Figure 11.4. OSHA mandates that each radiation area be clearly marked with the radiation sign shown in Figure 11.4 and the words:

CAUTION
RADIATION AREA

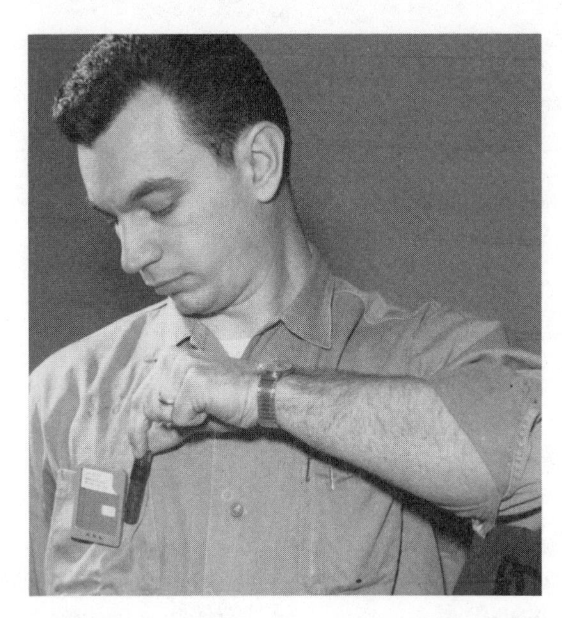

Figure 11.3. Example of personal radiation monitoring devices. (Source: Argonne National Laboratory)

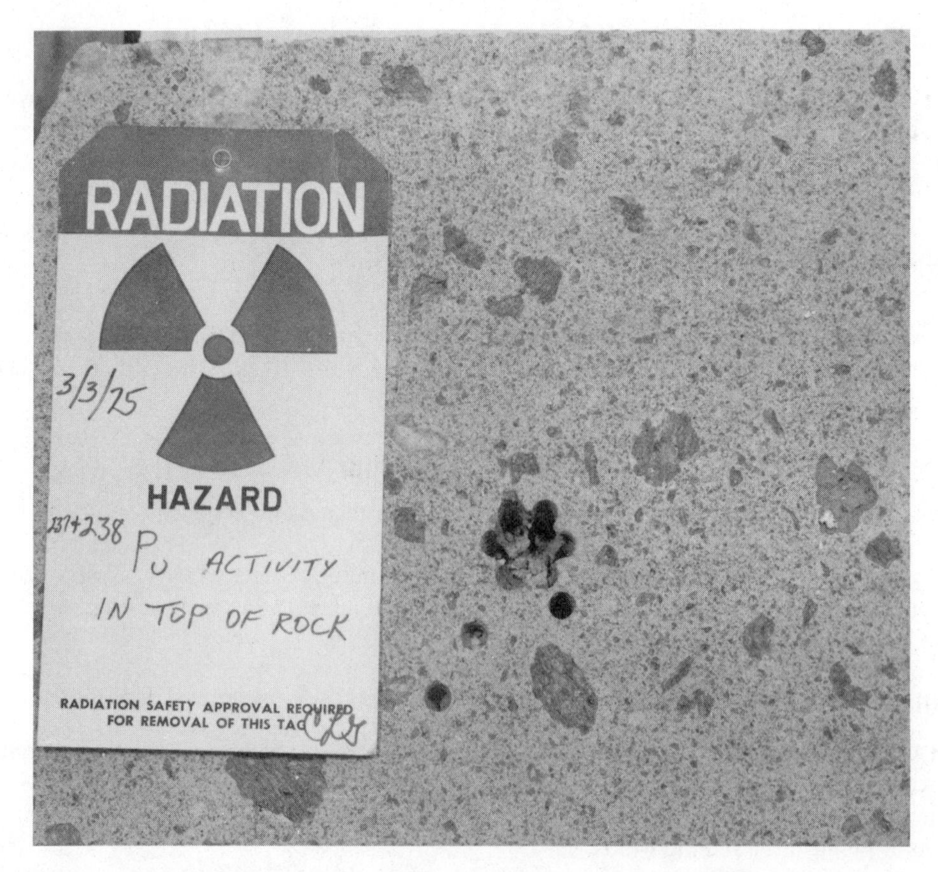

Figure 11.4. OSHA-prescribed, three-bladed design radiation sign. (Source: Argonne National Laboratory)

OSHA also requires that high radiation areas be clearly marked with the conventional radiation sign and the words:

CAUTION
HIGH RADIATION AREA

OSHA further requires that any high radiation area that is established for a period of 30 days or more be equipped with appropriate control devices either to reduce the level of radiation to below 100 millirems in one hour or to sound an audible alarm signal.

OSHA has specific regulations for signs in airborne radioactive areas and for containers of radioactive materials, as well as regulations for immediate evacuation warning signals.[14] The following areas and materials are excluded from OSHA radiation posting requirements:

- An area that contains a sealed source of radioactive material, provided that the amount of radiation 12 inches away from the surface of the sealed container does not exceed 5 millirems in one hour.
- Areas in medical facilities with patients who are treated with radioactive materials, provided that trained personnel are available to ensure that individuals entering the area are not overexposed to radiation.
- Areas that store radioactive materials for a period of less than eight hours provided that: (a) the area is constantly attended by a trained individual to ensure that overexposure to radiation will not occur, and (b) such area or room is subject to the employer's control.
- Radioactive materials packaged and labeled in accordance with regulations of the Department of Transportation.

Instruction of Personnel

OSHA requires that employers inform employees who work in radiation areas of the occurrence of radioactivity. OSHA also requires that employers provide for the instruction of employees on safety problems associated with overexposure. Employers also are required to post a current copy of their provisions and operating procedures in a conspicuous location.

Other Requirements

OSHA has set forth regulations in regard to storage of radioactive materials, radioactive waste disposal, Department of Labor notification of incidents involving radioactive materials, reports of overexposure, excessive levels and concentrations, and recordkeeping requirements.[14]

Nonionizing Radiation

Interest in the public health aspects of nonionizing radiation has increased manyfold because of the expanded production of electronic products that use or emit radiation (e.g., lasers, microwave ovens, radars, infrared inspection equipment, and high intensity light sources). In this section attention focuses mainly on ultraviolet radiation, lasers, and microwave radiation.[1,15]

Ultraviolet Radiation

Many items of equipment that emit ultraviolet radiation are in widespread use in laboratories. The overexposure of unprotected skin or eyes could result in tissue inflammation; also long-term effects may develop as a result of repeated overexposures.

Ultraviolet radiation has wavelengths ranging from 10 nanometers (nm) to 400 nm.[1] However, the rays that are most damaging to the eyes and skin have wavelengths between 200 nm and 315 nm.

Overexposure of unprotected skin to ultraviolet radiation may give rise to a condition called erythema which is similar to sunburn. Exposure of the unprotected eye can cause inflammation of the cornea, giving rise to conditions known as keratitis and conjuctivitis.

Infrared Radiation

Infrared radiation can damage the unprotected eye or skin by releasing energy and producing a heating effect. Overexposure of unprotected eyes to infrared radiation could give rise to cataracts. In order to protect against infrared radiation, every effort must be made to shield the emitting source by a heat-absorbing materials; also eye protection of an approved type must be used to prevent burns to the eyes.

Laser Radiation

The rate of development and manufacture of devices and systems based on the stimulated emission of radiation has been truly phenomenal. Lasers now are being used for a wide variety of purposes, including micromachining, welding, cutting, sealing, optical alignment, spectroscopy, and surgery, as well as in communications media.

Lasers can be produced in a variety of forms, including continuous and pulsed outputs of both visible and invisible radiation. High power pulsed laser manufacturing units could be extremely hazardous, as high power laser beams are capable of tissue destruction at the point of contact.

Although adequate eye and skin protection must be used in work with laser beams, the skin is much less sensitive to laser beams as compared to the eyes.[1,3,15]

Microwave Radiation

Microwave radiation wavelengths can vary from 10 meters to about 1 mm. Microwave-producing devices may pose the normal hazards associated with all electrical equipment as well as the danger of local burns caused by contact with the work coil or heat produced in conductors such as watches and rings. It is advisable to ground all radio frequency work coils and isolate them from high voltages. Any work conducted on microwave-producing equipment should be carried out only by well trained and authorized personnel.

In areas where high power microwave equipment is used, a warning sign must be posted to alert personnel of its radiation hazards.[1,15]

Radiation Protection Considerations

A basic concept in radiation protection practice[1,3,15] is the establishment of a controlled area. Access to such areas must be controlled, and supervision and control of occupational exposures must be provided in them. The emergence of beams and the escape of radioactive materials from these areas should be controlled. The areas must be identified by use of a standard radiation symbol with associated warning notices.

Because the useful beam of X-ray equipment may inflict a year's maximum permissible dose (MPD) in minutes or less, the design of industrial X-ray facilities must give proper consideration to the establishment of a suitably controlled area to assure proper radiation protection for two groups of individuals—those who operate the equipment (occupationally exposed) and those in the environs, either normally or casually (not occupationally exposed). Where possible, the X-ray equipment should be within a room or other enclosure arranged with controls outside and having interlocks to prevent entry when equipment is energized. Shielding can be provided so that the exposure rate outside the enclosure will be low enough to ensure that the applicable MPD or dose limit will not be exceeded. Small devices or instruments using X rays, such as laboratory equipment, usually can be totally enclosed with adequate shielding, but accessibility to the inside of the shield requires special consideration (interlocks, etc.). Where work requires truly mobile or portable equipment, exposure time and distance from the equipment become the basic means of limiting exposure rates to values that will ensure that no individual exceeds the applicable MPD or dose limit.

Gamma radiation, usually from a sealed source, is used for a variety of industrial purposes. The larger sources produce beams comparable in exposure rate to X-ray equipment. Gamma radiation cannot be turned off like X-rays, and this consideration imposes a severe requirement for retention of the sealed source (a) at a predetermined specific location where exposure control is assured or (b) within appropriate shielding at all times. Procedures and surveys must guarantee this control. The integrity of the encapsulation or bonded cover of the sealed source must be assured at all times to prevent release of the radioactive material into the environment, where it could be dispersed and inhaled or ingested. Periodic tests of the sealed source or its container should be made. Appropriate testing of radium sources is particularly important because any failure will release radon gas, which can contaminate the surrounding area.

Beta radiation sources, which frequently are built into some piece of equipment such as a thickness gage, must be shielded and arranged so that access to the beta radiation is prevented. Of particular concern is control of exposures during any maintenance procedures. To permit escape of the beta radiation, the encapsulating material must be relatively thin, at least over the useful area of the source. Damage to this encapsulation will permit release of the radionuclide to the environment, where it can be dispersed and inhaled or ingested.

Radioactive neutron sources commonly are small sealed sources of relatively substantial construction. The yield of neutrons is proportional to the activity in the source. Consideration must be given to gamma radiation as well as neutron radiation from these sources. If the radionuclide they contain is radium, the gamma dose equivalent rate is higher than that from neutrons, and the possibility of radon leakage must be recognized. Leakage of any of the radionuclides used in these sources presents a hazard of considerable magnitude, which necessitates care in use and periodic testing. Commonly used shielding materials are concrete, polyethylene, boronated polyethylene, or boron in other materials such as aluminum.

References

1. Freeman, N. T., and J. Whitehead, *Introduction to Safety in the Chemical Laboratory*, 1st Ed., Academic Press, New York (1982).
2. Pizzarello, D. J., *Basic Radiation Biology*, 1st Ed., Lea & Febiger, Philadelphia, Pa. (1967).
3. Barnes, E. C., *The Industrial Environment—Its Evaluation and Control*, U.S. Department of Health, Education, and Welfare, Washington, D.C. (1973).
4. Borowitz, S., and A. Beiser, *Essentials of Physics*, 1st Ed., Addison-Wesley Publishing Co., Reading, Mass. (1967).
5. Casarett, A. P., *Radiation Biology*, 1st Ed., Prentice-Hall, Englewood Cliffs, N.J. (1982).
6. National Research Council, *The Effects on Populations Exposure to Low Levels of Ionizing Radiation*, National Academy Press, Washington D.C. (1980).
7. Blot, W. J., and H. Sawada, Fertility among female survivors of the atomic bombs of Hiroshima and Nagasaki, *Amer. J. Hum. Genet.*, Vol. 24, pp. 613–622 (1977).
8. International Commission on Radiological Protection, *Radiosensitivity and Spatial Distribution of Dose*, Publication 14, Pergamon Press, Oxford (1969).
9. Seigel, D. G., Frequency of live births among survivors of Hiroshima and Nagasaki atomic bombs, *Radiat. Res.* Vol. 28, pp. 278–328 (1966).
10. Langham, W. H. Ed., *Radiobiological Factors in Manned Space Flight*; National Academy of Sciences, Washington, D.C. (1967).
11. Upton, A. C., Somatic and genetic effects of low-level radiation, pp. 1–40 in J. H. Lawrence, Ed., *Recent Advances in Nuclear Medicine*, Vol. 4, Grune & Stratton, New York (1974).
12. Casarett, G. W., Long-term effects of irradiation on sperm production of dogs, pp. 127–146 in W. D.

Carlson and F. X. Gassner, Eds., *Effects of Ionizing Radiation on the Reproductive System*, Macmillan, New York (1964).
13. Fabrikant, J. I., The effects of irradiation on the kinetics of proliferation and differentiation, *Amer. J. Roentgenol. Rad. Ther. Nucl. Med.*, Vol. 114, pp. 792–802 (1972).
14. Office of the Federal Register, *Code of Federal Regulations*, 29 CFR 1910, Office of the Federal Register, Washington, D.C. (1985).
15. Wilkening, G. M. *Nonionizing Radiation, the Industrial Environment—Its Evaluation and Control*, U.S. Department of Health, Education and Welfare, Washington, D.C. (1973).

Exercises

1. List the major differences between ionizing radiation and nonionizing radiation.
2. From an atomic level point of view, describe how ionizing radiation and nonionizing radiation are produced.
3. What is nuclear cohesive force, and how does it relate to radioactive decay?
4. What is the major difference between an isotope and a radioisotope?
5. Define the terms "ionization" and "excitation," and describe how they relate to the production of ionizing and nonionizing radiation.
6. List the major characteristics of alpha, beta, gamma, and X-rays.
7. List the symptoms of overexposure to ionizing radiation.
8. Describe the effects of ionizing radiation on human fertility.
9. Describe the effects of ionizing radiation on cancer and tumor development.
10. What are OSHA requirements for the exposure of individuals to radiation in restricted areas?
11. What are the radiation areas that are excluded from OSHA radiation posting requirements?
12. What is nonionizing radiation? List the biological effects of nonionizing radiation on humans.
13. List at least three general guidelines for protection against ionizing radiation.

Chapter

12

First Aid

First aid can be defined as the immediate care given to a person who has been injured or has fallen ill. First aid also includes self-help in the absence of medical facilities or personnel.

In many emergency situations, first aid could mean the difference between life and death, temporary or permanent disability, and short- or long-term hospitalization. Because accidents are the leading cause of death among people from age 1 to age 38, and in an emergency medical care delays of minutes can lead to disaster, a person with first aid knowledge may have the opportunity to prevent a fatal outcome.

A person with knowledge of first aid not only can save the lives of those who are stricken by an injury or illness, but also can provide instructions and generate a reasonable safety attitude among coworkers and family members.

With first aid training, self-care also can be exercised in an emergency. Even if first aiders become so incapacitated that they cannot provide self-help, they can instruct others in appropriate, and possibly lifesaving, actions.

First aid training can prove to be invaluable during a catastrophe, such as an earthquake or industrial accidents. Knowing what to do in an emergency alleviates confusion and disorganization. Knowledge of first aid is everyone's responsibility and should be considered an essential tool in preventing complications and saving lives.[1]

OSHA Requirements

First aid for employees who become sick or are injured on the job comes under the OSHA regulations in 29 CFR 1910 Subpart K Section 151. Such first aid may consist of attention to a minor injury that requires no further treatment or emergency help until professional medical personnel can take over. After evaluation of the hazards in the work area, appropriate provisions for on-site care can be made with the advice of a physician.

OSHA requires that medical personnel be readily available for consultation on matters of workplace safety and health. In situations where the workplace is not within a reasonable distance of a medical facility, OSHA also requires that an appropriate number of personnel be trained in first aid procedures and that approved first aid supplies be available on-site. In work situations where there is a potential for injury to the eye or skin due to contact with corrosive materials, OSHA requires that eyewash stations and safety showers be available for immediate emergency care.[2]

General Guidelines for Giving First Aid

After carrying out emergency measures to ensure the victim's safety, the first aider should observe the following guidelines prior to the actual administration of first aid:[1]

- Make no attempt to move the victim unless for safety reasons (such as a victim in contact with a live electrical conductor with no power shutoff mechanism).

- Determine the most appropriate position for the victim, and do not allow the victim to rise or walk.
- Do not disturb the victim unnecessarily (such as by asking questions that have no relevance to immediate medical care).
- Prevent chilling by means of covers or blankets.
- Examine the victim in a systematic manner, paying special attention to the nature of the accident or sudden illness and the needs of the situation. Have a reason for what you do.
- Administer the appropriate first aid procedures.

First Aid for Specific Injuries

This section is based on references 1 and 3.

Thermal and Chemical Burns

A burn can be defined as an injury resulting from heat, chemical agents, or radiation, and a burn can be classified according to the depth or degree of skin damage. Usually the degree of damage will differ in different parts of the same affected area.

First-degree burns are those that result from overexposure to sunlight, skin contact with hot objects, or scalding by hot water or steam. Redness or discoloration, some swelling and pain, and rapid healing are the usual signs.

Second-degree burns can result from very deep sunburns; contact with hot liquids; and flash burns from gasoline, kerosene, and other petroleum products. Second-degree burns usually are more painful than deeper burns in which the nerve endings in the skin are destroyed.

Third-degree burns can result from ignited clothing, immersion in hot water, contact with hot objects, and electricity. Important factors in determining the extent of tissue destruction are temperature and duration of contact.

First Aid for Burns

The primary objectives of first aid for burns are to minimize pain, prevent infection, and treat the victim for shock.

For first-degree burns, submerge the burned area in cold water and then cover it with a dry dressing.

For second-degree burns, immerse the burned area in cold water (not ice water) until the pain subsides and then apply freshly ironed or laundered cloths that have been wrung out in ice water. Gently blot the area dry and apply a dry, protective bandage, such as sterile gauze or a clean cloth. Do not break blisters or remove tissue. On a severe burn, do not use an antiseptic preparation, an ointment, a spray, or a home remedy.

For third-degree burns, do not remove particles of charred clothing. Cover the burns with thick, sterile dressings or a freshly ironed or laundered sheet or other household linen. If the hands are involved, keep them above the level of the victim's heart to minimize blood flow. Keep burned feet or legs elevated (do not allow the victim to walk).

A victim with a face burn needs to be observed continuously for breathing difficulty; the victim should be helped to sit up and should be kept under close observation. An open airway must be maintained if respiratory problems develop. The first aider should not immerse an extensively burned area or apply ice water over it because cold may intensify the shock reaction. However, a cold pack may be applied to the face or to the hands or feet.

As quickly as possible arrange transportation to the hospital. If medical help or trained ambulance personnel cannot come to help for an hour or more and the victim is conscious and not vomiting, administer a weak solution of salt and soda: one level teaspoonful salt and one-half level teaspoonful of baking soda to each quart of room-temperature water. Ask to victim to sip slowly. Give an adult about four ounces (a half glass) over a period of 15 minutes. A child from 1 to 12 years of age should be given about two ounces, and an infant under one year of age about one ounce. If the victim vomits, do not give any more of the solution. *Do not* give fluids if you suspect internal injuries.

Do not apply substances such as ointment, commercial preparations, grease, or other home remedy. (Such substances may cause further complications and interfere with treatment later.)

Chemical Burns

First aid steps for chemical burns of the skin are as follows:

- Flood the affected area with large quantities of water for at least 5 minutes.
- If the victim's clothing is contaminated with the chemicals, gently remove the clothing.
- The exact procedures for first aid after contact with a chemical depend, to a large extent, on the specific nature of the chemical. If appropriate procedures are available, apply them after washing the area with water. The material safety data sheet (MSDS) is a good source of first aid procedures for specific chemicals.
- Cover the affected area with a dressing bandage, and get medical help.

For acid burns to the eye, it is critical that first aid procedures be started immediately. The eyes must be flushed with running water for at least 15 minutes. If the victim is lying down, hold the eyelid open and pour water from the corner of the eyelid inward while exercising caution to prevent the contaminated water from entering the other eye. In order to neutralize the acid, it is advisable to wash the eyes with a weak solution of baking soda. After complete removal of any residual acid, the eye should be covered with a dry, clean dressing.

Alkali burns of the eye can result from eye contact with strong caustic solutions such as sodium hydroxide. These injuries are referred to as progressive injuries. An eye that might appear to have only minor irritation and redness can develop severe inflammation, which may result in the loss of sight. The immediate first aid for alkali burns of the eye is to flush the eye with running water for at least 15 minutes. Any residual dry chemical must be removed, and the eye should be covered with a dry, clean pad. Washing the eye with baking soda solution does not help and can make the situation worse, as alkali materials have basic properties and are not neutralized by baking soda, which is also a base.

Eye Injuries Due to Foreign Objects

Foreign objects such as dust and other solid particles harm the eye by scratching the cornea or becoming embedded in the eye, causing irritation. To remove a foreign object from the eye, follow the steps outlined below:

- Determine whether the object lies on the inner surface by gently pulling down the lower lid.
- Remove the object with a clean tissue if it lies on the inner surface. Cotton should not be used for removal of objects from the eye because it can stick to the surface of the eye and can result in further complications.
- If the foreign object is beneath the upper lid, grasp the lashes of the upper lid and gently pull forward and down over the lower lid; this may result in tears that dislodge the foreign object.
- If the foreign object has not been dislodged, lift the upper lid with a matchstick and remove the object with a clean tissue.
- Flush the eye with water.

External Bleeding

The rapid loss of as little as a quart of blood can result in shock and loss of consciousness. The first aider must stop any large rapid loss of blood immediately because it is possible for a victim to bleed to death in a very short period of time.

The preferred method for the control of severe bleeding is direct pressure by hand over a dressing. Direct pressure prevents loss of blood from the body without interference with normal blood circulation. In the control of severe bleeding, direct hand pressure over a thick pad of clothing causes the blood to clot, and on many occasions bleeding stops.

It should be noted that blood clots should not be disturbed or removed, as this action would cause bleeding to restart. On some occasions the blood soaks through the pad without clotting. This indicates that the pad is not thick enough, and an additional pad of cloth should be provided.

Head Injuries

For head injuries, first aid should be given as follows:

- Make no attempt to clean wounds, as this could result in severe bleeding and/or contamination of the brain.
- Control the bleeding by elevating the victim's head while exercising caution not to bend the neck, in case a neck injury has occurred.
- Place a clean dressing over the wound while avoiding excessive pressure, in case there is a fracture to the bone.
- After bleeding is under control, apply a bandage to hold the dressing in place.

Hand Injuries

For hand injuries, first aid should be given as follows:

- Elevate the victim's hand above the level of the heart. This is the most important first aid because it reduces further swelling of tissues due to gravity. (Only after snake bite and stings should the hand be kept hanging down after injury.)
- Do not attempt to cleanse a serious wound.
- To control bleeding, apply pressure over a sterile bandage.
- Separate the fingers by gauze or cloth dressing material, and cover the entire hand with a sterile towel.
- During transportation to receive medical care, elevate the victim's hand in a sling or on pillows.

Bone Fractures

Bone fractures, which are defined as a crack in the bone, can be classified into open fractures and closed fractures. Open fractures are directly related to an open wound on the surface of the body; closed fractures are not related to open wounds.

Fractures can be caused by a person's falling, a falling object's striking a person, or a sudden impact, as in a car accident. The first aider should exercise extreme caution when dealing with bone fractures, as inappropriate actions could result in further serious injuries. The following general guidelines should be used in trying to help someone suspected of having a bone fracture injury:

- Make no attempt to set (or reduce) a fracture or to push a protruding bone end back.
- Make no attempt to move a victim unless conditions are immediately dangerous to life or health.

- In lifting a victim who is unconscious, always assume that the person has an injury to the neck or spine.
- For a protruding bone, cover the entire wound with a large clean pad.
- Do not replace bone fragments.

Shock

Shock, which is a result of the depressed functioning of vital body organs, can be experienced by any person who has been involved in an accident. On many occasions victims have died because of shock created by their injuries.

The symptoms of shock include pale skin, cold sweat, weakness, rapid pulse (over 100), and rapid rate of breathing. The following guidelines should be used when trying to provide first aid to a shock victim:

- The victim should be asked to lie down and should be covered enough to prevent the loss of body heat.
- Although the appropriate position depends on the injuries, the victim should be placed in such a way as to improve blood circulation.
- The victim should not be moved in any way if injuries to the neck or lower spine are suspected.
- An unconscious victim should be positioned face down to ensure that airways remain open.
- Although administering fluids by mouth could be helpful, no attempt should be made to give any fluid to a victim who is unconscious, vomiting, or likely to have surgery.

Overexposure to Heat

In industrial situations, there are occasions when a worker may be overexposed to heat. Possible results are heat stroke, heat cramps, and heat exhaustion.

Heat stroke can be defined as the body's response to high temperatures, resulting in a high body temperature and interference with the body's sweating mechanism. Heat stroke can be immediately dangerous to life and requires urgent and proper medical attention.

Heat cramps can develop as a result of excessive sweating and substantial loss of body salt content. This condition involves muscular pain and spasms.

Heat exhaustion can be characterized as conditions of fatigue and weakness caused by inadequate intake of water to compensate for the loss of body fluids through sweating.

Heat Stroke

Heat stroke usually is accompanied by very high body temperatures (above 105°F); hot, red, and dry skin; rapid pulse; and a block of the body's sweating mechanism. Any first aid measures for heat stroke should be directed toward lowering the body temperature. The following steps should be followed for a victim suffering from heat stroke:

- The victim's clothes should be removed and the body temperature reduced by sponging the skin with water or rubbing alcohol, or by placing the victim in a cold water bathtub.
- The victim should be dried and placed in a cool area to promote body cooling. Monitor the body temperature, and if it goes up, repeat the cooling process. No attempt should be made to give stimulants to the victim of a heat stroke.

Heat Cramps

Heat cramp symptoms usually appear in the muscles of the legs and the abdomen.

The first aid procedure for heat cramps is to exert firm pressure or massage the affected muscles to relieve the spasm. The victim can be given sips of salt water at the rate of half a glass every 15 minutes for a period of one hour.

Heat Exhaustion

Although the victim of heat exhaustion may have a normal body temperature, this condition usually is accompanied by pale skin, profuse sweating, weakness, headache, and dizziness.

The appropriate first aid for victims of heat exhaustion is: make them comfortable by loosening their clothing; have them lie down and raise their feet, and apply cool cloths to their skin. A victim who is not vomiting can be given half a glass of salt water every 15 minutes for a period of one hour.

Cardiopulmonary Resuscitation (CPR)

The following paragraphs should not be considered an instruction course for cardiopulmonary resuscitation (CPR—a combination of mouth-to-mouth resuscitation and closed-chest heart massage); rather, they are meant to demonstrate the relative ease of learning and performing CPR. Classroom training, involving practice with special manikins, is essential for the prospective first aider. This is particularly true for closed-chest massage, which may cause cracked ribs or a punctured lung if not done properly.

CPR is an emergency first aid procedure that combines artificial respiration with artificial blood circulation—a combination of mouth-to-mouth breathing, which supplies air to the lungs, and chest compression, which provides blood circulation. In many cases where a person's breathing has stopped, the heart still beats. These cases require only mouth-to-mouth breathing. CPR should be used only in those situations where the victim's breathing and heart have stopped.[4]

Positioning the Victim

- Tap or gently shake the victim's shoulder. Ask "Are you O.K.?" If there is no response, get help from someone nearby.
- CPR must be performed with the victim lying on the back on a firm surface with the head lower than or level with the heart. If the victim is discovered lying face down or in an awkward position and you have determined that breathing has stopped, carefully roll the body as a unit onto the back. Take care not to worsen any possible injuries by jostling or twisting the victim.

Mouth-to-Mouth Breathing

- Place one hand on the victim's forehead and push it back firmly. Place the fingers of the other hand under the bony part of the lower jaw near the chin, and lift to bring the chin forward and the teeth almost to a closed position. This should open the airway. If foreign material or vomit is visible in the mouth, it should be removed quickly. Place your ear very close to the victim's mouth and nose. *Look* at the chest for breathing movements, *listen* for breaths, and *feel* for breathing against your cheek. This evaluation procedure should take only 3 to 5 seconds.
- If there is no breathing, give the victim two full breaths in a row. Keep the victim's head tipped, pinch the nose, and take a deep breath. Open your mouth wide and cover the victim's mouth, making a good seal. Now give two full, slow separate breaths. You should see the victim's chest rise and fall, and should hear and feel the air escape during exhalation.

- Check the victim's pulse and breathing for 5 to 10 seconds. To do this: Keep the head tipped with your hand on the victim's forehead. Place the fingertips of your other hand on the victim's adam's apple, sliding your fingers into the groove along the side of the neck nearest you. If there is a pulse but no breathing, give one breath every 5 seconds. If there is no pulse, immediately begin chest compression.

Chest Compression

In order to start chest compression, your hand should be in the proper position; to do this:

- Use your index and middle fingers to find the lower edge of the victim's rib cage on the side nearest you.
- Trace the edge of the ribs up to the notch where they meet the breastbone.
- Place your middle finger on the notch, pointing across the chest. Your index finger goes next to it, on the side closest to the victim's head.
- Put the heel of your other hand on the victim's breastbone right next to your index finger.
- Remove the two fingers from the notch, and place the heel of that hand on top of the other. Keep all your fingers *off* the chest.

In order to perform chest compression, with your elbows locked push straight down, keeping your shoulders directly over the heels of your hands. Keep your knees a shoulder-width apart.

Push down smoothly about $1\frac{1}{2}$ to 2 inches. Bend from the hip, not the knees. Keep your fingers *off* the chest.

Give 15 compressions at the rate of 80 to 100 times per minute. Then tip the victim's head and give two full, slow breaths in approximately 4 to 7 seconds. Continue, repeating the 15 compressions followed by two breaths. Check the pulse and breathing after the first minute, and every few minutes thereafter.

First Aid Summary

In emergencies, the nature of toxic exposure or hazardous situations that cause injuries and illnesses vary from site to site. Medical treatment may range from the bandaging of minor cuts and abrasions to lifesaving techniques. In many cases, essential medical help may not be immediately available. For this reason, it is vital to train on-site emergency personnel in on-the-spot treatment techniques, to establish and maintain telephone contact with medical experts, and to establish liaisons with local hospitals and ambulance services. The following essential points should be included in the design of this program:

- Train a cadre of personnel in emergency treatment such as first aid and CPR. Training should be thorough, repeated frequently, and geared to site-specific hazards.
- Establish liaisons with local medical personnel. Inform and educate these personnel about site-specific hazards so that they can be optimally helpful if an emergency occurs. Familiarize all on-site emergency personnel with these procedures.
- Set up on-site emergency first aid stations; see that they are well supplied and restocked immediately after each emergency.

Table 12.1 summarizes some of the more common injuries and appropriate first aid measures.[5,6]

Table 12.1. Common injuries and their first aid measures.

Injury	First Aid
Bleeding	Apply pressure over a clean pad. Elevate the wound unless there is a possibility of bone fractures. Do not disturb blood clots.
Thermal burns	If there is no blistering, immerse the burn in cold water; cover it with a clean cloth until help arrives.
Chemical burns	Flush the affected area with running water; remove contaminated clothing, and cover the affected area with a clean cloth.
Electric shock	Remove the victim from a live conductor or shut off the power; if victim is not breathing, administer mouth-to-mouth resuscitation; if victim's heart has stopped, administer CPR.
Shock	Place victim on back with feet higher than head unless ther is suspicion of bone fractures; check for breathing; apply artificial respiration if necessary; loosen victim's clothing.
Chemical eye injury	Flush victim's eyes with water for at least 15 minutes; remove any dry chemicals from the eyes, and cover them with a clean pad.
Eye injury due to a foreign object	Encourage the victim to tear without rubbing the eyes, which may dislodge the object; determine location of the object (upper or lower lid), and remove it with a clean tissue; flush the eyes with water.
Toxic gas poisoning	Get the victim to fresh air; check for breathing, and apply artificial respiration or CPR if necessary and you are properly trained.
Heat stroke	Lower victim's body temperature by applying water or rubbing alcohol to the skin or placing the victim in a cold water bathtub.

Source: References 5 and 6.

References

1. American Red Cross, (1979). *Standard First Aid and Personnel Safety*, 2nd Ed., American Red Cross (1979).
2. Office of the Federal Register, 1985, *Code of Federal Regulations*, 29 Parts 1900 to 1910, Office of the Federal Register, Washington, D.C. (1985).
3. Lefevre, M. J., and E. I. Becker, *First Aid Manual for Chemical Accidents*, 1st Ed., Van Nostrand Reinhold, New York (1980).
4. Bernstein, A. B., CPR A Step by Step Guide, Promotional Slide Guide Corp., Brooklyn, New York (1987).
5. Promotional Slide Guide Corp., Emergency Care until Help Arrives, Promotional Slide Guide Corp., Brooklyn, New York (1987).
6. Promotional Slide Guide Corp., First Aid Facts, Promotional Slide Guide Corp., Brooklyn, New York (1987).

Exercises

1. Summarize the OSHA requirements for first aid.
2. What general guidelines must be followed prior to the actual administration of first aid to a victim?
3. Define first-, second-, and third-degree burns, and list first aid measures for each.
4. What is meant by a progressive eye injury? What are the appropriate first aid measures for acid burns to the eye?
5. Why is the use of baking soda solution to wash the eyes effective for acid injuries to the eye and not effective for alkali burns of the eye?
6. List first aid measures for the control of external bleeding.
7. Why should wounds to the head not be cleaned by the first aider?
8. Explain why, in hand injuries that do not involve bone fracture, the hand should be elevated above the

heart level, and why, in hand injuries that involve snake bites or stings, the hand should be kept below the level of the heart.

9. What symptoms would you expect to observe in a victim suffering from a post-accident shock?
10. List the first aid measures that you would use for a coworker who is a victim of heat stroke.
11. Define cardiopulmonary resuscitation (CPR), list the conditions under which it must be used, and explain why someone without proper training should not attempt to use this procedure.

Chapter

13

Computer Systems and Statistical Methods for Occupational Safety and Health Management

Like other professionals, safety and health practitioners have seen tremendous increases in the amount of information and data they must analyze to keep up with rapidly growing and changing regulations. But even as safety and health professionals are experiencing an information explosion, computer technology is providing a means for better management and analysis of safety data. A computerized safety data system can improve the efficiency and effectiveness of a safety program by helping managers to easily search and manage safety data. Today's safety professionals also have experienced a rapid increase in their responsibility for gathering and analyzing safety data, on accident frequency, exposure to hazardous and toxic materials, employee training, and occupational illnesses and injuries. They must manage and analyze this information by using proper analytical and statistical methods, in order to provide necessary information to management and required reports to various local, state, and federal agencies of government.[1–4]

Today about 2,000 data bases are accessible by computer. The largest online data base service, DIALOG, maintains over 200 data bases with more than 75 million information records. Recently a few companies have begun to offer data files on CD-ROM disks, which are accessed from one's own CD-ROM disk drive.[2,4]

In the first part of this chapter, the mechanics of the development of a computerized safety data system (CSDS) are discussed. Attention then is focused on analytical tools and statistical techniques for the interpretation and analysis of safety data.

Development of a Computerized Safety Data System (CSDS)

Before developing a CSDS, the safety professional must conduct research into the organization's overall safety requirements and the type of safety information needed by management. Once this research has been completed, a decision regarding the type and format of output reports must be made. Efforts then must be directed toward developing a viable user's manual and designing an investigation form.[3,5,6]

Output Reports

The format of the output reports must be prepared according to management's needs; then the user's manual and the investigation forms can be developed to support the output reports. Because a CSDS is designed to produce information, the logic built into these reports must be compatible with the type of information that is to be produced. Development of the output reports prior to the user's manual and investigation forms offers a valuable advantage: no unnecessary rechecking of documents is required to ensure that all data needed for the output reports have been included.[1,3]

In developing these reports, one must ensure that they are readily understandable to the user, and present a good summary of accident cases and causes, along with statistical data as needed. Tables 13.1 through 13.3 show some typical output reports.

Table 13.1. Unsafe Condition Report, January 1, 19XX–December 31, 19XX.

Department	Unsafe Conditions Reported	Unsafe Conditions Corrected	Percent Corrected	Unsafe Conditions Existing
1	*A*	*F*	*K*	*P*
2	*B*	*G*	*L*	*Q*
3	*C*	*H*	*M*	*R*
4	*D*	*I*	*N*	*S*
5	*E*	*J*	*O*	*T*
Total	*U*	*V*	*W*	*X*

$K = F/A \times 100$
$P = A - F$

Table 13.2. Hazard Reporting.

		1989				1990			
Dept.	Hazard	1st Quarter	2nd Quarter	3rd Quarter	4th Quarter	1st Quarter	2nd Quarter	3rd Quarter	4th Quarter
A	Physical	21	6	15	9	19	31	17	5
	Health	14	26	11	7	10	9	18	21
B	Physical	11	5	2	14	5	7	9	15
	Health	2	0	1	3	2	1	5	2
C	Physical	1	0	2	4	1	0	1	2
	Health	0	1	2	0	0	1	1	0
Total	Physical	—	—	—	—	—	—	—	—
	Health	—	—	—	—	—	—	—	—

Table 13.3. Lost-Time Injuries and Their Associated Cost to the Company.

	1989		1990	
Dept.	No. of LTI	$ Cost	No. of LTI	$ Cost
A	X	X	X	X
B	X	X	X	X
C	X	X	X	X
D	X	X	X	X
E	X	X	X	X
F	X	X	X	X
Total	X	X	X	X

LTI: Lost time injury.

User's Manual

The user's manual, which is the key to the success of a CSDS, is a set of instructions on how to communicate information to the computer data base.[3,7] Because computers work with numbers, each fact or phrase about a case or accident must be assigned a code number, which is then placed on the input sheet for entry into the computer data base.

The user's manual must be designed so that it will satisfy the requirements of the output reports; thus it is important to decide on the format of the output reports before the development of a user's manual. This manual should be organized to allow for expansion or change without excessive reprogramming of the system.

A user's manual divides words and phrases related to an accident or case into files, each of which maintains specific information on each case. The nature and the organization of files in a CSDS largely depend on the type of industry involved and the desires of the system developer. For example, a chemical company's CSDS might divide the subject files into three general groups: identity, people, and analysis of management. Within each general group, the descriptive items are given special code numbers. For example, under the group code numbers for any fire or explosion can be assigned as follows:

IDENTITY

Code	*Description*
01	Source of ignition for fire or explosion (01–12)
02	Spontaneous ignition upon release
03	Runaway chemical reaction
04	Incompatible storage
05	Smoking around flammables
06	Electrical equipment malfunction
07	Static electricity
08	Vandalism, sabotage
09	Mechanical friction
10	Temperatures above autoignition
11	Oxidation overheating
12	Malfunction of pressure or temperature control devices

The identity group should cover other areas such as off-the-job incident, type of property, pipeline, production, type of property ownership, specific unit involved, specific component within each unit involved, types of explosives and/or flammables, nature of containment, and phase of operation at the time of incident.

Code numbers should be assigned to components of each one of the above-mentioned topics for entering the appropriate information into the computer data base.

The files under the people group can be organized into such topics as nature of injury or illness, part of body injured, severity of injury, and estimated days off work. Within each group additional classification must be carried out, along with the assignment of code numbers. For example, the file on nature of injury can be classified as follows:

Nature of Injury (30–39)	*Code*
Cuts	30
Heat disease	31
Asphyxiation	32
Bone fractures	33

Nature of Injury (30–39)	*Code*
Infection	34
Concussion	35
Chemical burns	36
Thermal burns	37
Crush	38
Abrasion	39

Files under analysis of management can be classified into such topics as human error, unsafe condition, design problems, construction problems, maintenance problems, operating problems, supervision problems, training problems, personnel problems, budget problems, purchasing problems, legal problems, medical problems, and security problems. Each topic mentioned should be further analyzed, with code numbers assigned to its components. For example, the unsafe condition file can be further classified as follows:

Unsafe Condition	*Code*
Defective equipment	50
Improper personal protective equipment	51
Improper placement	52
Temperature extremes	53
Hazardous noise	54
Inadequate illumination	55
Poor housekeeping	56
Lack of training	57
Poor visibility	58
Weather condition	59
Slippery surface	60
Improper labeling	61
Improper storage	62
Temperature or pressure control failure	63
Too heavy for handling	64
Failure of personal protective equipment	65
Bad roads	66

An explanation of the data files should be included in the user's manual to facilitate its use by persons who may not be familiar with the organization of the CSDS.

Designing an Investigation Form

The main objective of an investigation form is to provide the information required by the CSDS to satisfy the needs of its output reports. The investigation form should serve as a starting point for gathering data that will be further analyzed by the computer. It can be designed in a variety of formats and should be tailored to the specific needs of the CSDS. Table 13.4 highlights the components of a typical investigation form.[3]

The information provided by the investigation form can be transferred to the input form with appropriate code numbers for entry into the computer data base. Every effort must be made to ensure that the codes entered into the computer are the right codes for each case. Often the codes entered for a case are acceptable to the computer, but they are the wrong codes.

Table 13.4. Accident Investigation Form.

Employee's name ______________________________

Employee's SS number ______________________________

Accident location: Bldg. __________ Room __________ Other ______________________________

Time accident occurred ______________________________ Date ______________________________

Supervisor's name ______________________________

Type of Case:

First aid __________ Medical treatment ______________________________

Lost time __________ Fatality __________ Property damage ______________________________

Occupational illness ______________________________

Describe employee activity at time of accident:

Describe any tools or machinery involved:

Describe any personal protective equipment used by employee:

Describe condition of work atmosphere at time of accident:

Describe the accident (what happened):

In your opinion, what the probable causes of the accident are:

In your opinion, how this accident could have been prevented:

Changes in process, procedure, or equipment that you would recommend:

How you would classify the apparent causes of this accident:

Human error __________ Equipment __________

Material __________ Personal protective equipment ______________________________

Environmental ______________________________ Other ______________________________

Name and signature of person preparing this form ______________________________

Distribution:

Statistical Methods

Today's safety professional continually faces the problem of safety sampling and statistical analysis in order to draw meaningful conclusions for management decisions. Safety sampling and data analysis also provide a method of measuring worker safety and the overall safety performance of an organization. For example, these techniques, when used properly, enable the safety professional—without watching everybody and everything all the time—to determine the percentage of their time that workers are working safely. This, in turn, makes it possible to identify key points and areas requiring attention, such as the need for training, better procedures, improved material handling, or increased supervisor commitment.[7–12]

Sampling Methods

Safety practitioners are not always content with mere description of a set of data. Most of the time, however, they are interested in learning something about the safety performance of a larger group or an entire organization by conducting studies on a smaller group.[12]

A population can be defined as the set of individuals or units that an investigator wishes to study. A sample is a subset of a population on which studies are conducted and inferences concerning the entire population are made by statistical analysis of the data obtained.

In many instances the design and objectives of an evaluation project determine the most appropriate sampling technique and sample size. Some factors that affect both the sample size and the sampling technique are: accuracy of results desired, variability of units within a given population, population size, and frequency of occurrence.[10,12]

Random Sampling

One of the methods used for obtaining a representative sample from a given population is random sampling.[7,10] The important factor in selecting a random sample is to ensure that every unit within the population has the same probability of being selected. One procedure that is widely used in obtaining a random sample from a given population is to generate a table of random numbers and select every *n*th number (for example, every third number) from the population.

For example, suppose one needs to make inferences concerning the effectiveness of a safety training program by drawing a random sample from the test results (before and after training) of 5,000 students. A random sample for such study can be drawn by selecting, for example, every 100th number from a table of random numbers. It is also possible to determine the size of the sample required to achieve a given degree of accuracy.

Stratified Random Sampling

In situations where there is more than one way of classifying data (such as workers with different job assignments or male vs. female students), it is desirable to represent each category proportionately in the sample under study. In such circumstances the entire population can be divided into different categories, and then a proportionate random sample can be drawn from each category to form a representative random sample.

For example, in a study of safety test results for 3,000 male and 2,000 female students with a sample size of 50, a random sample of 30 of the male students' results and a random sample of 20 of the female students' results can be selected. This method of random sampling is referred to as stratified random sampling.[7,10]

As no two individuals or units within a given population are exactly alike, the random sample selected from a population does not have exactly the same makeup as the population itself. Therefore, some error is introduced into the statistical studies by chance alone every time a sample is selected from a population. This error is called the sampling error. The magnitude of the sampling error decreases as the sample size becomes larger, and it approaches zero as the sample size approaches the size of the entire population.

Descriptive and Inferential Statistics

Statistical methods enable investigators to draw reasonable conclusions about a large population by studying data obtained on a sample of the population. Statistical techniques fall into two broad categories: descriptive and inferential. In descriptive statistics, analysts summarize data, make calculations, and form graphs or tables that can be used to comprehend the data. Inferential statistics, on the other hand, is used to draw conclusions from data, and is based on the mathematical theory of probability.[12]

Frequency Distributions

In any study of statistics the main concern, either in fact or in theory, is with sets of numerical data. As mentioned earlier, the entire set is referred to as the population and the subset under study as the sample. Before analysis of the data, the members of a sample usually are arranged in an array to determine the frequency distribution of the population.

The frequency distribution that occurs most often, both in industry and in nature, is the so-called normal distribution. This distribution can be represented by a bell-shaped curve, with most of the values falling somewhere near the middle of the range of values, as shown in Figure 13.1.

It should be noted that many other distributions have been characterized and used; for example, the data obtained on the distribution of air pollutant particle size follow a log normal distribution. In this case, the logarithms of the actual measurements form a normal distribution.[11]

Mean, Median, Range, and Standard Deviation

In addition to the general information provided by a frequency distribution, there are quantitative characteristics of a population, called parameters.[7,11,12] The first of these, the mean, is a measure of central tendency, a number about which the data tend to concentrate.

The arithmetic mean is simply the sum of all individual values in a sample divided by the number of values. In mathematical form it can be expressed as follows:

$$\overline{Y} = \sum_{i=1}^{n} \frac{Y_i}{n}$$

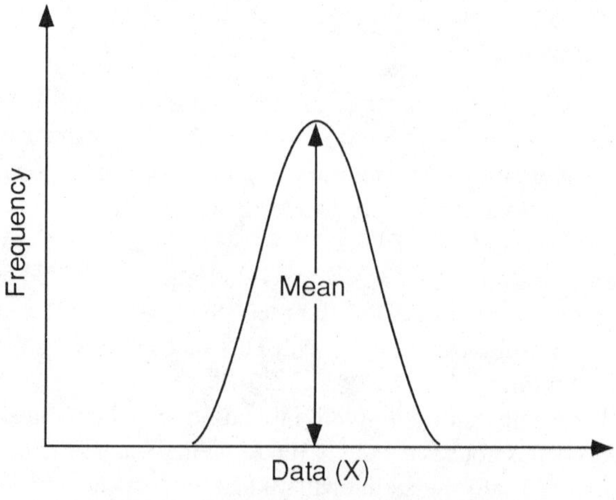

Figure 13.1. Normal distribution curve.

where:

$\overline{Y}$ = Arithmetic mean
Y_i = Individual values
n = Sample size

Another measure of central tendency is the median. The median of a distribution is the middle point, where half of the values fall above and half of the values fall below that point.

In addition to measures of average values, a description of the spread or dispersion of data is important in descriptive statistics. The two parameters that are commonly used to express the dispersion of data are the range and the standard deviation.

The range is simply the interval in which all the values fall; it is the difference between the high and the low values of the data set. Although range is a reasonable indication of dispersion when reliable data exist, its value can be greatly influenced by the existence of abnormally high or low values.

A better method than using the range for expressing data dispersion is to calculate the standard deviation. The difference between any value in the set and the arithmetic mean is called the deviation from the mean. If these deviations are squared, summed, and divided by $n - 1$ (n being the sample size), the resulting value is called the variance of the distribution; and the positive square root of the variance is called the standard deviation. Symbolically:

$$S^2 = \frac{(Y_i - \overline{Y})^2 + (Y_2 - \overline{Y})^2 + \cdots + (Y_n - \overline{Y})^2}{n - 1}$$

$$S^2 = \frac{\sum_{i=1}^{n} (Y_i - \overline{Y})^2}{n - 1}$$

$$S = \sqrt{\frac{\sum_{i=1}^{n} (Y_i - \overline{Y})^2}{n - 1}}$$

where:

S^2 = Variance of the distribution
S = Standard deviation
Y_i = Individual values
$\overline{Y}$ = Arithmetic mean
n = Sample size

Example. The safety department of company X reports the following data for lost time as a result of occupational injuries from 1980 through 1988. Calculate the mean, median, range, and standard deviation.

Year	*Lost time hours*	*Number of injuries*
1980	4,000	100
1981	9,000	180
1982	7,290	162

Year	*Lost time hours*	*Number of injuries*
1983	4,900	140
1984	3,450	115
1985	3,600	90
1986	2,490	83
1987	1,400	50
1988	1,525	61

$$\text{Lost time mean} = \frac{37{,}655}{9} = 4{,}184\ \frac{\text{hours}}{\text{year}}$$

Lost time median = 3,600 hours (half the data fall above and the other half below 3,600 hours)

Range, lost time = 9,000 − 1,400 = 7,600 hours

To calculate the standard deviation, it is convenient to construct the following table:

Y	$\lvert Y - \overline{Y} \rvert$	$(Y - \overline{Y})^2$
4,000	184	33,856
9,000	4,816	23,193,856
7,290	3,106	9,647,236
4,900	716	512,656
3,450	734	538,756
3,600	584	341,056
2,490	1,694	2,869,636
1,400	2,784	7,750,656
1,525	2,659	7,070,281
	Total:	51,957,989

$$\text{Standard deviation} = \sqrt{\frac{51{,}957{,}989}{9 - 1}} = 2{,}548 \text{ hours}$$

Testing Hypotheses

An investigator often has a particular hypothesis in mind about one or more of the parameters of the population being sampled. For example, an analyst might wonder whether he or she can conclude with a given confidence that all workers in a plant work safely 95% of the time based on information obtained from observing the work practices of 50 workers. Under these circumstances a test must be devised whereby the analyst can accept or reject the hypothesis once the sample is taken.[11,12] The determination of this test depends on certain characteristics of the population and will not be discussed here. It should be emphasized, however, that failure to reject a hypothesis does not necessarily mean that it is true. It simply means that, based on available information and with a given confidence limit, one is not in a position to reject the hypothesis. Similarly, failure to accept a hypothesis does not necessarily mean that it is false. For a detailed description of statistical significance tests, the reader is referred to any standard textbook on statistics, such as those listed at the end of this chapter.

Curve Fitting

In addition to characterizing a population, statistical methods are used for prediction purposes. This involves a consideration of the mathematical relationship between two or more variables of interest. This task usually is accomplished by first plotting the data on rectangular coordinates, which provides a visual image of the relationship between variables. In situations where the values of the dependent variable are approximately a linear function of the independent variable, a linear relationship is said to exist.

Correlation Analysis

A measure of the closeness of a set of data to linear correlation can be estimated by calculating the linear correlation coefficient as follows:[11]

$$r = \frac{\sum_{i=1}^{n} (X_i - \overline{X})(Y_i - \overline{Y})}{\sum_{i=1}^{n} (X_i - \overline{X})^2 \cdot \sum_{i=1}^{n} (Y_i - \overline{Y})^2}$$

where:

x and y = Independent and dependent variables, respectively
X_i = Individual values of independent variable
$\overline{X}$ = Arithmetic mean of independent variable
Y_i = Individual values of dependent variable
$\overline{Y}$ = Arithmetic mean of dependent variable
r = Linear correlation coefficient

The value of r can vary between -1 and $+1$. An r value close to 0 indicates that a linear relationship does not exist, and a value of r close to either $+1$ or -1 is an indication that a linear relationship exists.

Linear Regression Analysis

Linear regression analysis can be used to find the equation for the best fit line through a set of data points that indicate a linear relationship between variables. The mathematical equation for a straight line can be expressed as follows:[7,11]

$$y = m\,x + b$$

where (see Figure 13.2):

m = slope of the line
b = intercept with the y axis

The equations for finding the slope and the intercept can be summarized as follows:

$$m = \frac{n \sum_{i=1}^{n} X_i Y_i - \sum_{i=1}^{n} X_i \sum_{i=1}^{n} Y_i}{n \sum_{i=1}^{n} X_i^2 - \left(\sum X_i\right)^2}$$

$$b = \frac{\sum_{i=1}^{n} Y_i - m \sum_{i=1}^{n} X_i}{n}$$

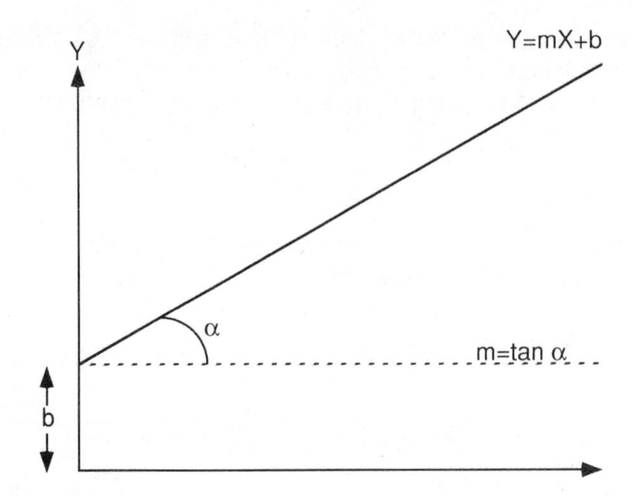

Figure 13.2. Representation of a straight line in rectangular coordinate system.

where n is the number of data points and Y_i and X_i are the dependent and independent variables, respectively. Linear regression can be used for predictive purposes by either interpolation or extrapolation when a linear relationship exist between two variables.

Example. The safety department of a company is interested in finding if a linear relationship exists between supervisor safety training and the number of unsafe conditions found in each supervisor's department. The work areas of five supervisors' were inspected, and the results were summarized in Table 13.5.[7,12]

In order to determine the possibility of a linear relationship we must calculate the linear correlation coefficient, r. To facilitate the calculations, Table 13.6 was constructed.

This analysis indicates a nearly 99% positive correlation in favor of a linear relationship between the hours of supervisor safety training and the number of unsafe conditions found in his or her department.

In order to find the mathematical relationship between dependent and independent variables, we proceed by way of the following tabulation and calculations:

X_i	Y_i	X_iY_i	X_i^2
40	10	400	1,600
5	16	80	25
80	2	160	6,400
20	14	280	400
16	14	224	256
$\Sigma = 161$	$\Sigma = 56$	$\Sigma = 1{,}144$	$\Sigma = 8{,}681$

$$m = \frac{5\,(1{,}144) - (161)\,(56)}{5\,(8{,}681) - (161)^2}$$

$$m = -0.188$$

$$b = \frac{56 - (-0.188)\,(161)}{5}$$

$$b = 17.25$$

Table 13.5. Relationship betwen Supervisor Safety Training and Unsafe Conditions.

Supervisor	Hours of Safety Training	Number of Unsafe Conditions
1	40	10
2	5	16
3	80	2
4	20	14
5	16	14

Table 13.6. Calculation of Linear Correlation Coefficient.

X_i	Y_i	$X_i - \overline{X}$	$Y_i - \overline{Y}$	$(X_i - \overline{X})^2$	$(Y_i - \overline{Y})^2$
40	10	8	−1	64	1
5	16	−27	6	729	36
80	2	48	−9	2,304	81
20	14	−12	3	144	9
16	14	−16	3	256	9
$\overline{X} = 32$	$\overline{Y} = 11$			$\Sigma = 3{,}497$	$\Sigma = 136$

$$r = \frac{(-8) + (-162) + (-432) + (-36) + (-48)}{\sqrt{3497 \times 136}}$$

$$= \frac{-686}{690} \rightarrow$$

$$r = -0.99$$

Therefore, the equation for the best fit straight line using linear regression is:

$$y = -0.188x + 17.25$$

The negative slope indicates that the number of unsafe condition in each department decreases as the amount of supervisor safety training increases.

The Method of Least Squares

One of the methods for fitting a straight line to a set of data points is the method of least squares.[12] The least square line has the property that the sum of the squared vertical distances from it to actual data points is smaller than the similar sum for any other line. Let's assume that X_i and Y_i denote a set of data points obtained from n individual observations or measurements. In the following analysis our notation will be:[7,11,12]

$$x_i = X_i - \overline{X}$$
$$y_i = Y_i - \overline{Y}$$

where $\overline{X}$ and $\overline{Y}$ are the arithmetic mean of n measurements. The equation for the least square line can be expressed as:

$$Y = m(X - \overline{X}) + b = b + mx$$

where:

$$m = \frac{\sum_{i=1}^{n} x_i y_i}{\sum_{i=1}^{n} x_i^2}$$

and $b = \overline{Y}$.

Let's rework the previous example by using the method of least squares. We proceed as follows:

X_i	Y_i	$X_i - \overline{X}$	$Y_i - \overline{Y}$	$x_i y_i$	x_i^2
40	10	8	−1	−8	64
5	16	−27	6	−162	729
80	2	48	−9	−432	2,304
20	14	−12	3	−36	144
16	14	−16	3	−48	256
		$\overline{X} = 32$	$\overline{Y} = 11$	$\Sigma = 3{,}497$	

$$m = \frac{(-8) + (-162) + (-432) + (-36) + (-48)}{3{,}497}$$

$$m = -0.196$$
$$b = 11$$

Therefore, the least square line relating the supervisor's safety training to the number of unsafe conditions in his or her area has a slope of −0.196 and an intercept of 11.

Other Useful Statistics

This discussion is based on reference 7.

OSHA Incidence Rate

The OSHA incidence rate (*IR*) is a measure of recordable occupational injuries. It is expressed as the ratio of the number of recordable injuries to the total man-hours worked, multiplied by 200,000 man-hours (100 employees, 40 hours a week, 50 weeks a year). This relationship can be expressed as follows:[11,12]

$$IR = \frac{I(200{,}000)}{H}$$

where:

IR = OSHA incidence rate
I = Number of injuries
H = Man-hours worked for the number of injuries, I

For example, if an organization experiences 10 recordable occupational injuries in a 100,000 man-hour period, its OSHA incidence rate is:

$$IR = \frac{(10)\,(200{,}000)}{100{,}000} = 20 \text{ lost time injuries per 100 employees annually}$$

Accident Severity Rate

The accident severity rate is defined as the number of lost days due to occupational injuries per specific number of annual worker-hours. The standard for accident severity rate is also 200,000 man-hours worked. The accident severity rate can be calculated from the following expression:

$$S = \frac{(LD)\,(200{,}000)}{H}$$

where:

S = Accident severity rate
LD = Number of lost days due to injuries
H = Total man-hours worked

For example, the accident severity rate for an organization that employs 50 workers with 100 days of lost time in one year can be calculated as follows:

$$S = \frac{(100)(200{,}000)}{(50)(40)(50)} = 200 \text{ lost days per OSHA standard for 100 employees}$$

References

1. ACGIH, *Microcomputer Applications in Occupational Health and Safety*, 1st Ed., Lewis Publishers, Chelsea, Mich. (1987).
2. American Society of Safety Engineers, *Directory of Safety Related Computer Resources*, Vol. II, 1st Ed., ASSE, Des Plaines, Ill. (1987).
3. Ross, C. W., *Computer Systems for Occupational Safety and Health Management*, 1st Ed., Marcel Dekker, New York (1984).
4. American Society of Safety Engineers, *Directory of Safety Related Computer Resources*, Vol. III, 1st Ed., ASSE, Des Plaines, Ill. (1987).
5. Murphy, J. D., *Secrets of Successful Selling*, 1st Ed., Dell Publishing, New York (1965).
6. Leterman, E. G., *The Sale Begins when the Customer Says "No,"* McFadden-Bartel Corp., New York (1963).
7. Revelle, J. B., *Safety Training Methods*, 1st Ed., John Wiley & Sons, New York (1980).
8. Allison, W. W., Safety statistics—A perspective for statistics in accident prevention, Professional Safety, p. 18, American Society of Safety Engineers (ASSE), Des Plaines, Ill. (Oct. 1988).
9. Tyler, W. W., Measuring unsafe behavior, Professional Safety, p. 20, ASSE, Des Plaines, Ill. (Nov. 1986).
10. Slote, L., *Handbook of Occupational Safety and Health*, 1st Ed., John Wiley & Sons, New York (1987).
11. National Institute for Occupational Safety and Health (NIOSH), *The Industrial Environment—Its Evaluation and Control*, NIOSH, Washington, D.C. (1973).
12. Dunn, O. J., and V. A. Clark, *Applied Statistics—Analysis of Variance and Regression*, 1st Ed., John Wiley & Sons, New York (1974).

Exercises

1. Describe the mechanics of creating a computerized safety data system (CSDS).
2. You are in the process of developing a CSDS for your company. How would you classify and code the maintenance problems that might contribute to an accident?
3. You are the safety director in a manufacturing plant that employs 200 male and 80 female workers. You are interested in conducting a study on 20 workers to find out what percentage of the time they work safely. Describe how you would select a stratified random sample, and at what time periods would you carry out your observations.

4. How do descriptive and inferential statistics differ?
5. The following data are obtained on the number of unsafe acts committed within a department of a manufacturing plant in a period of one month. Calculate the range, mean, median, and standard deviation for the data presented.

Day	*No. of unsafe acts*
1	12
3	4
6	8
9	15
12	6
15	3
18	7
21	9
24	8
27	2
30	5

6. The data obtained on noise production by a fin fan as a function of fan speed are summarized below:

Fan speed (rpm)	*Noise (dB)*
200	71
250	71
300	73
500	78
1,000	79
2,000	80

Use regression analysis to determine if a linear relationship exists between the fan rpm and the amount of noise produced. Determine the equation for the best fit straight line by regression analysis and by the method of least squares.

7. A chemical manufacturing facility that has 4,000 employees reports 15 injuries and 100 lost days for 1988. Determine the OSHA incidence rate and the accident severity rate for this facility.

APPENDIXES

Appendix

List of Extremely Hazardous Substances and Their Threshold Planning Quantities

CAS #	Chemical name	Notes	Reportable quantity* (pounds)	Threshold planning quantity (pounds)
75-86-5	Acetone cyanohydrin		10	1,000
1752-30-3	Acetone thiosemicarbazide	e	1	1,000/10,000
107-02-8	Acrolein		1	500
79-06-1	Acrylamide	d,l	5,000	1,000/10,000
107-13-1	Acrylonitrile	d,l	100	10,000
814-68-6	Acrylyl chloride	e,h	1	100
111-69-3	Adiponitrile	e,l	1	1,000
116-06-3	Aldicarb	c	1	100/10,000
309-00-2	Aldrin	d	1	500/10,000
107-18-6	Allyl alcohol		100	1,000
107-11-9	Allylamine	e	1	500
20859-73-8	Aluminum phosphide	b	100	500
54-62-6	Aminopterin	e	1	500/10,000
78-53-5	Amiton	e	1	500
3734-97-2	Amiton oxalate	e	1	100/10,000
7664-41-7	Ammonia	l	100	500
300-62-9	Amphetamine	e	1	1,000
62-53-3	Aniline	d,l	5,000	1,000
88-05-1	Aniline, 2,4,6-trimethyl	e	1	500
7783-70-2	Antimony pentafluoride	e	1	500
1397-94-0	Antimycin A	c,e	1	1,000/10,000
86-88-4	ANTU		100	500/10,000
1303-28-2	Arsenic pentoxide	d	5000	100/10,000
1327-53-3	Arsenous oxide	d,h	5000	100/10,000
7784-34-1	Arsenous trichloride	d	5000	500
7784-42-1	Arsine	e	1	100
2642-71-9	Azinphos-ethyl	e	1	100,10,000
86-50-0	Azinphos-methyl		1	10/10,000
98-87-3	Benzal chloride	d	5,000	500
98-16-8	Benzenamine, 3-(trifluoromethyl)	e	1	500
100-14-1	Benzene, 1-(chloromethyl)-4-nitro	e	1	500/10,000
98-05-5	Benzenearsonic acid	e	1	10/10,000
3615-21-2	Benzimidazole, 4,5-dichloro-2-(trifluoromethyl)	e,g	1	500/10,000
98-07-7	Benzotrichloride	d	1	100
100-44-7	Benzyl chloride	d	100	500
140-29-4	Benzyl cyanide	e,h	1	500
15271-41-7	Bicyclo[2.2.1]heptane-2-carbonitrile, 5-chloro-6-((((methylamino) carbonyl)oxy)imino)-, (1-alpha, 2-beta, 4-alpha, 5-alpha, 6E))	e	1	500/10,000
534-07-6	Bis(chloromethyl) ketone	e	1	10/10,000

CAS #	Chemical name	Notes	Reportable quantity* (pounds)	Threshold planning quantity (pounds)
4044-65-9	Bitoscanate	e	1	500/10,000
10294-34-5	Boron trichloride	e	1	500
7637-07-2	Boron trifluoride	e	1	500
353-42-4	Boron trifluoride compound with methyl ether (1:1)	e	1	1,000
28772-56-7	Bromadiolone	e	1	100/10,000
7726-95-6	Bromine	e,l	1	500
1306-?9-0	Cadmium oxide	e	1	100/10,000
2223-93-0	Cadmium stearate	c,e	1	1,000/10,000
7778-44-1	Calcium arsenate	d	1000	500/10,000
8001-35-2	Camphechlor	d	1	500/10,000
56-25-7	Cantharidin	e	1	100/10,000
51-83-2	Carbachol chloride	e	1	500/10,000
26419-73-8	Carbamic acid, methyl-, *O*-(((2,4-dimethyl-1, 3-dithiolan-2-Yl) methylene)amino)-	e	1	100/10,000
1563-66-2	Carbofuran		10	10/10,000
75-15-0	Carbon disulfide	l	100	10,000
786-19-6	Carbophenothion	e	1	500
57-74-9	Chlordane	d	1	1,000
470-90-6	Chlorfenvinfos	e	1	500
7782-50-5	Chlorine		10	100
24934-91-6	Chlormephos	e	1	500
999-81-5	Chlormequat chloride	e,h	1	100/10,000
79-11-8	Chloroacetic acid	e	1	100,10,000
107-07-3	Chloroethanol	e	1	500
627-11-2	Chloroethyl chloroformate	e	1	1,000
67-66-3	Chloroform	d,l	5,000	10,000
542-88-1	Chloromethyl ether	d,h	1	100
107-30-2	Chloromethyl methyl ether	c,d	1	100
3691-35-8	Chlorophacinone	e	1	100/10,000
1982-47-4	Chloroxuron	e	1	500/10,000
21923-23-9	Chlorthiophos	e,h	1	500
10025-73-7	Chromic chloride	e	1	1,10,000
62207-76-5	Cobalt, ((2,2′-(1,2-ethanediylbis (nitrilomethylidyne)) bis(6-fluorophenolato))(2-)-*N,N′O,O′*)	e	1	100/10,000
10210-68-1	Cobalt carbonyl	e,h	1	10,10,000
64-86-8	Colchicine	e,h	1	10/10,000
56-72-4	Coumaphos		10	100/10,000
5836-29-3	Coumatetralyl	e	1	500/10,000
95-48-7	Cresol, *o*-	d	1,000	1,000/10,000
535-89-7	Crimidine	e	1	100/10,000
4170-30-3	Crotonaldehyde		100	1,000
123-73-9	Crotonaldehyde, (E)-		100	1,000
506-68-3	Cyanogen bromide		1,000	500/10,000
506-78-5	Cyanogen iodide	e	1	1,000/10,000
2636-26-2	Cyanophos	e	1	1,000
675-14-9	Cyanuric fluoride	e	1	100
66-81-9	Cycloheximide	e	1	100/10,000
108-91-8	Cyclohexylamine	e,l	1	10,000
17702-41-9	Decaborane(14)	e	1	500/10,000
8065-48-3	Demeton	e	1	500
919-86-8	Demeton-S-methyl	e	1	500
10311-84-9	Dialifor	e	1	100/10,000
19287-45-7	Diborane	e	1	100
111-44-4	Dichloroethyl ether	d	1	10,000

CAS #	Chemical name	Notes	Reportable quantity* (pounds)	Threshold planning quantity (pounds)
149-74-6	Dichloromethylphenylsilane	e	1	1,000
62-73-7	Dichlorvos		10	1,000
141-66-2	Dicrotophos	e	1	100
1464-53-5	Diepoxybutane	d	1	500
814-49-3	Diethyl chlorophosphate	e,h	1	500
1642-54-2	Diethylcarbamazine citrate	e	1	100/10,000
71-63-6	Digitoxin	c,e	1	100/10,000
2238-07-5	Diglycidyl ether	e	1	1,000
20830-75-5	Digoxin	e,h	1	10/10,000
115-26-4	Dimefox	e	1	500
60-51-5	Dimethoate		10	500/10,000
2524-03-0	Dimethyl phosphorochloridothioate	e	1	500
77-78-1	Dimethyl sulfate	d	1	500
75-18-3	Dimethyl sulfide	e	1	100
75-78-5	Dimethyldichlorosilane	e,h	1	500
57-14-7	Dimethylhydrazine	d	1	1,000
99-98-9	Dimethyl-*p*-phenylenediamine	e	1	10/10,000
644-64-4	Dimetilan	e	1	500/10,000
534-52-1	Dinitrocresol		10	10/10,000
88-85-7	Dinoseb		1,000	100/10,000
1420-07-1	Dinoterb	e	1	500/10,000
78-34-2	Dioxathion	e	1	500
82-66-6	Diphacinone	e	1	10/10,000
152-16-9	Diphosphoramide, octamethyl-		100	100
298-04-4	Disulfoton		1	500
514-73-8	Dithiazanine iodide	e	1	500
541-53-7	Dithiobiuret		100	100/10,000
316-42-7	Emetine, dihydrochloride	e,h	1	1/10,000
115-29-7	Endosulfan		1	10/10,000
2778-04-3	Endothion	e	1	500/10,000
72-20-8	Endrin		1	500/10,000
106-89-8	Epichlorohydrin	d,l	1,000	1,000
2104-64-5	EPN	e	1	100/10,000
50-14-6	Ergocalciferol	c,e	1	1,000/10,000
379-79-3	Ergotamine tartrate	e	1	500/10,000
1622-32-8	Ethanesulfonyl chloride, 2-chloro-	e	1	500
10140-87-1	Ethanol, 1,2-dichloro-, acetate	4	1	1,000
563-12-2	Ethion		10	1,000
13194-48-4	Ethoprophos	e	1	1,000
538-07-8	Ethylbis(2-chloroethyl)amine	e,h	1	500
371-62-0	Ethylene fluorohydrin	c,e,h	1	10
75-21-8	Ethylene oxide	d,l	1	1,000
107-15-3	Ethylenediamine		5,000	10,000
151-56-4	Ethyleneimine	d	1	500
542-90-5	Ethylthiocyanate	e	1	10,000
22224-92-6	Fenamiphos	e	1	10/10,000
122-14-5	Fenithrothion	e	1	500
115-90-2	Fensulfothion	e,h	1	500
4301-50-2	Fluenetil	e	1	100/10,000
7782-41-4	Fluorine	k	10	500
640-19-7	Fluoroacetamide	j	100	100/10,000
144-49-0	Fluoroacetic acid	e	1	10/10,000
359-06-8	Fluoroacetyl chloride	c,e	1	10
51-21-8	Fluorouracil	e	1	500/10,000
944-22-9	Fonofos	e	1	500
50-00-0	Formaldehyde	d,l	1,000	500

CAS #	Chemical name	Notes	Reportable quantity* (pounds)	Threshold planning quantity (pounds)
107-16-4	Formaldehyde cyanohydrin	e,h	1	1,000
23422-53-9	Formetanate hydrochloride	e,h	1	500/10,000
2540-82-1	Formothion	e	1	100
17702-57-7	Formparanate	e	1	100/10,000
21548-32-3	Fosthietan	e	1	500
3878-19-1	Fuberidazole	e	1	100/10,000
110-00-9	Furan		100	500
13450-90-3	Gallium trichloride	e	1	500/10,000
77-47-4	Hexachlorocyclopentadiene	d,h	1	100
4835-11-4	Hexamethylenediamine, *N,N'*-dibuthyl-	e	1	500
302-01-2	Hydrazine	d	1	1,000
74-90-8	Hydrocyanic acid		10	100
7647-01-0	Hydrogen chloride (gas only)	e,l	1	500
7664-39-3	Hydrogen fluoride		100	100
7722-84-1	Hydrogen peroxide (conc. > 52%)	e,l	1	1,000
7783-07-5	Hydrogen selenide	e	1	10
7783-06-4	Hydrogen sulfide	l	100	500
123-31-9	Hydroquinone	e	1	500/10,000
13463-40-6	Iron, pentacarbonyl-	e	1	100
297-78-9	Isobenzan	e	1	100/10,000
78-82-0	Isobutyronitrile	e,h	1	1,000
102-36-3	Isocyanic acid, 3,4-dichlorophenyl ester	e	1	500/10,000
465-73-6	Isodrin		1	100/10,000
55-91-4	Isofluorphate	c	100	100
4098-71-9	Isophorone diisocyanate	b, e	1	100
108-23-6	Isopropyl chloroformate	e	1	1,000
625-55-8	Isopropyl formate	e	1	500
119-38-0	Isopropylmethylpyrazolyl dimethylcarbamate	e	1	500
78-97-7	Lactonitrile	e	1	1,000
21609-90-5	Leptophos	e	1	500/10,000
541-25-3	Lewisite	c,e,h	1	10
58-89-9	Lindane	d	1	1,000/10,000
7580-67-8	Lithium hydride	b,e	1	100
109-77-3	Malononitrile		1,000	500/10,000
12108-13-3	Manganese, tricarbonyl methylcyclopentadienyl	e,h	1	100
51-75-2	Mechlorethamine	c,e	1	10
950-10-7	Mephosfolan	e	1	500
1600-27-7	Mercuric acetate	e	1	500/10,000
7487-94-7	Mercuric chloride	e	1	500/10,000
21908-53-2	Mercuric oxide	e	1	500/10,000
10476-95-6	Methacrolein diacetate	e	1	1,000
760-93-0	Methacrylic anhydride	e	1	500
126-98-7	Methacrylonitrile	h	1	500
920-46-7	Methacryloyl chloride	e	1	100
30674-80-7	Methacryloyloxyethyl isocyanate	e,h	1	100
10265-92-6	Methamidophos	e	1	100/10,000
558-25-8	Methanesulfonyl fluoride	e	1	1,000
950-37-8	Methidathion	e	1	500/10,000
2032-65-7	Methiocarb		10	500/10,000
16752-77-5	Methomyl	h	100	500/10,000
151-38-2	Methoxyethylmercuric acetate	e	1	500/10,000
80-63-7	Methyl 2-chloroacrylate	e	1	500
74-83-9	Methyl bromide	l	1,000	1,000
79-22-1	Methyl chloroformate	d,h	1,000	500
624-92-0	Methyl disulfide	e	1	100
60-34-4	Methyl hydrazine		10	500

CAS #	Chemical name	Notes	Reportable quantity* (pounds)	Threshold planning quantity (pounds)
624-83-9	Methyl isocyanate	f	1	500
556-61-6	Methyl isothiocyanate	b,e	1	500
74-93-1	Methyl mercaptan		100	500
3735-23-7	Methyl phenkapton	e	1	500
676-97-1	Methyl phosphonic dichloride	b,e	1	100
556-64-9	Methyl thiocyanate	e	1	10,000
78-94-4	Methyl vinyl ketone	e	1	10
502-39-6	Methylmercuric dicyanamide	e	1	500/10,000
75-79-6	Methyltrichlorosilane	e,h	1	500
1129-41-5	Metolcarb	e	1	100/10,000
7786-34-7	Mevinphos		10	500
315-18-4	Mexacarbate		1,000	500/10,000
50-07-7	Mitomycin C	d	1	500/10,000
6923-22-4	Monocrotophos	e	1	10/10,000
2763-96-4	Muscimol	e,h	1,000	10,000
505-60-2	Mustard gas	e,h	1	500
13463-39-3	Nickel carbonyl	d	1	1
54-11-5	Nicotine	c	100	100
65-30-5	Nicotine sulfate	e	1	100/10,000
7697-37-2	Nitric acid		1,000	1,000
10102-43-9	Nitric oxide	c	10	100
98-95-3	Nitrobenzene	l	1,000	10,000
1122-60-7	Nitrocyclohexane	e	1	500
10102-44-0	Nitrogen dioxide		10	100
62-75-9	Nitrosodimethylamine	d,h	1	1,000
991-42-4	Norbormide	e	1	100/10,000
0	Organorhodium complex (PMN-82-147)	e	1	10/10,000
630-60-4	Ouabain	c,e	1	100/10,000
23135-22-0	Oxamyl	e	1	100/10,000
78-71-7	Oxetane, 3,3-bis(chloromethyl)-	e	1	500
2497-07-6	Oxydisulfoton	e,h	1	500
10028-15-6	Ozone	e	1	100
1910-42-5	Paraquat	e	1	10/10,000
2074-50-2	Paraquat methosulfate	e	1	10/10,000
56-38-2	Parathion	c,d	1	100
298-00-0	Parathion–methyl	c	100	100/10,000
12002-03-8	Paris green	d	100	500/10,000
19624-22-7	Pentaborane	e	1	500
2570-26-5	Pentadecylamine	e	1	100/10,000
79-21-0	Peracetic acid	e	1	500
594-42-3	Perchloromethylmercaptan		100	500
108-95-2	Phenol		1,000	500/10,000
97-18-7	Phenol, 2,2′-thiobis(4,6-dichloro)-	e	1	100/10,000
4418-66-0	Phenol, 2,2′-thiobis(4-chloro-6-methyl)-	e	1	100/10,000
64-00-6	Phenol, 3-(1-methylethyl)-, methylcarbamate	e	1	500/10,000
58-36-6	Phenoxarsine, 10,10′-oxydi-	e	1	500/10,000
696-28-6	Phenyl dichloroarsine	d,h	1	500
59-88-1	Phenylhydrazine hydrochloride	e	1	1,000/10,000
62-38-4	Phenylmercury acetate		100	500/10,000
2097-19-0	Phenylsilatrane	e,h	1	100/10,000
103-85-5	Phenylthiourea		100	100/10,000
298-02-2	Phorate		10	10
4104-14-7	Phosacetim	e	1	100/10,000
947-02-4	Phosfolan	e	1	100/10,000
75-44-5	Phosgene	l	10	10
732-11-6	Phosmet	e	1	10/10,000

CAS #	Chemical name	Notes	Reportable quantity* (pounds)	Threshold planning quantity (pounds)
13171-21-6	Phosphamidon	e	1	100
7803-51-2	Phosphine		100	500
2703-13-1	Phosphonothioic acid, methyl-, *O*-ethyl *O*-(4-(methylthio)phenyl)ester	e	1	500
50782-69-9	Phosphonothioic acid, methyl-, *S*-(2-(bis(1-methylethyl)amino)ethyl)	e	1	100
2665-30-7	Phosphonothioic acid, methyl-, *O*-(4-nitrophenyl) *O*-phenyl ester	e	1	500
3254-63-5	Phosphoric acid, dimethyl 4-(methylthio) phenyl ester	e	1	500
2587-90-8	Phosphorothioic acid, *O,O*-dimethyl-*S*-(2-methylthio) ethyl ester	c,e,g	1	500
7723-14-0	Phosphorus	b,h	1	100
10025-87-3	Phosphorus oxychloride	d	1,000	500
10026-13-8	Phosphorus pentachloride	b,e	1	500
1314-56-3	Phosphorus pentoxide	b,e	1	10
7719-12-2	Phosphorus trichloride		1,000	1,000
57-47-6	Physostigmine	e	1	100/10,000
57-64-7	Physotigmine, salicylate (1 : 1)	e	1	100/10,000
124-87-8	Picrotoxin	e	1	500/10,000
110-89-4	Piperidine	e	1	1,000
5281-13-0	Piprotal	e	1	100/10,000
23505-41-1	Pirimifos-ethyl	e	1	1,000
10124-50-2	Potassium arsenite	d	1,000	500/10,000
151-50-8	Potassium cyanide	b	10	100
506-61-6	Potassium silver cyanide	b	1	500
2631-37-0	Promecarb	e,h	1	500/10,000
106-96-7	Propargyl bromide	e	1	10
57-57-8	Propiolactone, beta-	e	1	500
107-12-0	Propionitrile		10	500
542-76-7	Propionitrile, 3-chloro-		1,000	1,000
70-69-9	Propiophenone, 4-amino-	e,g	1	100/10,000
109-61-5	Propyl chloroformate	e	1	500
75-56-9	Propylene oxide	l	100	10,000
75-55-8	Propyleneimine	d	1	10,000
2275-18-5	Prothoate	e	1	100
129-00-0	Pyrene	c	5,000	1,000/10,000
140-76-1	Pyridine, 2-methyl-5-vinyl-	e	1	500
504-24-5	Pyrdine, 4-amino-	h	1,000	500/10,000
1124-33-0	Pyridine, 4-nitro-, 1-oxide	e	1	500/10,000
53558-25-1	Pyriminil	e,h	1	100/10,000
14167-18-1	Salcomine	e	1	500/10,000
107-44-8	Sarin	e,h	1	10
7783-00-8	Selenious acid		10	1,000/10,000
7791-23-3	Selenium oxychloride	e	1	500
563-41-7	Semicarbazide hydrochloride	e	1	1,000/10,000
3037-72-7	Silane, (4-aminobutyl)diethoxymethyl-	e	1	1,000
7631-89-2	Sodium arsenate	d	1,000	1,000/10,000
7784-46-5	Sodium arsenite	d	1,000	500/10,000
26628-22-8	Sodium azide (Na(N3))	b	1,000	500
124-65-2	Sodium cacodylate	e	1	100/10,000
143-33-9	Sodium cyanide (Na(CN))	b	10	100
62-74-8	Sodium fluoroacetate		10	10/10,000
131-52-2	Sodium pentachlorophenate	e	1	100/10,000
13410-01-0	Sodium selenate	e	1	100/10,000

CAS #	Chemical name	Notes	Reportable quantity* (pounds)	Threshold planning quantity (pounds)
10102-18-8	Sodium selenite	h	100	100/10,000
10102-20-2	Sodium tellurite	e	1	500/10,000
900-95-8	Stannane, acetoxytripheynl-	e,g	1	500/10,000
57-24-9	Strychnine	c	10	100/10,000
60-41-3	Strychnine, sulfate	e	1	100/10,000
3689-24-5	Sulfotep		100	500
3569-57-1	Sulfoxide, 3-chloropropyl octyl	e	1	500
7446-09-5	Sulfur dioxide	e,l	1	500
7783-60-0	Sulfur tetrafluoride	e	1	100
7446-11-9	Sulfur trioxide	b,e	1	100
7664-93-9	Sulfuric acid		1,000	1,000
77-81-6	Tabun	c,e,h	1	10
13494-80-9	Tellurium	e	1	500/10,000
7783-80-4	Tellurium hexafluoride	e,k	1	100
107-49-3	TEPP		10	100
13071-79-9	Terbufos	e,h	1	100
78-00-2	Tetraethyllead	c,d	10	100
597-64-8	Tetraethyltin	c,e	1	100
75-74-1	Tetramethyllead	c,e,l	1	100
509-14-8	Tetranitromethane		10	500
10031-59-1	Thallium sulfate	h	100	100/10,000
6533-73-9	Thallous carbonate	c,h	100	100/10,000
7791-12-0	Thallous chloride	c,h	100	100/10,000
2757-18-8	Thallous malonate	c,e,h	1	100/10,000
7446-18-6	Thallous sulfate		100	100/10,000
2231-57-4	Thiocarbazide	e	1	1,000/10,000
39196-18-4	Thiofanox		100	100/10,000
297-97-2	Thionazin		100	500
108-98-5	Thiophenol		100	500
79-19-6	Thiosemicarbazide		100	100/10,000
5344-82-1	Thiourea, (2-chlorophenyl)-		100	100/10,000
614-78-8	Thiourea, (2-methylphenyl)-	e	1	500/10,000
7550-45-0	Titanium tetrachloride	e	1	100
584-84-9	Toluene 2,4-diisocyanate		100	500
91-08-7	Toluene 2,6-diisocyanate		100	100
110-57-6	Trans-1,4-dichlorobutene	e	1	500
1031-47-6	Triamiphos	e	1	500/10,000
24017-47-8	Triazofos	e	1	500
76-02-8	Trichloroacetyl chloride	e	1	500
115-21-9	Trichloroethylsilane	e,h	1	500
327-98-0	Trichloronate	e,k	1	500
98-13-5	Trichlorophenylsilane	e,h	1	500
1558-25-4	Trichloro(chloromethyl)silane	e	1	100
27137-85-5	Trichloro(dichlorophenyl)silane	e	1	500
998-30-1	Triethoxysilane	e	1	500
75-77-4	Trimethylchlorosilane	e	1	1,000
824-11-3	Trimethylolpropane phosphite	e,h	1	100/10,000
1066-45-1	Trimethyltin chloride	e	1	500/10,000
639-58-7	Triphenyltin chloride	e	1	500/10,000
555-77-1	Tris(2-chloroethyl)amine	e,h	1	100
2001-95-8	Valinomycin	c,e	1	1,000/10,000
1314-62-1	Vanadium pentoxide		1,000	100/10,000
108-05-4	Vinyl acetate monomer	d,l	5,000	1,000
81-81-2	Warfarin		100	500/10,000
129-06-6	Warfarin sodium	e,h	1	100/10,000

CAS #	Chemical name	Notes	Reportable quantity* (pounds)	Threshold planning quantity (pounds)
28347-13-9	Xylylene dichloride	e	1	100/10,000
58270-08-9	Zinc, dichloro(4,4-dimethyl-5((((methylamino) carbonyl)oxy)imino)pentanenitrile)-, (T-4)-	e	1	100/10,000
1314-84-7	Zinc phosphide	b	100	500

*Only the statutory or final reportable quantity (RQ) is shown. For more information, see 40CFR Table 302.4.

Notes: b This material is a reactive solid. The threshold planning quantity (TPQ) does not default to 10,000 pounds for nonpowder, nonmolten, nonsolution form.

c The calculated TPQ changed after technical review as described in the technical support document.

d Indicates that the RQ is subject to change when the assessment of potential carcinogeneity and/or other toxicity is completed.

e Statutory reportable quantity for purposes of notification under SARA sect. 304(a)(2).

f The statutory one-pound reportable quantity for methyl isocyanate may be adjusted in a future rulemaking action.

g New chemicals added that were not part of the original list of 402 substances.

h Revised TPQ based on new or reevaluated toxicity data.

j TPQ is revised to its calculated value and does not change due to technical review as in proposed rule.

k The TPQ was revised after proposal due to calculation error.

l Chemicals on the original list that do not meet the toxicity criteria but because of their production volume and recognized toxicity are considered chemicals of concern (''Other chemicals'').

Source: Federal Register, Vol. 52, No. 77/Wed; April 22, 1987/Rules and Regulations.

Appendix

B

Known Carcinogens

4-Aminobiphenyl
Analgesic mixtures containing phenacetin*
Arsenic and certain arsenic compounds
Asbestos
Azathioprine
Benzene
Benzidine
N,*N*-bis(2-chloroethyl)2-naphthylamine (chlornaphthazine)
Bis(chloromethyl)ether and technical grade chloromethyl methyl ether
1,4-Butanediol dimethylsulfonate (myleran)
Certain combined chemotherapy for lymphomas
Chlorambucil
Chromium and certain chromium compounds
Coke oven emissions
Conjugated estrogens
Cyclophosphamide
Diethylstilbestrol
Hematite underground mining
Isopropyl alcohol manufacturing (strong-acid process)
Manufacture of auramine
Melphalan
Methoxsalen with ultraviolet A therapy (PUVA)
Mustard gas
2-Naphthylamine
Nickel refining
Rubber industry (certain occupations)
Soots, tars, and mineral oils
Thorium dioxide
Vinyl chloride

Source: Fourth Annual Report on Carcinogens, National Toxicology Program, U.S. Department of Health and Human Services (1985).

Appendix

C

Suspected Carcinogens

2-Acetylaminofluorene
Acrylonitrile
Adriamycin
Aflatoxins
2-Aminoanthraquinone
1-Amino-2-methylanthraquinone
Amitrole
o-Anisidine and *o*-anisidine hydrochloride
Aramite®
Benz(a)anthracene
Benzo(b)fluoranthene
Benzo(a)pyrene
Benzotrichloride
Beryllium and certain beryllium compounds
Bischloroethyl nitrosourea
Cadmium and certain cadmium compounds
Carbon tetrachloride
1-(2-Chloroethyl)-3-cyclohexyl-1-nitrosourea (CCNU)
Chloroform
4-Chloro-*o*-phenylenediamine
p-Cresidine
Cupferron
Cycasin
Dacarbazine
DDT
2,4-Diaminoanisole sulfate
2,4-Diaminotoluene
Dibenz(a,h)acridine
Dibenz(a,j)acridine
Dibenz(a,h)anthracene
7H-Dibenzo(c,g)carbazole
Dibenzo(a,h)pyrene
Dibenzo(a,i)pyrene
1,2-Dibromo-3-chloropropane
1,2-Dibromoethane (EDB)
3,3′-Dichlorobenzidine
1,2-Dichloroethane
Diepoxybutane
Di(2-ethylhexyl)phthalate
Diethyl sulfate
3,3′-Dimethoxybenzidine
4-Dimethylaminoazobenzene
3,3′-Dimethylbenzidine
Dimethylcarbamoyl chloride
1,1-Dimethylhydrazine
Dimethyl sulfate
1,4-Dioxane
Direct Black 38
Direct Blue 6
Epichlorohydrin
Estrogens (not conjugated): 1. Estradiol 17β
Estrogens (not conjugated): 2. Estrone
Estrogens (not conjugated): 3. Ethinylestradiol
Estrogens (not conjugated): 4. Mestranol
Ethylene oxide
Ethylene thiourea
Formaldehyde (gas)
Hexachlorobenzene
Hexamethylphosphoramide
Hydrazine and hydrazine sulfate
Hydrazobenzene
Indeno (1,2,3-cd)pyrene
Iron dextran complex
Kepone® (Chlordecone)
Lead acetate and lead phosphate
Lindane and other hexachlorocyclohexane isomers
2-Methylaziridine (propyleneimine)
4,4′-Methylenebis(2-chloroaniline) (MBOCA)
4,4′-Methylenebis(*N,N*-dimethyl)benzenamine
4,4′-Methylenedianiline and its dihydrochloride
Methyl iodide
Metronidazole
Michler's ketone
Mirex
Nickel and certain nickel compounds
Nitrilotriacetic acid
5-Nitro-*o*-anisidine
Nitrofen
Nitrogen mustard
2-Nitropropane
N-Nitrosodi-*n*-butylamine
N-Nitrosodiethanolamine
N-Nitrosodiethylamine
N-Nitrosodimethylamine
p-Nitrosodiphenylamine
N-Nitrosodi-*n*-propylamine
N-Nitroso-*N*-ethylurea
N-Nitroso-*N*-methylurea
N-Nitrosomethylvinylamine
N-Nitrosomorpholine
N-Nitrosodiethylamine
N-Nitrosopiperidine
N-Nitrosopyrrolidine
N-Nitrososarcosine
Norethisterone
Oxymetholone
Phenacetin
Phenazopyridine hydrochloride
Phenytoin and sodium salt or phenytoin
Polybrominated biphenyls

Polychlorinated biphenyls
Procarbazine and procarbazine hydrochloride
Progesterone
1,3-Propane sultone
Prophylthiouracil
β-Propiolactone
Reserpine
Saccharin
Safrole
Selenium sulfide
Streptozotocin
Sulfallate
2,3,7,8-Tetrachlorodibenzo-*p*-dioxin (TCDD)
Thioacetamide
Thiourea
Toluene diisocyanate
o-Toluidine and *o*-toluidine hydrochloride
Toxaphene
2,4,6-Trichlorophenol
Tris(1-aziridinyl)phosphine sulfide
Tris(2,3-dibromopropyl)phosphate
Urethane

Source: Fourth Annual Report on Carcinogens, National Toxicology Program, U.S. Department of Health and Human Services (1985).

Appendix

D

OSHA Permissible Exposure Limits

Chemical name	Permissible exposure limit (8-hr TWA)	Target organ
Acetaldehyde	200 ppm (360 mg/m^3)	RRespiratorysystem, skin, kidneys
Acetic acid	10 ppm (25 mg/m^3)	Respiratory system, skin, eyes, teeth
Acetic anhydride	5 ppm (20 mg/m^3)	Respiratory system, eyes, skin
Acetone	1,000 ppm (2,400 mg/m^3)	Respiratory system, skin
Acetonitrile	40 ppm (70 mg/m^3)	Kidneys, liver, CVS, CNS, lungs, skin, eyes
Acetylene tetrabromide	1 ppm (14 mg/m^3)	Eyes, upper respiratory system, liver
Acrolein	0.1 ppm (0.25 mg/m^3)	Heart, eyes, skin, respiratory system
Acrylamide	0.3 mg/m^3	CNS, PNS, skin, eyes
Acrylonitrile	2 ppm; 10 ppm ceiling, 15 min	CVS, liver, kidneys, CNS, skin, brain tumor, lung and bowel cancer
Aldrin	0.25 mg/m^3	Cancer, CNS, liver, kidneys, skin
Allyl alcohol	2 ppm (5 mg/m^3)	Eyes, skin, respiratory system
Allyl chloride	1 ppm (3 mg/m^3)	Respiratory system, skin, eyes, liver, kidneys
Allyl glycidyl ether	10 ppm/ceiling (45 mg/m^3)	Respiratory system, skin
2-Aminopyridine	0.5 ppm (2 mg/m^3)	CNS, respiratory system
Ammonia	50 ppm (35 mg/m^3)	Respiratory system, eyes
Ammonium sulfamate	15 mg/m^3	None known
n-Amyl acetate	100 ppm (525 mg/m^3)	Eyes, skin, respiratory system
sec-Amyl acetate	125 ppm (650 mg/m^3)	Respiratory system, eyes, skin
Aniline	5 ppm (19 mg/m^3)	Blood, CVS, liver, kidneys
Anisidine (*o*-, *p*-isomers)	0.5 mg/m^3	Blood, kidneys, liver, CVS
Antimony and compounds (as Sb)	0.5 mg/m^3	Respiratory system, CVS, skin, eyes
ANTU	0.3 mg/m^3	Respiratory system
Arsenic and compounds (as As)	10 μg/m^3	Liver, kidneys, skin, lungs, lymphatic system
Arsine	0.05 ppm (0.2 mg/m^3)	Blood, kidneys, liver
Asbestos	0.2 fiber/cc	Lungs
Azinphos-methyl	0.2 mg/m^3	Respiratory system, CNS, CVS, blood cholinesterase
Barium (soluble compounds as Ba)	0.5 mg/m^3	Heart, CNS, skin, respiratory system, eyes
Benzene	10 ppm; 50 ppm ceiling, 10 min	Blood, CNS, skin, bone marrow, eyes, respiratory system
Benzoyl peroxide	5 mg/m^3	Skin, respiratory system, eyes
Benzyl chloride	1 ppm (5 mg/m^3)	Eyes, respiratory system, skin

Chemical name	Permissible exposure limit (8-hr TWA)	Target organ
Beryllium and compounds (as Be)	20 μg/m^3; 5.0 μg/m^3 ceiling; 25 μg/m^3 (30-min ceiling)	Lung, skin, eyes, mucous membrances
Boron oxide	15 mg/m^3	Skin, eyes
Boron trifluoride	1 ppm ceiling (3 mg/m^3)	Respiratory system, kidneys, eyes, skin
Bromine	0.1 ppm (0.7 mg/m^3)	Respiratory system, eyes, CNS
Bromoform	0.5 ppm (5 mg/m^3)	Skin, liver, kidneys, respiratory system, CNS
Butadiene	1,000 ppm (2,200 mg/m^3)	Eyes, respiratory system, CNS
2-Butanone	200 ppm (590 mg/m^3)	CNS, lungs
2-Butoxy ethanol	50 ppm (240 mg/m^3)	Liver, kidneys, lymphoid system, skin, blood, eyes, respiratory system
Butyl acetate	150 ppm (710 mg/m^3)	Eyes, skin, respiratory system
sec-Butyl acetate	200 ppm (950 mg/m^3)	Eyes, skin, respiratory system
tert-Butyl acetate	200 ppm (950 mg/m^3)	Respiratory system, eyes, skin
Butyl alcohol	100 ppm (300 mg/m^3)	Skin, eyes, respiratory system
sec-Butyl alcohol	150 ppm (450 mg/m^3)	Eyes, skin, CNS
tert-Butyl alcohol	100 ppm (300 mg/m^3)	Eyes, skin
Butylamine	5 ppm ceiling (15 mg/m^3)	Respiratory system, skin, eyes
tert-Butyl chromate (as CrO_3)	0.1 mg/m^3 ceiling	Respiratory system, skin, eyes, CNS
n-Butyl glycidyl	50 ppm (270 mg/m^3)	Eyes, skin, respiratory system, CNS
Butyl mercaptan	10 ppm (35 mg/m^3)	Respiratory system; in animals: CNS, liver, kidneys
p-tert-Butyltoluene	10 ppm (60 mg/m^3)	CVS, CNS, skin, bone marrow, eyes, upper respiratory system
Cadmium dust (as Cd)	0.2 mg/m^3; 0.6 mg/m^3 ceiling	Respiratory system, kidneys, prostate, blood
Cadmium fume (as Cd)	0.1 mg/m^3; 0.3 mg/m^3 ceiling	Respiratory system, kidneys, blood
Calcium arsenate	10 μg/m^3	Eyes, respiratory system, liver, skin, lymphatics, CNS
Calcium oxide	5 mg/m^3	Respiratory system, skin, eyes
Camphor	2 ppm (12 mg/m^3)	CNS, eyes, skin, respiratory system
Carbaryl (Sevin)	5 mg/m^3	Respiratory system, CNS, CVS, skin
Carbon black	3.5 mg/m^3	None known
Carbon dioxide	5,000 ppm	Lungs, skin, CVS
Carbon disulfide	20 ppm, 30 ppm ceiling, 100 ppm; 30-min ceiling	CNS, PNS, CVS, eyes, kidneys, skin
Carbon monoxide	50 ppm (55 mg/m^3)	CVS, lungs, blood, CNS
Carbon tetrachloride	10 ppm; 25 ppm ceiling; 200 ppm, 5-min/4-hr peak	CNS, eyes, lungs, liver, kidneys, skin
Chlordane	0.5 mg/m^3	CNS, eyes, lungs, liver, kidneys, skin
Chlorinated camphene	0.5 mg/m^3	CNS, skin
Chlorinated diphenyl oxide	0.5 mg/m^3	Skin, liver
Chlorine	1 ppm ceiling (3 mg/m^3)	Respiratory system
Chlorine dioxide	0.1 ppm (0.3 mg/m^3)	Respiratory system, eyes
Chlorine trifluoride	0.1 ppm ceiling (0.4 mg/m^3)	Skin, eyes
Chloro-acetaldehyde	1 ppm ceiling (3 mg/m^3)	Eyes, skin, respiratory system
alpha-Chloro-acetophenone	0.05 ppm (0.3 mg/m^3)	Eyes, skin, respiratory system
Chlorobenzene	75 ppm (350 mg/m^3)	Respiratory system, eyes, skin, CNS, liver

Chemical name	Permissible exposure limit (8-hr TWA)	Target organ
o-Chloro-benzylidene malonitrile	0.05 ppm (0.4 mg/m^3)	Respiratory system, skin, eyes
Chlorobromomethane	200 ppm (1,050 mg/m^3)	Skin, liver, kidneys, respiratory system, CNS
Chlorodiphenyl (42% chlorine)	1 mg/m^3	Skin, eyes, liver
Chlorodipheynl (54% chlorine)	0.5 mg/m^3	Skin, eyes, liver
Chloroform	50 ppm (240 mg/m^3)	Liver, kidneys, heart, eyes, skin
1-Chloro-1-nitropropane	20 ppm (100 mg/m^3)	In animals: respiratory system, liver, kidneys, CVS
Chloropicrin	0.1 ppm (0.7 mg/m^3)	Respiratory system, skin, eyes
Chloroprene	25 ppm (90 mg/m^3)	Respiratory system, skin, eyes
Chromic acid and chromates (as CrO_3)	0.1 mg/m^3 ceiling	Blood, respiratory system, liver, kidneys, eyes, skin
Chromium, metal and insoluble salts (as Cr)	1 mg/m^3	Respiratory system
Chromium, soluble chromic, chromous salts (as Cr)	0.5 mg/m^3 (NIOSH)	Skin
Coal tar pitch volatiles	0.2 mg/m^3 (benzene-soluble fraction)	Respiratory system, bladder, kidneys, skin
Cobalt metal, fume, and dust (as Co)	0.1 mg/m^3	Respiratory system, skin
Copper dust and mist (as Cu)	1 mg/m^3	Respiratory system, skin, liver, increased risk with Wilson's disease, kidneys
Copper fume (as Cu)	0.1 mg/m^3	Respiratory system, skin, eyes, increased risk with Wilson's disease
Crag herbicide	15 mg/m^3	None known
Cresol	5 ppm (22 mg/m^3)	CNS, respiratory system, liver, kidneys, skin, eyes
Crotonaldehyde	2 ppm (6 mg/m^3)	Respiratory system, eyes, skin
Cumene	50 ppm (245 mg/m^3)	Eyes, upper respiratory system, skin, CNS
Cyanides (as CN)	5 mg/m^3	CVS, CNS, liver, kidneys, skin
Cyclohexane	300 ppm (1,050 mg/m^3)	Eyes, respiratory system, skin, CNS
Cyclohexanol	500 ppm (200 mg/m^3)	Eyes, respiratory system, skin
Cyclohexanone	50 ppm (200 mg/m^3)	Respiratory system, eyes, skin, CNS
Cyclohexene	300 ppm (1,015 mg/m^3)	Skin, eyes, respiratory system
Cyclopentadiene	75 ppm (200 mg/m^3)	Eyes, respiratory system
2,4-D	10 mg/m^3	Skin, CNS
DDT	1 mg/m^3	CNS, kidneys, liver, skin, PNS
Decaborane	0.05 ppm (0.3 mg/m^3)	CNS
Demeton	0.1 mg/m^3	Respiratory system, CVS, CNS, skin, eyes, blood cholinesterase
Diacetone alcohol	50 ppm (240 mg/m^3)	Eyes, skin, respiratory system
Diazomethane	0.2 ppm (0.4 mg/m^3)	Respiratory system, eyes, skin
Diborane	0.1 ppm (0.1 mg/m^3)	Respiratory system, CNS
Dibromochloropropane	1 ppb	CNS, skin, liver, kidney, spleen, reproductive system, digestive system
Dibutyl phosphate	1 ppm (5 mg/m^3)	Respiratory system, skin
Dibutylphthalate	5 mg/m^3	Respiratory system, GI tract
o-Dichlorobenzenc	50 ppm ceiling (300 mg/m^3)	Liver, kidneys, skin, eyes
p-Dichlorobenzene	75 ppm (450 mg/m^3)	Liver, respiratory system, eyes, kidneys, skin

Chemical name	Permissible exposure limit (8-hr TWA)	Target organ
Dichlorodifluoromethane	1,000 ppm (4,950 mg/m^3)	CVS, PNS
1,3-Dichloro-5,5-dimethyl hydantoin	0.2 mg/m^3	Respiratory system, eyes
1,1-Dichloroethane	100 ppm (400 mg/m^3)	Skin, liver, kidneys
1,2-Dichloroethylene	200 ppm (790 mg/m^3)	Respiratory system, eyes, CNS
Dichloroethyl ether	15 ppm ceiling (90 mg/m^3)	Respiratory system, skin, eyes
Dichloromonofluoromethane	1,000 ppm (4,200 mg/m^3)	Respiratory system, CVS
1,1-Dichloro-1-nitroethane	10 ppm ceiling (60 mg/m^3)	Lungs
Dichlorotetrafluoroethane	1,000 ppm (7,000 mg/m^3)	Respiratory system, CVS
Dichlorvos	1 mg/m^3	Respiratory system, CVS, CNS, eyes, skin, blood cholinesterase
Dieldrin	0.25 mg/m^3	CNS, liver, kidneys, skin
Diethylamine	25 ppm (75 mg/m^3)	Respiratory system, skin, eyes
Diethylaminoethanol	10 ppm (50 mg/m^3)	Respiratory system, skin, eyes
Difluorodibromomethane	100 ppm (860 mg/m^3)	Skin, respiratory system
Diglycidyl ether	0.5 ppm (2.8 mg/m^3)	Skin, eyes, respiratory system
Diisobutyl ketone	50 ppm (290 mg/m^3)	Respiratory system, skin, eyes
Diisopropylamine	5 ppm (20 mg/m^3)	Respiratory system, skin, eyes
Dimethyl acetamide	10 ppm (35 mg/m^3)	Liver, skin
Dimethylamine	10 ppm (18 mg/m^3)	Respiratory system, skin, eyes
Dimethylaniline	5 ppm (25 mg/m^3)	Blood, kidneys, liver, CVS
Dimethyl-1,2-dibromo-2,2-dichlorethyl phosphate	3 mg/m^3	Respiratory system, CNS, CVS, skin, eyes, blood cholinesterase
Dimethyl formamide	10 ppm (30 mg/m^3)	Liver, kidneys, CVS, skin
1,1-Dimethylhydrazine	0.5 ppm (1 mg/m^3)	CNS, liver, GI tract, blood, respiratory system, eyes, skin
Dimethylphthalate	5 mg/m^3	Respiratory system, GI tract
Dimethylsulfate	1 ppm (5 mg/m^3)	Eyes, respiratory system, liver, kidneys, CNS, skin
Dinitrobenzene (all isomers)	1 mg/m^3	Blood, liver, CVS, eyes, CNS
Dinitro-*o*-cresol	0.2 mg/m^3	CVS, endocrine system, eyes
Dinitrotoluene	1.5 mg/m^3	Blood, liver, CVS
Di-*sec*-octyl phthalate	5 mg/m^3	Eyes, upper respiratory system, GI tract
Dioxane	100 ppm (360 mg/m^3)	Liver, kidneys, skin, eyes
Diphenyl	0.2 ppm (1 mg/m^3)	Liver, skin, CNS, upper respiratory system, eyes
Dipropylene glycol methyl ether	100 ppm (600 mg/m^3)	Respiratory system, eyes
Endrin	0.1 mg/m^3	CNS, liver
Epichlorohydrin	5 ppm (19 mg/m^3)	Respiratory system, lungs, skin, kidneys
EPN	0.5 mg/m^3	Respiratory system, CVS, CNS, eyes, skin, blood cholinesterase
Ethanolamine	3 ppm (6 mg/m^3)	Skin, eyes, respiratory system
2-Ethoxyethanol	200 ppm (740 mg/m^3)	In animals: lungs, eyes, blood, kidneys, liver
2-Ethoxyethylacetate	100 ppm (540 mg/m^3)	Respiratory system, eyes, GI tract
Ethyl acetate	400 ppm (1,400 mg/m^3)	Eyes, skin, respiratory system
Ethylamine	10 ppm (18 mg/m^3)	Respiratory system, eyes, skin
Ethyl acrylate	25 ppm (100 mg/m^3)	Respiratory system, eyes, skin
Ethyl benzene	100 ppm (435 mg/m^3)	Eyes, upper respiratory system, skin, CNS
Ethyl bromide	200 ppm (890 mg/m^3)	Skin, liver, kidneys, respiratory system, CVS, CNS
Ethyl butyl ketone	50 ppm (230 mg/m^3)	Eyes, skin, respiratory system
Ethyl chloride	1,000 ppm (2,600 mg/m^3)	Liver, kidneys, respiratory system, CVS

Chemical name	Permissible exposure limit (8-hr TWA)	Target organ
Ethylene chlorohydrin	5 ppm (16 mg/m^3)	Respiratory system, liver, kidneys, CNS, skin, CVS
Ethylenediamine	10 ppm (25 mg/m^3)	Respiratory system, liver, kidneys, skin
Ethylene dibromide	10 ppm; 30 ppm ceiling; 50 ppm, 5-min peak	Respiratory system, liver, kidneys, skin, eyes
Ethylene dichloride	50 ppm; 100 ppm ceiling; 200 ppm peak	Kidneys, liver, eyes, skin, CNS
Ethylene glycol dinitrate	1 mg/m^3 ceiling	CVS, blood, skin
Ethylene oxide	1 ppm (1.8 mg/m^3)	Eyes, blood, respiratory system, liver, CNS, kidneys
Ethyl ether	400 ppm (1,200 mg/m^3)	CNS, skin, respiratory system, eyes
Ethyl formate	100 ppm (300 mg/m^3)	Eyes, respiratory system
Ethyl mercaptan	10 ppm ceiling (25 mg/m^3)	Respiratory system; in animals: liver, kidneys
n-Ethylmorpholine	20 ppm (94 mg/m^3)	Respiratory system, eyes, skin
Ethyl silicate	100 ppm (850 mg/m^3)	Respiratory system, liver, kidneys, blood, skin
Ferbam	15 mg/m^3	Respiratory system, skin, GI tract
Ferrovanadium dust	1 mg/m^3	Respiratory system, eyes
Fluorides (as F)	2.5 mg/m^3	Eyes, respiratory system, CNS, skeleton, kidneys, skin
Fluorine	0.1 ppm (0.2 mg/m^3)	Respiratory system, eyes, skin; in animals: liver, kidneys
Fluorotrichloromethane	1000 ppm (5600 mg/m^3)	CVS, skin
Formaldehyde	3 ppm; 5 ppm ceiling; 10 ppm, 30-min ceiling	Respiratory system, eyes, skin
Formic acid	5 ppm (9 mg/m^3)	Respiratory system, skin, kidneys, liver, eyes
Furfural	5 ppm (20 mg/m^3)	Eyes, respiratory system, skin
Furfuryl alcohol	50 ppm (200 mg/m^3)	Respiratory system
Glycidol	50 ppm (150 mg/m^3)	Eyes, skin, respiratory system, CNS
Hafnium and compounds (as Hf)	0.5 mg/m^3	Eyes, skin, mucous membranes
Heptachlor	0.5 mg/m^3	In animals: CNS, liver
Heptane	500 ppm (2,000 mg/m^3)	Skin, respiratory system, PNS
Hexachloroethane	1 ppm (10 mg/m^3)	Eyes
Hexachloronaphthalene	0.2 mg/m^3	Liver, skin
Hexane	500 ppm (1,800 mg/m^3)	Skin, eyes, respiratory system, lungs
2-Hexanone	100 ppm (410 mg/m^3)	CNS, skin, respiratory system
Hexone	100 ppm (410 mg/m^3)	Respiratory system, eyes, skin, CNS
sec-Hexyl acetate	50 ppm (300 mg/m^3)	CNS, eyes
Hydrazine	1 ppm (1.3 mg/m^3)	CNS, respiratory system, skin, eyes
Hydrogen bromide	3 ppm (10 mg/m^3)	Respiratory system, eyes, skin
Hydrogen chloride	5 ppm ceiling (7 mg/m^3)	Respiratory system, skin, eyes
Hydrogen cyanide (as CN)	5 mg/m^3	CNS, CVS, liver, kidneys
Hydrogen fluoride	3 ppm (2 mg/m^3)	Eyes, respiratory system, skin
Hydrogen peroxide	1 ppm (1.4 mg/m^3)	Eyes, skin, respiratory system
Hydrogen selenide	0.05 ppm (0.2 mg/m^3)	Respiratory system, eyes
Hydrogen sulfide	20 ppm ceiling; 50 ppm, 10-min peak	Respiratory system, eyes
Hydroquinone	2 mg/m^3	Eyes, respiratory system, skin, CNS

Chemical name	Permissible exposure limit (8-hr TWA)	Target organ
Iodine	0.1 ppm ceiling (1 mg/m^3)	Respiratory system, eyes, skin, CNS, CVS
Iron oxide fume	10 mg/m^3	Respiratory system
Isoamyl acetate	100 ppm (525 mg/m^3)	Eyes, skin, respiratory system
Isoamyl alcohol	100 ppm (360 mg/m^3)	Eyes, skin, respiratory system
Isobutyl acetate	150 ppm (700 mg/m^3)	Skin, eyes, respiratory system
Isophorone	25 ppm (140 mg/m^3)	Respiratory system
Isopropyl acetate	250 ppm (950 mg/m^3)	Eyes, skin, respiratory system
Isopropyl alcohol	400 ppm (980 mg/m^3)	Eyes, skin, respiratory system
Isopropylamine	5 ppm (12 mg/m^3)	Respiratory system, skin, eyes
Isopropyl ether	500 ppm (2,100 mg/m^3)	Respiratory system, skin
Isopropyl glycidyl ether	50 ppm (240 mg/m^3)	Eyes, skin, respiratory system
Ketene	0.5 ppm (0.9 mg/m^3)	Respiratory system, eyes, skin
Lead, inorganic fumes and dusts (as Pb)	0.05 mg/m^3	GI tract, CNS, kidneys, blood, gingival tissue
Lead arsenate	0.05 mg/m^3 (as lead)	GI tract, CNS, kidneys, blood, gingival tissue, lymphatics, skin
Lindane	0.5 mg/m^3	Eyes, CNS, blood, liver, kidneys, skin
Lithium hydride	0.025 mg/m^3	Respiratory system, skin, eyes
LPG	1,000 ppm (1,800 mg/m^3)	Respiratory system, CNS
Magnesium oxide fume	15 mg/m^3	Respiratory system, eyes
Malathion	15 mg/m^3	Respiratory system, liver, blood cholinesterase, CNS, CVS, GI tract
Maleic anhydride	0.25 ppm (1 mg/m^3)	Eyes, respiratory system, skin
Manganese and compounds (as Mn)	5 mg/m^3 ceiling	Respiratory system, CNS, blood, kidneys
Mercury and inorganic compounds (as Hg)	0.1 mg/m^3 ceiling	Skin, respiratory system, CNS, kidneys, eyes
Mercury (organo) alkyl compounds (as Hg)	0.01 mg/m^3; 0.04 mg/m^3	CNS, kidneys, eyes, skin
Mesityl oxide	25 ppm (100 mg/m^3)	Eyes, skin, respiratory system, CNS
Methoxychlor	15 mg/m^3	None known
Methyl acetate	200 ppm (610 mg/m^3)	Respiratory system, skin, eyes
Methyl acetylene	1,000 ppm (1,650 mg/m^3)	CNS
Methyl acetylene–propadiene mixture	1,000 ppm (1,800 mg/m^3)	CNS, skin, eyes
Methyl acrylate	10 ppm (35 mg/m^3)	Respiratory system, eyes, skin
Methylal	1,000 ppm (3,100 mg/m^3)	Skin, respiratory system, CNS
Methyl alcohol	200 ppm (260 mg/m^3)	Eyes, skin, CNS, GI tract
Methylamine	10 ppm (12 mg/m^3)	Respiratory system, eyes, skin
Methyl (*n*-amyl) ketone	100 ppm (465 mg/m^3)	Eyes, skin, respiratory system, CNS, PNS
Methyl bromide	20 ppm (80 mg/m^3)	CNS, respiratory system, skin, eyes
Methyl cellosolve	25 ppm (80 mg/m^3)	CNS, blood, skin, eyes kidneys
Methyl cellosolve acetate	25 ppm; 120 mg/m^3)	Kidneys, brain, CNS, PNS
Methyl chloride	100 ppm; 200 ppm ceiling; 300 ppm, 5-min/3-hr peak	CNS, liver, kidneys, skin
Methyl chloroform	350 ppm (1,900 mg/m^3)	Skin, CNS, CVS, eyes
Methylcyclohexane	500 ppm (2,000 mg/m^3)	Respiratory system, skin
Methylcyclohexanol	100 ppm (470 mg/m^3)	Respiratory system, skin, eyes; in animals: CNS, liver, kidneys
o-Methylcyclohexanone	100 ppm (460 mg/m^3)	In animals: lungs, liver, kidneys, skin
Methylene bisphenyl isocyanate	0.02 ppm ceiling (0.2 mg/m^3)	Respiratory system, eyes

Chemical name	Permissible exposure limit (8-hr TWA)	Target organ
Methylene chloride	500 ppm; 1,000 ppm ceiling; 2,000 ppm, 5-min/2-hr peak	Skin, CVS, eyes, CNS
Methyl formate	100 ppm (250 mg/m^3)	Eyes, respiratory system, CNS
5-Methyl-3-heptanone	25 ppm (130 mg/m^3)	Eyes, skin, respiratory system, CNS
Methyl iodide	5 ppm (28 mg/m^3)	CNS, skin, eyes
Methyl isobutyl carbinol	25 ppm (100 mg/m^3)	Eyes, skin
Methyl isocyanate	0.02 ppm (0.05 mg/m^3)	Respiratory system, eyes, skin
Methyl mercaptan	10 ppm, 15-min ceiling (20 mg/m^3)	Respiratory system, CNS
Methyl methacrylate	100 ppm (410 mg/m^3)	Eyes, upper respiratory system, skin
alpha-Methyl styrene	100 ppm ceiling (480 mg/m^3)	Eyes, respiratory system, skin
Mica (less than 1% quartz)	20 mppcf	Lungs
Molybdenum soluble compounds (as Mo)	5 mg/m^3	Respiratory system; in animals: kidneys, blood
Molybdenum insoluble compounds (as Mo)	15 mg/m^3	None known
Monomethyl aniline	2 ppm (9 mg/m^3)	Respiratory system, liver, kidneys, blood
Monomethyl hydrazine	0.2 ppm ceiling (0.35 mg/m^3)	CNS, respiratory system, liver, blood, CVS, eyes
Morpholine	20 ppm (70 mg/m^3)	Respiratory system, eyes, skin
Naphtha (coal tar)	100 ppm (400 mg/m^3)	Respiratory system, eyes, skin
Naphthalene	10 ppm (50 mg/m^3)	Eyes, blood, liver, kidneys, skin, RBC, CNS
Nickel, metal and soluble compounds (as Ni)	1 mg/m^3	Nasal cavities, lungs, skin
Nickel carbonyl	0.001 ppm (0.007 mg/m^3)	Lungs, paranasal sinus, CNS
Nicotine	0.5 mg/m^3	CNS, CVS, lungs, GI tract
Nitric acid	2 ppm (5 mg/m^3)	Eyes, respiratory system, skin, teeth
Nitric oxide	25 ppm (30 mg/m^3)	Respiratory system
p-Nitroaniline	1 ppm (6 mg/m^3)	Blood, heart, lungs, liver
Nitrobenzene	1 ppm (5 mg/m^3)	Blood, liver, kidneys, CVS, skin
p-Nitrochlorobenzene	1 mg/m^3	Blood, liver, kidneys, CVS
Nitroethane	100 ppm (310 mg/m^3)	Skin
Nitrogen dioxide	5 ppm ceiling (9 mg/m^3)	Respiratory system, CVS
Nitrogen trifluoride	10 ppm (29 mg/m^3)	In animals: blood
Nitromethane	100 ppm (250 mg/m^3)	Skin
1-Nitropropane	25 ppm (90 mg/m^3)	Eyes, CNS
2-Nitropropane	25 ppm (90 mg/m^3)	Respiratory system, CNS
Nitrotoluene	5 ppm (30 mg/m^3)	Blood, CNS, CVS, skin, GI tract
Octachloronaphthalene	0.1 mg/m^3	Skin, liver
Octane	500 ppm (2,350 mg/m^3)	Skin, eyes, respiratory system
Oil mist (mineral)	5 mg/m^3	Respiratory system, skin
Osmium tetroxide	0.002 mg/m^3	Eyes, respiratory system, skin
Oxalic acid	1 mg/m^3	Respiratory system, skin, kidneys, eyes
Oxygen difluoride	0.05 ppm (0.1 mg/m^3)	Lungs, eyes
Ozone	0.1 ppm (0.2 mg/m^3)	Eyes, respiratory system
Paraquat compounds	0.5 mg/m^3	Eyes, respiratory system, heart, liver, kidneys, GI tract
Parathion	0.1 mg/m^3	Respiratory system, CNS, CVS, eyes, skin, blood cholinesterase
Pentaborane	0.005 ppm (0.01 mg/m^3)	CNS, eyes, skin

Chemical name	Permissible exposure limit (8-hr TWA)	Target organ
Pentachloronaphthalene	0.5 mg/m^3	Skin, liver, CNS
Pentachlorophenol	0.5 mg/m^3	CVS, respiratory system, eyes, liver, kidneys, skin, CNS
Pentane	1,000 ppm (2,950 mg/m^3)	Skin, eyes, respiratory system
2-Pentanone	200 ppm (700 mg/m^3)	Respiratory system, eyes, skin, CNS
Perchloromethyl mercaptan	0.1 ppm (0.8 mg/m^3)	Eyes, respiratory system, liver, kidneys, skin
Perchloryl fluoride	3 ppm (13.5 mg/m^3)	Respiratory system, skin, blood
Petroleum distillates (naphtha)	500 ppm (2,000 mg/m^3)	Skin, eyes, respiratory system, CNS
Phenol	5 ppm (19 mg/m^3)	Liver, kidneys, skin
p-Phenylene diamine	0.1 mg/m^3	Respiratory system, skin
Phenyl ether	1 ppm (7 mg/m^3)	Eyes, skin, respiratory system
Phenyl ether-biphenyl mixture	1 ppm (7 mg/m^3)	Eyes, skin, respiratory system
Phenyl glycidyl ether	10 ppm (60 mg/m^3)	Skin, eyes, CNS
Phenylhydrazine	5 ppm (22 mg/m^3)	Blood, respiratory system, liver, kidneys, skin
Phosdrin	0.1 mg/m^3	Respiratory system, CNS, CVS, skin, blood cholinesterase
Phosgene	0.1 ppm (0.4 mg/m^3)	Respiratory system, skin, eyes
Phosphine	0.3 ppm (0.4 mg/m^3)	Respiratory system
Phosphoric acid	1 mg/m^3	Respiratory system, eyes, skin
Phosphorus (yellow)	0.1 mg/m^3	Respiratory system, liver, kidneys, jaw, teeth, blood, eyes, skin
Phosphorus pentachloride	1 mg/m^3	Respiratory system, eyes, skin
Phosphorus pentasulfide	1 mg/m^3	Respiratory system, CNS, eyes, skin
Phosphorus trichloride	0.5 ppm (3 mg/m^3)	Respiratory system, eyes, skin
Phthalic anhydride	2 ppm (12 mg/m^3)	Respiratory system, eyes, skin, liver, kidneys
Picric acid	0.1 mg/m^3	Kidneys, liver, blood, skin, eyes
Pival	0.1 mg/m^3	Blood prothrombin
Platinum (soluble salts as Pt)	0.002 mg/m^3	Respiratory system, skin, eyes
Portland cement (less than 1% quartz)	50 mppcf	Respiratory system, eyes,
Propane	1,000 ppm (1,800 mg/m^3)	CNS
n-Propyl acetate	200 ppm (840 mg/m^3)	Respiratory system, eyes, skin, CNS
Propyl alcohol	200 ppm (500 mg/m^3)	Skin, eyes, respiratory system, GI tract
Propylene dichloride	75 ppm (350 mg/m^3)	Skin, eyes, respiratory system, liver, kidneys
Propyleneimine	2 ppm (5 mg/m^3)	Eyes, skin
Propylene oxide	100 ppm (240 mg/m^3)	Eyes, skin, respiratory system
N-Propyl nitrate	25 ppm (110 mg/m^3)	None known
Pyrethrum	5 mg/m^3	Respiratory system, skin, CNS
Pyridine	5 ppm (15 mg/m^3)	CNS, liver, kidneys, skin, GI tract
Quinone	0.1 ppm (0.4 mg/m^3)	Eyes, skin
Rhodium, metal fume and dust (as Rh)	0.1 mg/m^3	None known
Rhodium, soluble salts (as Rh)	0.001 mg/m^3	Eyes
Ronnel	15 mg/m^3	Skin, liver, kidneys, blood plasma

Chemical name	Permissible exposure limit (8-hr TWA)	Target organ
Rotenone	5 mg/m^3	CNS, eyes, respiratory system
Selenium and compounds (as Se)	0.2 mg/m^3	Upper respiratory system, eyes, skin, liver, kidneys, blood
Selenium hexafluoride (as Se)	0.05 ppm (0.4 mg/m^3)	None known
Silica (amorphous)	20 mppcf	Respiratory system
Silica (crystalline)	10 mg/m^3/%SiO_2 + 2	Respiratory system
Silver, metal and soluble compounds (as Ag)	0.01 mg/m^3	Nasal septum, skin, eyes
Soapstone	20 mppcf	Lungs, CVS
Sodium fluoroacetate	0.5 mg/m^3	CVS, lungs, kidneys, CNS
Sodium hydroxide	2 mg/m^3	Eyes, respiratory system, skin
Stibine	0.1 pm (0.5 mg/m^3)	Blood, liver, kidneys, lungs
Stoddard solvent	500 ppm (2,900 mg/m^3)	Skin, eyes, respiratory system, CNS
Strychnine	0.15 mg/m^3	CNS
Styrene	100 ppm; 200 ppm ceiling; 600 ppm, 5-min/3-hr peak	CNS, respiratory system, eyes, skin
Sulfur dioxide	5 ppm (13 mg/m^3)	Respiratory system, skin, eyes
Sulfuric acid	1 mg/m^3	Respiratory system, eyes, skin, teeth
Sulfur monochloride	1 ppm (6 mg/m^3)	Respiratory system, skin, eyes
Sulfur pentafluoride	0.025 ppm (0.25 mg/m^3)	Respiratory system, CNS
Sulfuryl fluoride	5 ppm (920 mg/m^3)	Respiratory system, CNS
2,4,4-T	10 mg/m^3	Skin, liver, GI tract
Talc (nonasbestiform)	20 mppcf	Lungs, CVS
Tantalum metal, oxide dusts (as Ta)	5 mg/m^3	None known in humans
TEDP	0.2 mg/m^3	CNS, respiratory system, CVS
Tellurium compounds (as Te)	0.1 mg/m^3	Skin, CNS
Tellurium hexafluoride (as Te)	0.02 ppm (0.2 mg/m^3)	Respiratory system
TEPP	0.05 mg/m^3	CNS, respiratory system, CVS, GI tract
Terphenyls	1 ppm ceiling (9 mg/m^3)	Skin, respiratory system
1,1,2,2-Tetrachloro-1,2-difluoroethane	500 ppm (4,170 mg/m^3)	Lungs, skin
1,1,1,2-Tetrachloro-2,2-difluoroethane	500 ppm (4,170 mg/m^3)	Respiratory system, skin
1,1,2,2-Tetrachloro-ethane	5 ppm (35 mg/m^3)	Liver, kidneys, CNS
Tetrachloroethylene	100 ppm; 200 ppm ceiling; 300 ppm, 5-min/3-hr peak	Liver, kidneys, eyes, upper respiratory system, CNS
Tetrachloronaphthalene	2 mg/m^3	Liver, skin
Tetraethyl lead (as Pb)	0.075 mg/m^3	CNS, CVS, kidneys, eyes
Tetrahydrofuran	200 ppm (590 mg/m^3)	Eyes, skin, respiratory system, CNS
Tetramethyl lead	0.075 mg/m^3	CNS, CVS, kidneys
Tetramethyl succinonitrile	0.5 ppm (3 mg/m^3)	CNS
Tetranitromethane	1 ppm (8 mg/m^3)	Respiratory system, eyes, skin, blood, CNS
Tetryl	1.5 mg/m^3	Respiratory system, eyes, CNS, skin; in animals: liver, kidneys
Thallium, soluble compounds (as T1)	0.1 mg/m^3	Eyes, CNS, lung, liver, kidneys, GI tract, body hair
Thiram	5 mg/m^3	Respiratory system, skin
Tin, inorganic compounds except oxides (as Sn)	2 mg/m^3	Eyes, skin, respiratory system
Tin, organic compounds (as Sn)	0.1 mg/m^3	CNS, eyes, liver, urinary tract, skin, blood

Chemical name	Permissible exposure limit (8-hr TWA)	Target organ
Titanium oxide	15 mg/m^3	Lungs
Toluene	200 ppm; 300 ppm ceiling; 500 ppm, 10-min peak	CNS, liver, kidneys, skin
Toluene-2,4-diisocyanate	0.02 ppm ceiling (0.14 mg/m^3)	Respiratory system, skin
o-Toluidine	5 ppm (22 mg/m^3)	Blood, kidneys, liver, CVS, skin, eyes
Tributyl phosphate	5 mg/m^3	Respiratory system, skin, eyes
1,1,2-Trichloroethane	10 ppm (45 mg/m^3)	CNS, eyes, nose, liver, kidneys
Trichloroethylene	100 ppm; 200 ppm ceiling; 300 ppm peak	Respiratory system, heart, liver, kidneys, CNS, skin
Trichloronaphthalene	5 mg/m^3	Skin, liver
1,2,3-Trichloropropane	50 ppm (300 mg/m^3)	Eyes, respiratory system, skin, CNS, liver
1,1,2-Trichloro-1,2,2-trifluoroethane	1,000 ppm (7,600 mg/m^3)	Skin, heart
Triethylamine	25 ppm (100 mg/m^3)	Respiratory system, eyes, skin
Trifluoromonobromomethane	1,000 ppm (6,100 mg/m^3)	Heart, CNS
Trinitrotoluene	1.5 mg/m^3	Blood, liver, eyes, CVS, CNS, kidneys, skin
Triorthocresyl phosphate	0.1 mg/m^3	PNS, CNS
Triphenyl phosphate	3 mg/m^3	Blood
Turpentine	100 ppm (560 mg/m^3)	Skin, eyes, kidneys, respiratory system
Uranium, insoluble compounds (as U)	0.25 mg/m^3	Skin, bone marrow, lymphatics
Uranium, soluble compounds (as U)	0.05 mg/m^3	Respiratory system, blood, liver, lymphatics, kidneys, skin, bone marrow
Vanadium pentoxide dust (as V)	0.5 mg/m^3 ceiling	Respiratory system, skin, eyes
Vanadium pentoxide fume (as V)	0.1 mg/m^3 ceiling	Respiratory system, skin, eyes
Vinyl chloride	1 ppm; 5 ppm, 15-min ceiling	Liver, CNS, blood, respiratory system, lymphatic system
Vinyltoluene	100 ppm (480 mg/m^3)	Eyes, skin, respiratory system
Warfarin	0.1 mg/m^3	Blood, CVS
Xylene (*o*-, *m*-, and *p*-isomers)	100 ppm (435 mg/m^3)	CNS, eyes, GI tract, blood, liver, kidneys, skin
Xylidine	5 ppm (25 mg/m^3)	Blood, lungs, liver, kidneys, CVS
Yttrium compounds (as Y)	1 mg/m^3	Eyes, lungs
Zinc chloride fume	1 mg/m^3	Respiratory system, skin, eyes
Zinc oxide fume	5 mg/m^3	Respiratory system
Zirconium compounds (as Z)	5 mg/m^3	Respiratory system, skin

Source: NIOSH Pocket Guide to Chemical Hazards, Sept. 1985.
CNS = central nervous system; PNS = peripheral nervous system; CVS = cardiovascular system; GI = gastrointestinal; TWA = time-weighted average.

Index

Index